Fundamentals of Nuclear Physics

Noboru Takigawa · Kouhei Washiyama

Fundamentals of Nuclear Physics

Springer

Noboru Takigawa
Department of Physics
Graduate School of Science
Tohoku University
Sendai
Japan

Kouhei Washiyama
Center for Computational Sciences
University of Tsukuba
Tsukuba, Ibaraki
Japan

ISBN 978-4-431-55377-9 ISBN 978-4-431-55378-6 (eBook)
DOI 10.1007/978-4-431-55378-6

Library of Congress Control Number: 2016958483

Translation from the Japanese language edition: *Genshikaku Butsurigaku* by Noboru Takigawa,

Printed on acid-free paper

This Springer imprint is published by Springer Nature
The registered company is Springer Japan KK
The registered company address is: Chiyoda First Bldg. East, 3-8-1 Nishi-Kanda, Chiyoda-ku, Tokyo 101-0065, Japan

To Noriko, Akiko, Tsuyoshi
and to our parents.

Preface

This book provides an introduction to nuclear physics. Research in nuclear physics covers a wide variety of subjects, and one can list many key words: nuclear structure and reactions of stable and unstable nuclei, fission and decay of a nucleus, extreme states such as the limits of existence and high-spin states, properties at high temperature and high density, hypernuclei, neutron stars, and nucleosynthesis, among others. All of these are the subjects of nuclear physics. In addition to these rather static properties, nuclear reactions such as heavy-ion collisions introduce new aspects of research, i.e., dynamical properties of nuclei or reaction mechanisms, such as heavy-ion fusion reactions, dissipation phenomena and liquid–gas phase transition. Many of these phenomena can be understood from the point of view that a nucleus is a quantum many-body system of nucleons stabilized by nuclear force. On the other hand, phenomena at higher energies, driven by, e.g., high-energy heavy-ion collisions, require a different approach: the approach based on the quantum chromodynamics (QCD). The study of quark–gluon plasma and of the QCD phase diagram is representative and forms a large stream of current nuclear physics.

In this book, we largely restrict our subjects and describe basic features of nuclear structure and of nuclear decays. β-decay and most excitation motions are left to other books. Also, nuclear reactions and current subjects such as physics of unstable nuclei, hypernuclei, and nuclear physics based on QCD are untouched except for occasional very brief references. Even with these limitations, we could only briefly mention recent developments. However, we have tried to convey part of them through columns on the QCD phase diagram of nuclear matter, superheavy elements, superdeformed states, and overview of the synthesis of elements. We hope that together with the main text they help readers to grasp our current knowledge of the nucleus and some recent research trends in nuclear physics. By restricting the subjects, our aim was to contain many experimental data of basic nuclear properties or suitable illustrations and explain the main structural features of nuclei in some detail. This will be useful because most of the phenomena listed in the first paragraph, but omitted from the book, are intimately related to those basic properties. We also have attempted to explain how the nuclear model has developed from the original phenomenological level of a shell model to the modern understanding based on a many-body theory such as the Hartree–Fock calculations. We also have described nuclear force, the basis of nuclear physics, in some detail. We sometimes introduce semi-classical approximation to the original quantum mechanical formalism. We hope that it helps the readers to intuitively grasp the underlying physics of complicated nuclear phenomena.

In writing the book our intention was to create not only a good introduction to nuclear physics, but also a good reference book for physicists to learn the application of quantum mechanics and mathematics. Toward that aim, in addition to describing the basic phenomena of a nucleus, we attempted to convey how the basic subjects of modern physics such as quantum mechanics, statistical physics, mathematics for physics, e.g., complex integrals, are used to describe or interpret various phenomena of nucleus. We also considered the standard level of knowledge of junior and senior students, and gave a detailed description to enable them to derive each equation. Finally, we added the appendix to prove a number of important formulae in the main text and to show some fundamental formulae.

The contents of the book are based on the lectures that one of the authors, N.T., has delivered at Tohoku University, Sendai, Japan, for a long time to junior and senior students in the undergraduate physics course and also to beginning graduate students. We have included as sidebars some additional material that was presented in the class in order to keep the atmosphere of the lecture. Many textbooks and original papers and figures therein have helped in preparing the lectures and this book. Several of the figures are taken from them. Using this opportunity, we wish to thank the authors. The papers cited at various places are not at all complete. Moreover, it does not mean that they are necessarily the representative papers on each citation. Nevertheless, we hope that they can help readers to do further study. This book is an English translation of the Japanese edition, which one of the authors, N.T., published in 2013. The appendix contains the errata to the original Japanese edition.

We would like to thank Akif Baha Balantekin for many useful comments for the writing of this English edition. We thank D.M. Brink, A.B. Balantekin, N. Rowley, F. Michel, S.Y. Lee, P. Fröbrich, S. Landowne, K. McVoy, W.A. Friedman, G.F. Bertsch, A. Brown, H. Weidenmüller, H. Friedrich, H. Esbensen, M.S. Hussein, L.F. Canto, C. Bertulani, P.R.S. Gomes, D. Hinde, M. Dasgupta, G. Pollarolo, A. Bonasera, M. Di Toro, C. Spitaleri, C. Rolfs, I. Thompson, S. Ayik, K. Hara, Y. Abe, H. Sagawa, A. Iwamoto, T. Tazawa, M. Ohta, J. Kasagi, and many other colleagues and friends around the world; and K. Hagino and A. Ono in the Nuclear Theory Group of Tohoku University, Sendai, for useful discussions. We are grateful to students of Tohoku University, especially foreign students, for stimulating us to write this English book. We would like to thank F. Minato, S. Yusa, S. Iwasaki and his wife, H. Tamura, and H. Koura for preparing figures, and K. Morita, T. Hirano, Y. Aritomo, T. Kajino, and Y. Motizuki in preparing the columns; and P. Möller, T. Wada, K. Matsuyanagi, T. Tamae, and K. Kato for kindly reading sections of the Japanese edition and making many helpful suggestions. We also thank all the mentors and collaborators who provided us opportunities to work in their institutions. We are grateful to H. Niko and R. Takizawa for their help as the editors of this English version. Above all N.T. would like to thank his wife, Noriko, and his family for their support over the years.

Sendai, Japan
September 2016

Noboru Takigawa
Kouhei Washiyama

Contents

Chapter 1
Introduction

Abstract It would be useful to have an overview of some fundamental aspects of nuclei before discussing each subject in detail. In this connection, we briefly describe in this chapter the constituents and basic structure of atomic nuclei, properties of particles which are closely related to nuclear physics, the role of the four fundamental forces in nature in nuclear physics, nuclear species, the abundance of elements and the phase diagram of nuclear matter.

1.1 The Constituents and Basic Structure of Atomic Nuclei

The atomic nuclei are self-bound many-body systems of protons (p) and neutrons (n) by strong interaction. Although other baryons such as $\Delta(1232)$ are also contained, their amounts are small. For example, the percentage of $\Delta(1232)\Delta(1232)$ contained in the lightest nucleus deuteron (d) is about 1%. π-mesons in virtual states mediate the interaction between the constituent particles and affect the electromagnetic properties of protons and neutrons. Furthermore, each proton and neutron is also a composite particle consisting of three quarks. The other hadrons also consist of quarks. One can therefore take also the view that atomic nuclei are many-body systems of quarks.

The picture of nuclei and of nuclear phenomena, hence the appropriate way to describe them, depend on the object and method of observation and the related energy scale, and lead to various models for nuclei. This book restricts to low-energy phenomena and discusses the nuclear structure and properties primarily from the point of view that nuclei are many-body systems of protons and neutrons. The governing law is quantum mechanics. This contrasts to quantum chromodynamics for high-energy phenomena. Among various quantum many-body systems, nuclei have characteristics that the number of constituents is small and also that the leading forces are strong interactions.

Each nucleus is represented, for example, as $^{16}_{8}$O. O is the symbol of the chemical element. It represents oxygen in this example, hence the number of protons is 8. This number is called the atomic number, and is given at the left lower side. It is often omitted, because it has a one to one correspondence to the symbol of element. The number at the left upper side is called mass number and is given by the sum

N. Takigawa and K. Washiyama, *Fundamentals of Nuclear Physics*,
DOI 10.1007/978-4-431-55378-6_1

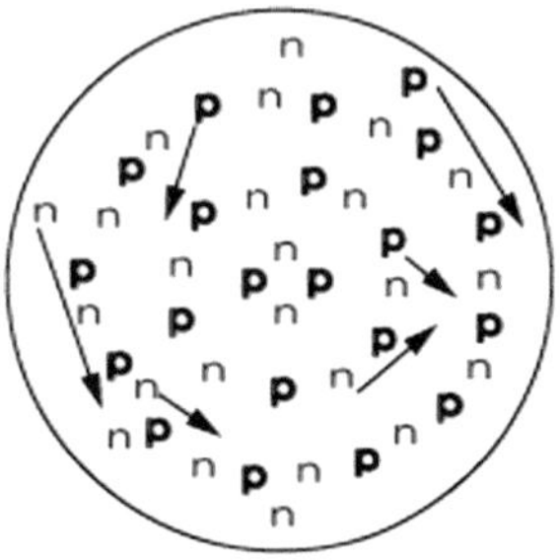

Fig. 1.1 Conceptual illustration of nuclear structure: example of ${}^{45}_{21}$Sc

of the atomic number and the neutron number. They are denoted by A, Z and N, respectively, and $A = Z + N$.

Figure 1.1 is a conceptual illustration of nuclear structure exemplified by ${}^{45}_{21}$Sc. The enclosing circle has been drawn to indicate the finiteness of the nuclear size. In reality, it is absent, because nuclei are not given by any external boundary conditions, but are self-bound systems. The arrows indicate that protons p and neutrons n inside a nucleus are not fixed at lattice points like atoms in solids, but are moving around with finite velocities. We learn later that nuclei behave like either liquid or gas depending on the observables or phenomena we are interested in.

1.2 Properties of Particles Relevant to Nuclear Physics

Table 1.1 gives the properties of particles which are closely related to this book. As the table shows, proton and neutron resemble each other in many properties such as the mass and the spin except for electric properties, and are jointly called *nucleons*. In order to distinguish them, one introduces the concept of *isospin space* related to charge, and considers proton and neutron to be two different states in the isospin space. The operators and states in the isospin space obey the same law as that of angular momentum, and are called isospin operators and isospin states, respectively.

Nucleon has two states in the isospin space. Similarly to the spin operators for electron $\hat{\mathbf{s}}$, one therefore introduces the isospin operators $\hat{\mathbf{t}}$ by

$$\hat{t}_x = \frac{1}{2}\begin{pmatrix} 0 & 1 \\ 1 & 0 \end{pmatrix}, \quad \hat{t}_y = \frac{1}{2}\begin{pmatrix} 0 & -\mathrm{i} \\ \mathrm{i} & 0 \end{pmatrix}, \quad \hat{t}_z = \frac{1}{2}\begin{pmatrix} 1 & 0 \\ 0 & -1 \end{pmatrix}, \tag{1.1}$$

and, analogously to the Pauli spin operators $\hat{\boldsymbol{\sigma}}$, $\hat{\boldsymbol{\tau}} = 2\hat{\mathbf{t}}$ by

$$\hat{\tau}_x = \begin{pmatrix} 0 & 1 \\ 1 & 0 \end{pmatrix}, \quad \hat{\tau}_y = \begin{pmatrix} 0 & -\mathrm{i} \\ \mathrm{i} & 0 \end{pmatrix}, \quad \hat{\tau}_z = \begin{pmatrix} 1 & 0 \\ 0 & -1 \end{pmatrix}, \tag{1.2}$$

Table 1.1 Properties of particles relevant to nuclear physics. I: isospin, S: strangeness, R_c: radius of charge distribution, μ: magnetic dipole moment, m in the magneton $e\hbar/2mc$ is m_e for electron, m_μ for μ particle, m_p for proton and neutron, and m_p for Λ and Σ particles. The number represents the mean value if error is not given. The lifetime of proton depends on methods. The mass of ν is from the decay of tritium. The lifetime of ν is from nuclear reactor with m_{ν_e} in units of eV. The quark structure for $\rho^{\pm,0}$ is the same as that for $\pi^{\pm,0}$. Taken from [1]

Name	I	I_3	J^π	S	mc^2 (MeV)	R_c (fm)	μ $(\frac{e\hbar}{2mc})$	τ(mean life) (s)	Quark model
p	$\frac{1}{2}$	$-\frac{1}{2}$	$\frac{1}{2}^+$	0	938.3	0.88 ± 0.01	2.79	$>1.9 \times 10^{29}$ y	uud
n	$\frac{1}{2}$	$\frac{1}{2}$	$\frac{1}{2}^+$	0	939.6	0.34[a]	−1.91	885.7 ± 0.8	udd
γ			1^-		$<6 \times 10^{-23}$			stable	
$W^\pm$			1		80.4×10^3			3.1×10^{-25}	
Z^0			1		91.2×10^3			2.7×10^{-25}	
ν_e			$\frac{1}{2}$		$<2 \times 10^{-6}$		[b]	$>300 m_{\nu_e}$	
e^-			$\frac{1}{2}$		0.511		1.00	$>4.6 \times 10^{26}$ y	
μ^-			$\frac{1}{2}$		105.7		1.00	2.2×10^{-6}	
π^+	1	1	0^-	0	139.6			2.6×10^{-8}	$u\bar{d}$
π^-	1	−1	0^-	0	139.6			2.6×10^{-8}	$d\bar{u}$
π^0	1	0	0^-	0	135.0			0.84×10^{-16}	$\frac{1}{\sqrt{2}}(u\bar{u}-d\bar{d})$
$\rho^{\pm,0}$	1		1^-		775.5			4.5×10^{-24}	$(u\bar{d}, d\bar{u})$
ω	0		1^-		782.7			7.9×10^{-23}	[c]
K^+	$\frac{1}{2}$	$\frac{1}{2}$	0^-	1	493.7			1.24×10^{-8}	$u\bar{s}$
K^-	$\frac{1}{2}$	$-\frac{1}{2}$	0^-	−1	493.7			1.24×10^{-8}	$s\bar{u}$
K^0	$\frac{1}{2}$	$-\frac{1}{2}$	0^-	1	497.6				$d\,\bar{s}$
$\bar{K}^0$	$\frac{1}{2}$	$\frac{1}{2}$	0^-	−1	497.6				$s\,\bar{d}$
Λ	0	0	$\frac{1}{2}^+$	−1	1115.7		−0.61	2.63×10^{-10}	uds
Σ^+	1	1	$\frac{1}{2}^+$	−1	1189.4		2.46	0.80×10^{-10}	uus
Σ^0	1	0	$\frac{1}{2}^+$	−1	1192.6			$(7.4 \pm 0.7) \times 10^{-20}$	uds
Σ^-	1	−1	$\frac{1}{2}^+$	−1	1197.4			1.5×10^{-10}	dds
Ξ^0	$\frac{1}{2}$	$\frac{1}{2}$	$\frac{1}{2}^+$	−2	1314.8			2.9×10^{-10}	uss
Ξ^-	$\frac{1}{2}$	$-\frac{1}{2}$	$\frac{1}{2}^+$	−2	1321.3			1.6×10^{-10}	dss
Δ	$\frac{3}{2}$		$\frac{3}{2}^+$	0	~ 1232			$\sim 6 \times 10^{-24}$	[d]

[a]The mean square charge radius of neutron is $\langle r_n^2 \rangle = -0.1161 \pm 0.0022\,\mathrm{fm}^2$

[b]$\mu_\nu < 0.54 \times 10^{-10} \mu_B$

[c]$c_1(u\bar{u}+d\bar{d})+c_2 s\bar{s}$

[d]Δ^{++} =uuu, Δ^+ =uud, Δ^0 =udd, Δ^- =ddd

and considers proton and neutron to be simultaneous eigenstates of $\hat{\mathbf{t}}^2$ and $\hat{t}_z$ in such a way that[1]

$$|\mathrm{n}\rangle = \left|\frac{1}{2}\ \frac{1}{2}\right\rangle = \begin{pmatrix}1\\0\end{pmatrix}, \quad |\mathrm{p}\rangle = \left|\frac{1}{2}\ -\frac{1}{2}\right\rangle = \begin{pmatrix}0\\1\end{pmatrix}. \tag{1.3}$$

As Table 1.1 shows, the isospin is one of the important quantum numbers to specify the property of each particle. Its magnitude is assigned to be I when there exist $2I+1$ particles which have common properties for all aspects such as the mass and spin but the electric charge. For example, the isospin of π-mesons is 1, since there exist three particles which differ only in the electric charge. In nuclei, the isospin quantum number of each state and the symmetry concerning the isospin play important roles reflecting the symmetry properties of nuclear force in the isospin space.

If a particle is a structureless fermion, one can deduce from the Dirac equation that its magnetic dipole moment, which is often simply called magnetic moment, is given by $\mu = e\hbar/2mc$, where m is the mass of the particle. In fact, the magnetic moment of an electron is 1 in units of the *Bohr magneton* $\mu_{\mathrm{B}} = e\hbar/2m_{\mathrm{e}}c$.[2] However, Table 1.1 shows that the magnetic moment of a proton significantly deviates from the *nuclear magneton* $\mu_N = e\hbar/2m_{\mathrm{p}}c$. Also, the magnetic moment of a neutron is not zero, but is nearly comparable in magnitude and opposite in sign to that of a proton. They are called *anomalous magnetic moments* and imply that both the proton and the neutron are composite particles with intrinsic structure.[3]

Exercise 1.1 Derive the approximate equation for two large components φ in the four-component spinor ψ starting from the Dirac equation in the presence of electromagnetic fields and assuming that the velocity v of the particle is much smaller than the speed of light in vacuum c, i.e., $v \ll c$, and show that the term which describes the interaction with the magnetic field in the effective Hamiltonian is given by $H = -\frac{e\hbar}{2mc}\boldsymbol{\sigma}\cdot\mathbf{B}$, where $\mathbf{B}$ is the magnetic field. One can thus prove that the magnetic moment of a Dirac particle is given by $e\hbar/2mc$.

The fact that both proton and neutron are not point particles, but have intrinsic structure, can be seen also directly from the data of charge distribution. The radius of the charge distribution of a proton is about 1 fm.[4]

---

[1]There exists an alternative definition, where proton and neutron are inverted such that $|\mathrm{p}\rangle = |\frac{1}{2}\ \frac{1}{2}\rangle$, $|\mathrm{n}\rangle = |\frac{1}{2}\ -\frac{1}{2}\rangle$. Since $N \geq Z$ for most stable nuclei, we adopt the Definition (1.3) in this book.

[2]Precisely speaking, the experimental value of the magnetic moment of an electron is larger than the prediction of the Dirac theory by about 0.1%, and can be explained by quantum electrodynamics.

[3]It is Otto Stern who experimentally determined the magnetic moment of a proton for the first time. There remains an episode that Pauli visited Stern while he was conducting the experiment and denied the significance of the experiment based on the Dirac theory. Despite the criticism, Stern continued his experiment, and discovered the anomalous magnetic moment of a proton, and consequently was awarded the Nobel Prize in Physics in 1943.

[4]Recent experiments of the scattering of high-energy electrons, and also of polarized electrons, are shedding new lights on the intrinsic structure of nucleons. For example, it is getting uncovered that the electric and magnetic charge distributions are different.

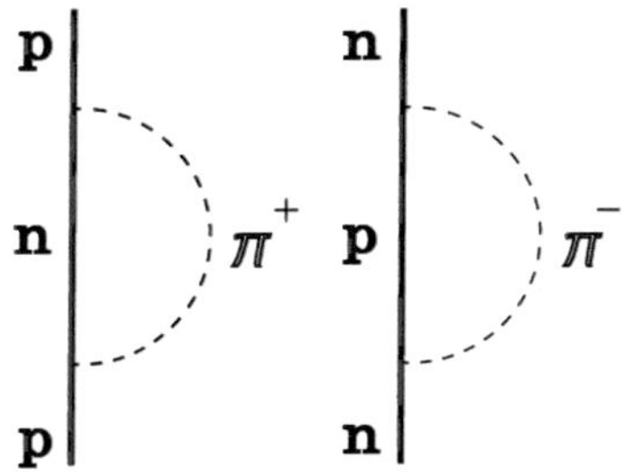

Fig. 1.2 A physical nucleon dressed with a virtual π-meson cloud

The anomalous magnetic moments of nucleons can be understood by either the meson theory or the quark model. The latter will be explained in Sect. 5.4.3. Here, we learn the former.

Let us assume that the proton observed in experiments is a superposition of a bare proton as a structureless Dirac particle and a bare neutron as a Dirac particle surrounded by a virtual π^+-meson (the left part of Fig. 1.2),

$$|\mathrm{p}\uparrow) = \sqrt{1-C_\mathrm{p}}|\mathrm{p}\uparrow\rangle + \sqrt{C_\mathrm{p}}|(\mathrm{n}\times\pi^+)\uparrow\rangle\ . \tag{1.4}$$

Since the π-meson is a pseudoscalar particle, the orbit of the π-meson around the neutron is p-orbit. Corresponding to Eq. (1.4), the magnetic moment of a proton will be given by

$$\mu_\mathrm{p} = (1-C_\mathrm{p})\mu_N + C_\mathrm{p}\frac{e\hbar}{2m_\pi c}\ . \tag{1.5}$$

The first and the second terms on the right-hand side represent the contribution of the interaction of the electromagnetic field with a bare proton and with π-meson, respectively. Because of the difference between the masses of a nucleon and a π-meson, one can explain the anomalous magnetic moment of a proton by assuming the admixture of the (neutron$\times\pi^+$-meson)-component in a proton to be about 30%.

Similarly, if one assumes that a neutron is not a genuine Dirac particle, but contains the component, where π^--meson is moving around a proton as a Dirac particle, by C_n in proportion (the right part of Fig. 1.2), one obtains

$$\mu_\mathrm{n} = C_\mathrm{n}\left(\mu_N - \frac{e\hbar}{2m_\pi c}\right) \tag{1.6}$$

for the magnetic moment of a neutron. Assuming $C_\mathrm{p} = C_\mathrm{n}$, one obtains

$$\mu_\mathrm{p} + \mu_\mathrm{n} = \mu_N\ . \tag{1.7}$$

This agrees well with the experimental data $(\mu_\mathrm{p}+\mu_\mathrm{n})_\mathrm{exp} \sim 0.88\mu_N$.

1.3 The Role of Various Forces

It is known that four types of forces exist in nature. In this section, we briefly survey the role of four forces in nuclear physics.

The strong interaction is principally responsible for the stability and the structure of nuclei. *The electromagnetic interaction* provides a powerful probe of nuclear structure through the electron scattering from nuclei as well as electromagnetic transitions and moments thanks to its well understood nature and the weakness of the force. Furthermore, it governs the lifetime of excited states of nuclei through the electromagnetic transition by γ-ray emission. *The weak interaction* governs the stability of nuclei through $\beta^{\pm}$-decay. The representative example is the beta decay of neutron (Fig. 1.3), which is given by

$$\mathrm{n} \to \mathrm{p} + \mathrm{e}^- + \bar{\nu}_\mathrm{e} \,. \tag{1.8}$$

As shown in Table 1.1, the mean life of the neutron in free space is $\tau \sim 14.8$ min, and the corresponding half-life is $T_{1/2} \sim 10.2$ min. The weak interaction plays an important role also in the synthesis of elements beyond Fe.

Exercise 1.2 Explain the reason why the third particle besides the proton and the electron is needed in the final state of the β-decay of the neutron. Also, discuss the properties of that particle.

All of the strong, the electromagnetic and the weak interactions are relevant to the decay of nuclei. Though there are a large variety of decays, the time scale of the lifetime associated with them is of the order of 10^{-21} s, 1 ps $= 10^{-12}$ s, and 1 min, respectively, reflecting the difference among their strengths.

The gravitational force plays almost no role in the structure of nuclei. However, it plays a crucial role in the synthesis of elements. Also, as we learn later, although there exists no stable nucleus of dineutrons, there exist *neutron stars* because of the gravitational force.

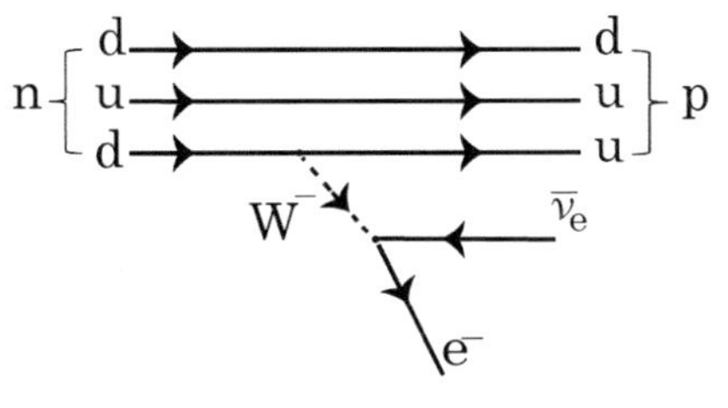

Fig. 1.3 The β-decay of a neutron

1.4 Useful Physical Quantities

It is sometimes worth making order-of-magnitude estimates of various physical quantities. In that connection. it is useful to remember the following approximate values related to the fundamental physical constants c (the speed of light in vacuum), $\hbar$ (the Planck constant divided by 2π), e (electron charge magnitude), k_B (the Boltzmann constant) as

$$c = 2.99792458 \times 10^8 \,\text{m/s} \approx 3.00 \times 10^8 \,\text{m/s}, \tag{1.9}$$

$$\hbar c = 197.326968 \,\text{MeV fm} \approx 200 \,\text{MeV fm}, \tag{1.10}$$

$$\frac{e^2}{\hbar c} = \frac{1}{137.035999074} \approx \frac{1}{137} \quad \text{(fine structure constant)}, \tag{1.11}$$

$$k_B T = 0.02482 \,\text{eV} \approx \frac{1}{40} \,\text{eV at } T = 288 \,\text{K}. \tag{1.12}$$

The quantity $e^2/\hbar c$ is called the fine structure constant. Equation (1.11) holds when the proportional coefficient in the static electric force between two particles with electric charge q_1, q_2 is determined such that the force is given by $F(r) = q_1 q_2 / r^2$ when the two particles are apart from each other by the distance r. Equation (1.12) represents the kinetic energy of thermal neutrons. It is useful to convert the temperature given in units of Kelvin to the corresponding energy in units of MeV.

Exercise 1.3 The range of force is given by the Compton wave length $\hbar/mc$ of the corresponding gauge particle. Estimate the range of the strong interaction and of the weak interaction.

Exercise 1.4 As Fig. 1.4 shows, the force between two protons is dominated by a repulsive Coulomb interaction at large distances, and turns attractive in the region inside their touching radius due to the nuclear force, i.e., due to the strong interaction. Estimate the height of the Coulomb barrier $V_{\text{CB}} = V(r_B)$ by assuming that the touching radius is given by $r = r_B \sim 2 \times R_{\text{p}} \sim 2\,\text{fm}$, where R_{p} is the radius of the proton.

Exercise 1.5 The temperature of the Sun at the core is about 16 million K. Estimate the collision energy E at the core of the Sun.

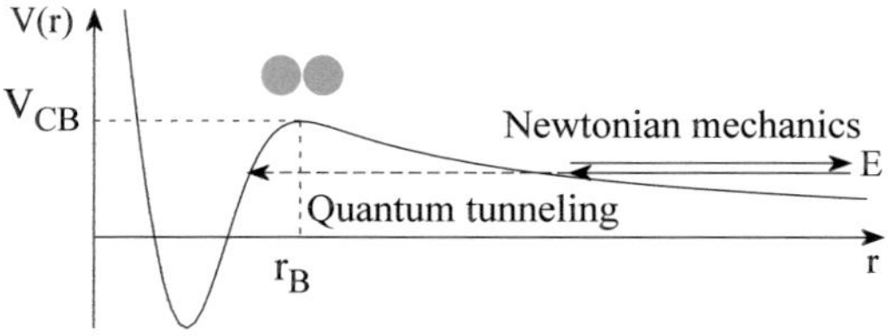

Fig. 1.4 Illustration of the collision between two protons in the Sun

Nuclear reactions in the Sun occur very slowly, because they take place by quantum tunneling as indicated by Exercises 1.4 and 1.5. In reality, there exists another important hindrance factor. As we learn later, there exists no stable bound state in the diproton system. The only stable dinucleon system is deuteron consisting of one proton and one neutron. In order for the fusion of two protons to take place, the inverse reaction of Eq. (1.8), where a proton is converted into a neutron by weak interaction, must therefore be involved. Because of the superposition of the quantum tunneling and weak interaction, the nuclear reaction between two protons is doubly hindered. Consequently, the Sun burns very slowly. It has burnt already for 4.6 billion years and is expected to continue to shine for another almost same period.

1.5 Species of Nuclei

The display of nuclei on the two-dimensional plane, where one axis, say the abscissa, represents the neutron number and the other, say the ordinate, the proton number, is called *Nuclear chart* or *Segré chart*. There are 256 stable nuclei if one includes those nuclei which have long lifetimes of the order comparable to the lifetime of the Sun such as U. They lie in the vicinity of the diagonal line of the nuclear chart for the reason we learn later. The reason why there exist more stable nuclei than the number of stable elements about 92 in nature is because there exist about three stable nuclei for each element on average. For example, there exist two stable nuclei, called proton p and deuteron d, for the element hydrogen.

Incidentally, the nuclei which have the same number of protons (i.e., the same atomic number), but differ in the neutron number, hence in the mass number as well, are called *isotopes* to each other, and those whose neutron numbers are the same, but the proton numbers are different, the *isotones*, and those which have same mass numbers the *isobars*, respectively.

The study of unstable nuclei with short lifetimes is currently one of the hot subjects of nuclear physics. If one includes unstable nuclei whose lifetime is longer than 1 μs, about 7000 nuclei are theoretically predicted to exist, among which about 3000 nuclei have already been discovered experimentally. Also, a group of nuclei stabilized by shell effects (see Chap. 5) are predicted to exist in the region far beyond U. They are called *superheavy nuclei* or *superheavy elements*, on which extensive studies are going on both experimentally and theoretically.

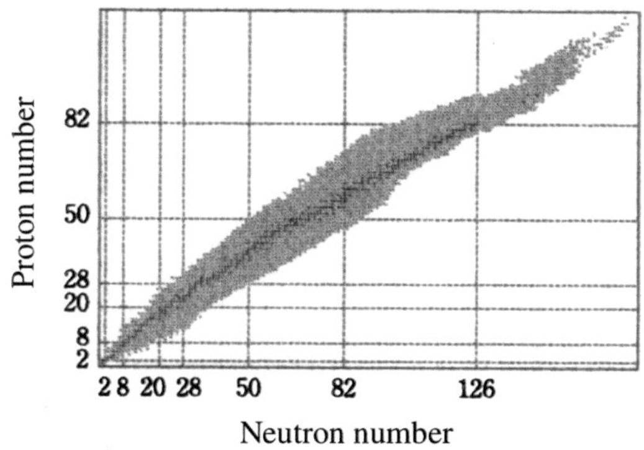

Fig. 1.5 Nuclear chart. Made from the 2010 version of Japan Atomic Energy Agency (JAEA)

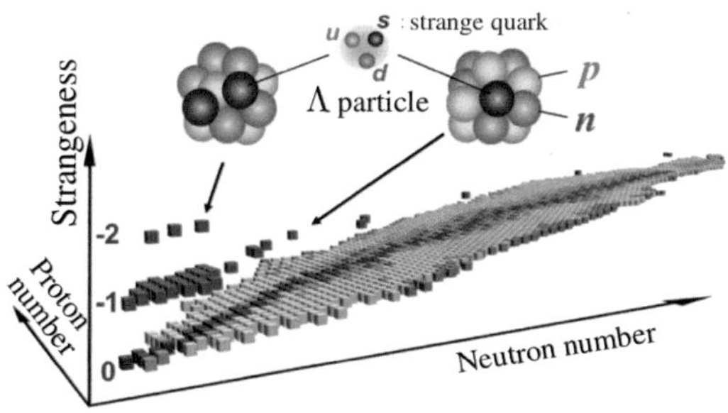

Fig. 1.6 Three-dimensional nuclear chart. Made by the experimental group of nuclear physics in the graduate school of science of Tohoku University

This book deals with nuclei made from the first-generation quarks, i.e., u and d quarks, from the point of view of the quark model. In recent years, however, nuclei containing Λ or Σ or Ξ particles, which include one or two second-generation s quark (s) as constituents, are also under extensive studies. These novel nuclei are called *hypernuclei*. A big progress of their study is expected to be stimulated by the operation of, e.g., J-PARC. Figure 1.5 shows the ordinary nuclear chart, while Fig. 1.6 a multi-layers nuclear chart, where the ordinate represents the strangeness number. The numbers appearing in Fig. 1.5 are the magic numbers, which we learn in Chap. 5.

To end this chapter, we mention the abundance of elements. Figure 1.7 shows the relative abundance of elements[5] in the conventional unit, i.e., by taking the abundance

---

[5]See also [4], where the relative abundance of even–even nuclear species with $A \geq 50$ in the solar system is given.

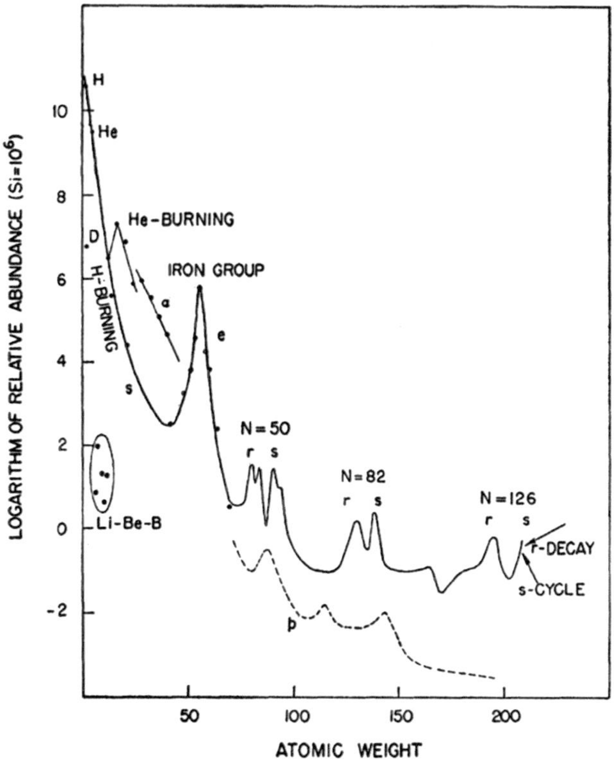

Fig. 1.7 Schematic curve of atomic abundances as a function of atomic weight based on the data of Suess and Urey [2]. Taken from [3]

of Si to be 10^6, i.e., $H(Si) = 10^6$. The hydrogen dominates by far, and the general trend is that the abundance sharply decreases with increasing mass number up to $A \sim 100$, then decreases much more slowly. In addition, it is noticeable that there appears a sharp peak of Fe group and that the abundances of nuclei with particular neutron numbers are large. Also, one notices that each of the latter abundances has twin peaks. These features originate from the stability of nuclei, and the magic numbers, and the way of nucleosynthesis. We will gradually learn them in this book.

1.6 Column: QCD Phase Diagram of Nuclear Matter

Water changes its state from ice (*solid phase*) to water (*liquid phase*), then to vapour (*gas phase*) when the temperature and the pressure vary. Similarly, nuclear matter changes its state or phase with temperature and density.

Figure 1.8 conceptually represents the phase diagram of nuclear matter from the point of view of QCD by taking the temperature and the chemical potential, which corresponds to the baryon density, for the ordinate and the abscissa, respectively. The region marked as nuclear matter is the state of nucleus treated in detail in this book.

As the figure shows, it is conjectured that a phase of quark–gluon plasma is achieved irrespective of the density if the temperature becomes high. Related experimental as well as theoretical studies are extensively performed including the high-energy heavy-ion collisions (Au + Au) using the RHIC (Relativistic Heavy Ion Collider) at the Brookhaven National Laboratory (BNL) in USA, and the Pb + Pb Collision experiments by the Large Hadron Collider (LHC) at the European Organization for Nuclear Research (CERN). The stars and the arrows in the diagram indicate the regions expected to be achieved by collision experiments and the expected paths of time evolution, respectively. There still remain, however, many uncertainties, including the phase diagram itself, and studies from various aspects are under way.

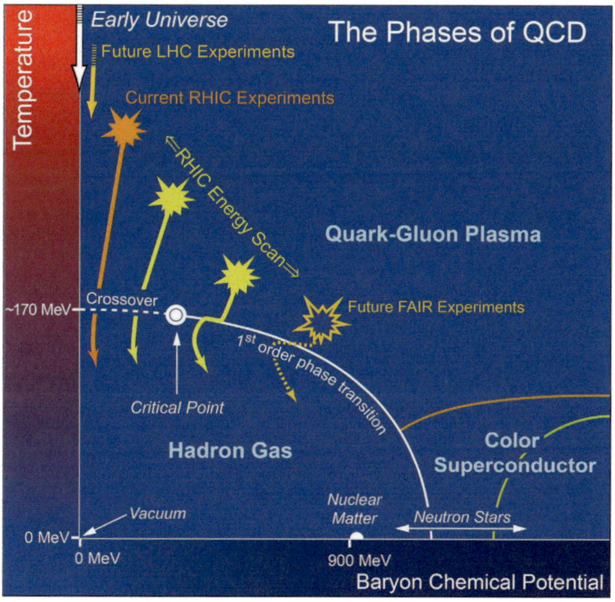

Fig. 1.8 QCD phase diagram. Taken from [5], courtesy of Brookhaven National Laboratory

References

1. W.-M. Yao et al. (Particle Data Group), J. Phys. G **33**, 1 (2006)
2. H.E. Suess, H.C. Urey, Rev. Mod. Phys. **28**, 53 (1956)
3. E.M. Burbidge, G.R. Burbidge, W.A. Fowler, F. Hoyle, Rev. Mod. Phys. **29**, 547 (1957)
4. Aage Bohr, Ben R. Mottelson, *Nuclear Structure*, vol. I (Benjamin, New York, 1969)
5. The Nuclear Science Advisory Committee, The Frontiers of Nuclear Science – A Long Range Plan (2007). http://science.energy.gov/np/nsac/

Chapter 2
Bulk Properties of Nuclei

Abstract Their sizes and masses are the most fundamental properties of nuclei. They have simple mass number dependences which suggest that the nucleus behaves like a liquid and lead to the liquid-drop model for the nucleus. In this chapter we learn these bulk properties of nuclei and their applications to discussing nuclear stability, muon-catalysed fusion and the structure of heavy stars. As an example of the applications we discuss somewhat in detail the basic features of fission and nuclear reactors. We also mention deviations from what are expected from the liquid-drop model which suggest the pairing correlation and shell effects. We also discuss the velocity and the density distributions of nucleons inside a nucleus.

2.1 Nuclear Sizes

2.1.1 Rutherford Scattering

At the beginning of the twentieth century when quantum mechanics was born, various models were proposed for the structure of atoms such as the *plum pudding* or *raisin bread model* of J.J. Thompson, which assumes that the plus charge distributes over whole atom together with electrons, and the *Saturn model* by Hantaro Nagaoka. Rutherford led these debates to conclusion through the study of scattering of alpha particles on atom. He proposed the existence of a central part of the atom, i.e., the core part, which bears all the positive charge that cancels out the total negative charge of electrons and also carries the dominant part of the mass of the atom. Rutherford named this core part nucleus, and gave the limiting value to its size.[1] At that time, Rutherford had been engaged in the detailed studies of the properties of alpha particles emitted from radioactive materials, and knew that the alpha particle is the ionized He. What Rutherford remarked in the experimental results of Marsden is that alpha

[1] It was 1911 when Rutherford submitted his article on the atomic model to a science journal. The idea and the formula of Rutherford were derived by stimulation of experimental results of his coworker Marsden who had been engaged in the study of scattering of alpha particles emitted from natural radioactive elements on matter. Furthermore, they have been confirmed experimentally to be correct by his collaborators Geiger and Marsden.

N. Takigawa and K. Washiyama, *Fundamentals of Nuclear Physics*,
DOI 10.1007/978-4-431-55378-6_2

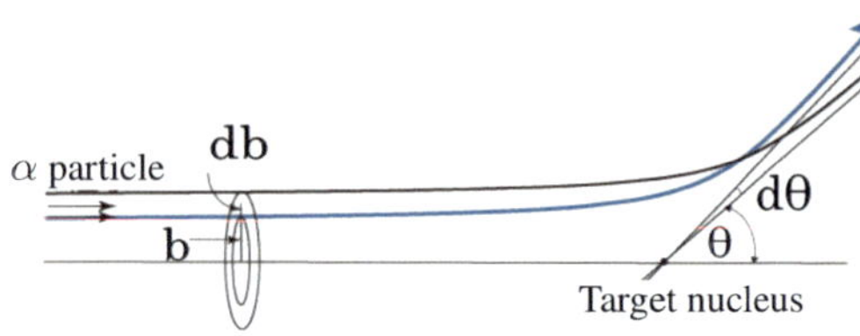

Fig. 2.1 Connection between the classical orbits of the Rutherford scattering and differential cross section

particles passing through matter sometimes make a large angle scattering, although most of them go nearly straight. This experimental feature cannot be explained by the Thompson model which assumes that positive charges distribute over the whole atom.[2]

It is established today that alpha particle is the nucleus of He. Here, we learn how the Rutherford model for atom was born, and how the information on the nuclear size is obtained through the scattering experiments of alpha particles.

Let us now consider the scattering of two structureless charged particles by Coulomb interaction, which is called Coulomb scattering or Rutherford scattering. Here, the two charged particles represent the alpha particle and the nucleus of the target atom. Since the mass of electrons is extremely small, one can ignore the scattering by electrons. The correct expression of the differential cross section for the Coulomb scattering can be obtained by classical mechanics, although, correctly speaking, one should use quantum mechanics. One important thing in this connection is that one to one correspondence holds between the impact parameter b and the scattering angle θ as illustrated in Fig. 2.1. The alpha particles which pass the area $2\pi b\mathrm{d}b$ of the impact parameter between b and $b + \mathrm{d}b$ in Fig. 2.1 are scattered to the region of solid angle $\mathrm{d}\Omega = 2\pi \sin\theta d\theta$ around the scattering angle θ. On the other hand, the differential cross section is defined as the number of particles scattered to the region of solid angle $\mathrm{d}\Omega$ when there exists one incident particle per unit time and unit area. Hence by definition

$$\frac{\mathrm{d}\sigma}{\mathrm{d}\theta} = \frac{2\pi b}{|\mathrm{d}\theta/\mathrm{d}b|} . \tag{2.1}$$

Using the relation

$$b = a \cot\left(\frac{\theta}{2}\right) \tag{2.2}$$

with

$$a = \frac{Z_1 Z_2 e^2}{\mu v^2} \tag{2.3}$$

[2]The history of the progress and discoveries of modern physics in the period from late in the nineteenth century to the beginning of the twentieth century is vividly described in the book by E. Segré [1].

which holds between the impact parameter b and the scattering angle θ in the case of Coulomb scattering, we obtain

$$\frac{d\sigma}{d\Omega} = \frac{d\sigma_R}{d\Omega} \equiv \frac{1}{2\pi \sin\theta} \frac{d\sigma}{d\theta} = \frac{a^2}{4} \frac{1}{\sin^4(\theta/2)}, \tag{2.4}$$

where the index R in σ_R means the Rutherford scattering. In Eq. (2.3), μ is the reduced mass, and v is the speed of the relative motion in the asymptotic region, i.e., at the beginning of scattering. Note that Eq. (2.4) exactly agrees with the formula of the differential cross section $d\sigma_R/d\Omega$ obtained quantum mechanically for the Rutherford scattering. The characteristics of the Coulomb scattering given by Eq. (2.4) are that the forward scattering is strong, but also that backward scattering takes place with a certain probability as well. These match the experimental results of Marsden.

The ground state of the natural radioactive nucleus $^{210}_{84}$Po decays with the half-life of 138.4 days by emitting an alpha particle ($^{210}_{84}\text{Po} \rightarrow {}^{206}_{82}\text{Pb} + \alpha$). The kinetic energy of the alpha particle is about 5.3 MeV corresponding to the Q-value of the decay 5.4 MeV. The differential cross section for the scattering, where this α particle is used to bombard the Au target of atomic number 79, agrees with that for the Rutherford scattering right up to the backward angle $\theta = \pi$. This suggests that the sum of the radii of Au and α ($R(\text{Au}) + R(\alpha)$) is smaller than the distance of closest approach $d(\theta = \pi)$ for the scattering with the impact parameter $b = 0$ leading to the backward scattering $\theta = 180°$ in the case of Rutherford scattering. Since the distance of closest approach d and the scattering angle θ or the impact parameter b is related by

$$d = a\left[1 + \csc\left(\frac{\theta}{2}\right)\right] = a + \sqrt{a^2 + b^2} \tag{2.5}$$

for the Rutherford scattering, the above mentioned experimental results give the upper boundary of the radius as $R(\text{Au}) + R(\alpha) < 4.3 \times 10^{-12}$ cm. This upper boundary is much smaller than the radii of atoms, which are of the order of 10^{-8} cm. Rutherford was thus guided to his atomic model.

Figure 2.2 shows the ratio of the experimental differential cross section for the scattering of α particles of 48.2 MeV by Pb target to that for the corresponding Rutherford scattering $d\sigma_R/d\Omega$ as a function of the scattering angle. The experimental cross section gets rapidly smaller than that of the Rutherford scattering beyond $\theta = 30°$. This can be understood as the consequence of that the distance of closest approach is small for the scattering corresponding to large angle scattering as Eq. (2.5) shows, and hence the overlap between α particle and Pb becomes large, and consequently, those phenomena such as inelastic scattering which are excluded in the Rutherford scattering take place. Figure 2.3 shows classical trajectories obtained by fixing Pb at the origin and by making the incident energy of α particles to 48.2 MeV so as to match with Fig. 2.2. The circle shows the region corresponding to the sum of the radii of Pb and α particle, i.e., about 9.1 fm. The figure confirms that the distance of closest approach for the trajectories corresponding to large angle scattering becomes indeed smaller than the sum of the radii of Pb and α particle.

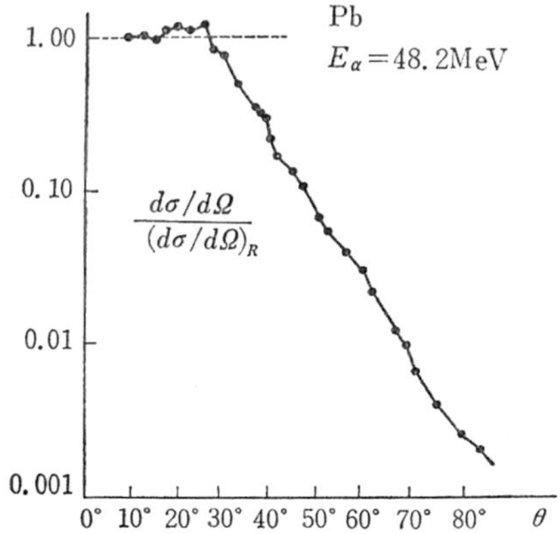

Fig. 2.2 Differential cross section of the elastic scattering of α particles by Pb. Taken from [2]

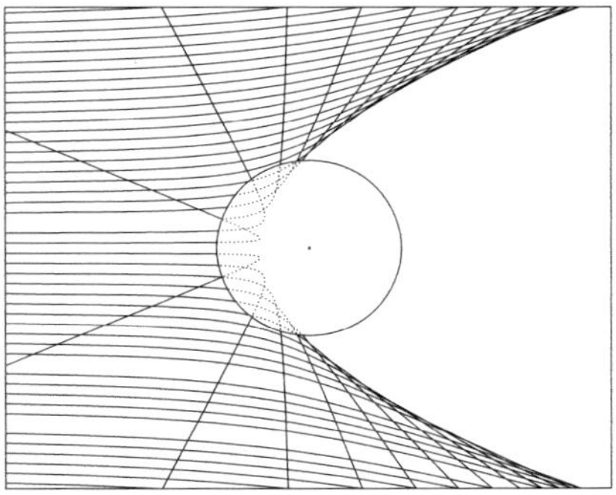

Fig. 2.3 Trajectories of Rutherford scattering

Exercise 2.1 Estimate the sum of the radii of α and Pb from the result of Fig. 2.2 based on the idea mentioned above.

The ratio of the differential cross section shown in Fig. 2.2 resembles the Fresnel diffraction of classical optics. The Fresnel diffraction occurs in the case where the source of light is located in the vicinity of the object that causes diffraction, e.g., an absorber. The scattering of α particles by a nucleus with large atomic number behaves like a Fresnel scattering, because the large Coulomb repulsion strongly bends the trajectory and works to make the source of light effectively locate in the vicinity of the scatterer. It is also because the partial waves corresponding to small impact parameters whose distance of closest approach is small are removed from the elastic scattering due to, e.g., inelastic scattering. Concerning the elastic scattering, this plays effectively the same role as that of an absorber.

This picture holds when the Coulomb interaction dominates the scattering process. The strength of the Coulomb interaction increases in proportion to the product of the charges of the projectile and target nuclei. On the other hand, roughly speaking, the strength of the nuclear interaction increases in proportion to the reduced mass $A_1 A_2/(A_1 + A_2)$. Hence the refraction effect due to the nuclear interaction becomes non-negligible for the scattering by a target nucleus with small atomic number. In fact, a similar differential cross section appears not by diffraction effect, but by refraction effect. The interpretation and analysis then get complicated. Accordingly, one can safely estimate the nuclear size based on the consideration of the Coulomb trajectory together with the strong absorption due to inelastic scattering as described in the present section when the atomic number of the target nucleus is large.

2.1.2 Electron Scattering

The scattering of α particles by a nucleus had led to the Rutherford atomic model, and provided a way to estimate the nuclear size. It thus played an important historical role. However, as stated at the end of the last section, it has a limitation regarding applicability. In contrast, the scattering of electrons by a nucleus, which we learn in this section, is a powerful method to study the nuclear size, more exactly, the distribution of protons inside a nucleus, because only well understood electromagnetic force is involved.[3,4]

2.1.2.1 The de Broglie Wavelength of Electron

In the experiments of electron scattering, electrons are injected on the target nucleus after they are accelerated by, e.g., a linear accelerator. In order to deduce the information on the nuclear size from electron scattering, the de Broglie wavelength of the electron must be of the same order of magnitude as that of the nuclear size or smaller. In this connection, let us first study the relation between the de Broglie wavelength of electron and the kinetic energy of electron supplied by the accelerator.

The de Broglie wavelength of electron λ_e is given by

$$\lambda_e = \frac{h}{p_e} \tag{2.6}$$

in terms of the momentum of the electron p_e and the Planck constant h. On the other hand, it holds that

$$E_{\text{total}} = \sqrt{m_e^2 c^4 + p_e^2 c^2} = m_e c^2 + E_{\text{kin}} = m_e c^2 + E_e \tag{2.7}$$

---

[3]R. Hofstadter was awarded the Nobel Prize in Physics 1961 for the study of high energy electron scattering with linear accelerator and the discovery of the structure of nucleon. He performed also systematic studies of nuclei by electron scattering.

[4]The electron scattering is a powerful method to learn the structure of nucleons mentioned in Chap. 1, and also to study nuclear excitations such as *giant resonances* and hypernuclei as well.

Table 2.1 The de Broglie wavelength of electron for some representative acceleration energies

Acceleration energy E_e (MeV)	100	200	300	1000	4000
de Broglie wavelength λ_e (fm)	12.4	6.2	4.1	1.2	0.31

by using the relation of relativity between the momentum and the energy of an electron, since the total energy of electron is given by adding the energy E_e supplied by the accelerator, i.e., the kinetic energy E_{kin}, to the rest energy. Hence,

$$p_e^2 = 2m_e E_e + E_e^2/c^2 . \tag{2.8}$$

By inserting Eq. (2.8) into (2.6), we obtain

$$\frac{\lambda_e}{2\pi} = \frac{\hbar c}{E_e(1 + 2m_e c^2/E_e)^{1/2}} \approx \frac{\hbar c}{E_e} \approx \frac{200}{E_e/\mathrm{MeV}}\ \mathrm{fm} . \tag{2.9}$$

We ignored the rest energy of electron $m_e c^2$ in the third and fourth terms of Eq. (2.9) by assuming that it is much smaller than the acceleration energy E_e. We have given not the wavelength itself, but the quantity which is obtained by dividing the wavelength by 2π to facilitate the estimate of the order of magnitude.

Table 2.1 gives the de Broglie wavelength of electron estimated by Eq. (2.9) for some representative acceleration energies.

2.1.2.2 Form Factor

As Table 2.1 shows, one has to inject electrons on a nucleus after having accelerated them to much higher energy than the rest energy of electron $m_e c^2 \approx 0.51$ MeV in order to learn the nuclear size, which is of the order of fm. Hence one needs to use the Dirac equation, to which a relativistic Fermi particle obeys, in order to theoretically derive the proper expression of the differential cross section [3].

Exercise 2.2 Evaluate the ratio v/c of the velocity of the electron v to the speed of light in the vacuum c when the acceleration energy of the electron E_e is 100 MeV.

However, here let us simplify the problem in the following way, and learn how the information on nuclei can be obtained from the analyses of the electron scattering:

1. Treat the scattering of electrons by the electromagnetic field made by a nucleus, instead of considering the scattering of electrons by the nucleons inside the nucleus.
2. Consider only the Coulomb force (electric force) and ignore the magnetic force.
3. Use non-relativistic Schrödinger equation.

We express the Coulomb potential as $V(\mathbf{r})$. The scattering amplitude to the angle θ is then given by

$$f^{(1)}(\theta) = -\frac{1}{4\pi}\frac{2\mu}{\hbar^2}\int e^{-i\mathbf{q}\cdot\mathbf{r}}V(\mathbf{r})d\mathbf{r} \tag{2.10}$$

following the first order Born approximation, which is valid because of the high energy scattering, and also because the involved electric force is weak compared to the kinetic energy. The μ is the reduced mass. It can be identified with the mass of electron m_e to a high degree of accuracy. The $\mathbf{q}$ is the momentum transfer divided by $\hbar$ and is given by

$$\mathbf{q} = \mathbf{k}_f - \mathbf{k}_i\,, \tag{2.11}$$

$$q = |\mathbf{q}| = 2k\sin(\theta/2)\,. \tag{2.12}$$

where $\mathbf{k}_i$ is the wave-vector of the incident electron, $\mathbf{k}_f$ is the wave-vector of the electron scattered to the direction of angle θ and k is the wave number of the electron corresponding to the incident energy.

Exercise 2.3 Derive Eqs. (2.10)–(2.12).

We remark that the potential $V(\mathbf{r})$ obeys the following Poisson equation,

$$\Delta V = 4\pi Ze^2\rho_C(\mathbf{r})\,, \tag{2.13}$$

in order to relate the scattering amplitude to the distribution of protons inside a nucleus. The Z is the atomic number of the nucleus to be studied, and $\rho_C(\mathbf{r})$ is the charge density at the position $\mathbf{r}$ measured from the center of the nucleus.[5] It is normalized as

$$\int \rho_C(\mathbf{r})d\mathbf{r} = 1\,. \tag{2.14}$$

By repeating the integration by parts twice in Eq. (2.10), and by using Eq. (2.13), we obtain,

$$f^{(1)}(\theta) = \frac{Ze^2}{2\mu v^2}\frac{1}{\sin^2(\theta/2)}F(\mathbf{q})\,, \tag{2.15}$$

where $F(\mathbf{q})$ is defined as

$$F(\mathbf{q}) \equiv \int e^{-i\mathbf{q}\cdot\mathbf{r}}\rho_C(\mathbf{r})d\mathbf{r}\,. \tag{2.16}$$

The differential cross section is therefore given by

$$\frac{d\sigma^{(1)}}{d\Omega} = |f^{(1)}(\theta)|^2 = \frac{d\sigma_R}{d\Omega}|F(\mathbf{q})|^2\,. \tag{2.17}$$

[5]The distribution of protons ρ_p can be derived from ρ_C by taking into account the intrinsic structure of proton.

Correctly, the expression of the differential cross section is obtained as

$$\frac{d\sigma^{(1)}}{d\Omega} = \frac{d\sigma_M}{d\Omega}|F(\mathbf{q})|^2, \tag{2.18}$$

$$\frac{d\sigma_M}{d\Omega} = \left[\frac{Ze^2}{2E_e \sin^2(\theta/2)}\right]^2 \left[1 - \frac{v^2}{c^2}\sin^2(\theta/2)\right]$$
$$\approx \left[\frac{Ze^2 \cos(\theta/2)}{2E_e \sin^2(\theta/2)}\right]^2, \tag{2.19}$$

by replacing the differential cross section of the Rutherford scattering $d\sigma_R/d\Omega$ by that for the Mott scattering $d\sigma_M/d\Omega$ which takes into account relativistic effects for electrons. The $d\sigma_M/d\Omega$ gives the differential cross section of the scattering of electrons by the Coulomb force made by a point charge. Equation (2.18) shows that the information on the density distribution of protons inside a nucleus can be obtained through the ratio of the experimental differential cross section to that for the Mott scattering. The $F(\mathbf{q})$ defined by Eq. (2.16) is the factor which represents the effects of the finiteness of the nuclear size and is called the *form factor*.

Especially, if the nucleus is spherical, i.e., if the charge distribution is spherical, the form factor is given by

$$F(q) = 4\pi \int_0^\infty \rho_C(r) j_0(qr) r^2 dr = 4\pi \int_0^\infty \rho_C(r) \frac{\sin(qr)}{qr} r^2 dr, \tag{2.20}$$

where $j_0(x)$ is the spherical Bessel function of the first kind.

Exercise 2.4 Prove Eq. (2.20) in the following two methods:

1. Perform directly the integration over the angular part of $\mathbf{r}$ ($d\Omega_r$) in Eq. (2.16).
2. Expand at first the $e^{-i\mathbf{q}\cdot\mathbf{r}}$ in terms of Legendre functions, and then use the orthonormal property of the spherical harmonics.

2.1.2.3 Density Distribution

(1) Estimate of the Nuclear Radius from Diffraction Pattern As Table 2.1 implies, the experiments of electron scattering are performed with high energy in order to study nuclear size and the density distribution of protons inside a nucleus. The corresponding differential cross section is expected to show a diffraction pattern similar to that of the Fraunhofer diffraction in optics. Figure 2.4 shows the differential cross section for the elastic scattering of electrons from Au target at 153 and 183 MeV. The monotonically decreasing solid line is the Mott scattering cross section for 183 MeV. The figure shows that the observed cross section is smaller than that for the Mott scattering, and that it has indeed the diffraction pattern, i.e., oscillation, of the Fraunhofer type.

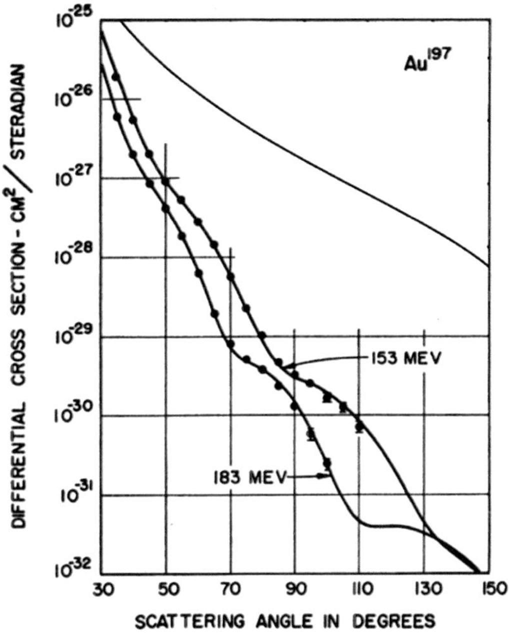

Fig. 2.4 The elastic electron-scattering differential cross section from Au at energies of 153 and 183 MeV. Taken from [4]

Let us assume that the charge distributes uniformly inside a nucleus with a spherical shape of radius R in order to see how the nuclear size is estimated from the diffraction pattern. The resulting form factor reads

$$F(q) = \frac{3}{qR} j_1(qR) = \frac{3}{qR}(qR)^{-2}[\sin(qR) - (qR)\cos(qR)] . \tag{2.21}$$

We then find that the zeros of $j_1(x)/x$ correspond to the angles where the differential cross section shows local minima. We denote the magnitude of the transferred momenta corresponding to those scattering angles $\theta_1, \theta_2, \ldots$ by $q_1, q_2, \ldots$. The first zero of $j_1(x)/x$ is $x_1 \approx 4.49$, and the interval between the successive zeros Δx is about π afterwards. One can therefore estimate the radius either by $R \sim x_1/q_1$ or by $R \sim \pi/(q_2 - q_1)$ using the $q_1, q_2, \ldots$ evaluated from $\theta_1, \theta_2, \ldots$, which are extracted from the experimental data.

Exercise 2.5 Estimate the radius of the nucleus of Au from the experimental data shown in Fig. 2.4.[6]

[6]The dips in diffraction pattern are buried by the distortion effects due to Coulomb force in the case of target nuclei with a large mass number such as Au. The diffraction pattern appears more clearly for the target nuclei with small mass number.

(2) The Woods–Saxon Representation of the Density Distribution Accurate information on the density distribution[7] of nucleons inside a nucleus can be obtained by making the Fourier transformation of the form factor given by the experiments of electron scattering. However, one usually takes the method to first postulate a plausible functional form, then determine the parameters therein to reproduce the experimental data. In that case, a widely adopted choice is to assume the following functional form called the *Woods–Saxon type*,[8]

$$\rho(r) = \frac{\rho_0}{1 + \mathrm{e}^{(r-R)/a}} , \tag{2.22}$$

where R is the parameter to represent the radius. a is the parameter to represent the thickness of the surface area and is called the surface diffuseness parameter. The density falls from 90 to 10% of the central density over the region of thickness of 4.4 times a around R. The ρ_0 is the central density, and is given as a function of R and a through the normalization condition,

$$\int \rho d\mathbf{r} \sim \rho_0 \frac{4\pi}{3} R^3 \left[1 + \pi^2 \left(\frac{a}{R} \right)^2 \right] = A . \tag{2.23}$$

The two solid curves in Fig. 2.4 which reproduce fairly well the experimental differential cross section have been calculated with the best fit parameters for the data at 183 MeV by assuming the Woods–Saxon representation for the charge distribution.

The following values are obtained from the analyses of experimental data for a large number of stable target nuclei,

$$R \sim (1.1\text{–}1.2) A^{1/3}\,\text{fm} , \quad a \sim 0.6\,\text{fm} , \quad \rho_0 \sim (0.14\text{–}0.17)\,\text{fm}^{-3} , \tag{2.24}$$

as parameters in the Woods–Saxon parametrization.[9] The fact that the radius is proportional to the $1/3$ power of the mass number, i.e., the number of nucleons composing the nucleus, and that the density is independent of the mass number

[7]In stable nuclei, the protons and neutrons distribute inside the nucleus almost in the same way. Here, we therefore treat the density distribution of protons and of nucleons as the same except for the absolute value. In these days, extensive studies are performed on the nuclei far from the β-stability line, which are called unstable nuclei. It is then getting known that some of them have very different distributions for protons and for neutrons. For example, the region where there exists only neutrons largely extends over the surface region in some nuclei such as ^{11}Li in the vicinity of the neutron drip line. Such layer is called the *neutron halo*. Recently, it is reported from the inelastic scattering of polarized protons (see [5]) and also from the elastic scattering of polarized electrons that even a typical stable nucleus ^{208}Pb has a larger radius of the neutron distribution than that of the proton distribution by 0.15–0.33 fm. Relatedly, the study of the existence of the region consisting of only neutrons, which is called *neutron skin*, is an active area of research.

[8]There exist deviations from the Woods–Saxon type for individual nucleus. They are explained by shell model.

[9]The value $\rho_0 \sim 0.17\,\text{fm}^{-3}$ is widely accepted as the density of nuclear matter (see [6] for the argument about the detailed mass number dependence of the central density of nuclei with large mass number).

suggest that the nucleus has a property as a *liquid*.[10] The fact that the density is independent of the mass number is called the *saturation property of nuclear density*.

Incidentally, the surface area dominates in light nuclei. The *Gaussian type* is therefore often used, instead of the Woods–Saxon type, as a more realistic functional form.

2.1.2.4 The Root-Mean-Square Radius

The radius of a nucleus is the average of the radii of the region of spatial occupation of the nucleons composing the nucleus. Hence the concept of root-mean-square radius is often used in discussing nuclear size. We first define the mean-square radius by

$$\langle r^2 \rangle \equiv \int_0^\infty r^2 \rho(r) 4\pi r^2 \mathrm{d}r \Big/ \int_0^\infty \rho(r) 4\pi r^2 \mathrm{d}r \,. \tag{2.25}$$

The root-mean-square radius $R_{\mathrm{r.m.s.}}$ is then defined by taking the square root as $R_{\mathrm{r.m.s.}} \equiv \sqrt{\langle r^2 \rangle}$.

We assumed a sharp density distribution without a surface diffuseness, i.e., a step function, in a simple analysis of the diffraction pattern of the electron scattering, e.g., as shown in Eq. (2.21). We name the resulting radius the *equivalent radius* or the *effective sharp radius* and designate it as R_{eq}. By definition, $R_{\mathrm{r.m.s.}}$ and R_{eq} are related to each other as

$$R_{\mathrm{eq}} = \sqrt{\frac{5}{3}\langle r^2 \rangle} = \sqrt{\frac{5}{3}} R_{\mathrm{r.m.s.}} \,. \tag{2.26}$$

Since R_{eq} is proportional to the $1/3$ power of the mass number A, let us represent it as

$$R_{\mathrm{eq}} = r_0 A^{1/3} \,. \tag{2.27}$$

The proportional coefficient r_0, called *radius parameter*, is then independent of individual nucleus, and takes experimentally the value[11] of

$$r_0 \sim 1.1\text{–}1.2\ \mathrm{fm} \,. \tag{2.28}$$

---

[10] Although the nucleus appears like a drop of liquid, it is not a classical liquid, but a drop of *quantum liquid*, since commutation relations govern the nucleus. In the classical liquid, the momentum space is not deformed even if it is spatially deformed. On the other hand, in the case of nuclei, a spatial deformation leads to the deformation in the momentum space through the uncertainty relation (see the footnote concerning quantum liquid in Sect. 8.3.4).

[11] Equations (2.24), (2.27) and (2.28) hold for stable nuclei. It has been found that the radii of unstable nuclei, especially of nuclei in the vicinity of neutron drip line, significantly deviate from these equations. For example, in the case of Li isotopes, the radius of ^{11}Li which has a neutron halo as mentioned before and thus called a halo nucleus or a neutron halo nucleus is much larger than what we expect from Eq. (2.27), although the radii of stable isotopes ^{6,7}Li and of the isotopes ^{8,9}Li which lie close to the stability line vary with A almost following Eq. (2.27).

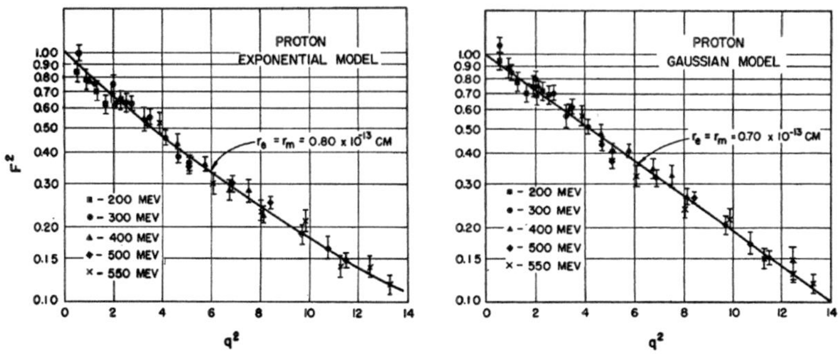

Fig. 2.5 The form factor of the electron scattering by proton. Taken from [7]

If we assume the density distribution of the Woods–Saxon type given by Eq. (2.22), Eq. (2.26) is replaced by

$$\langle r^n \rangle \equiv \frac{\int r^n \rho(r) \mathrm{d}\mathbf{r}}{\int \rho(r) \mathrm{d}\mathbf{r}} = \frac{3}{n+3} R^n \left[1 + \frac{n(n+5)}{6} \pi^2 \left(\frac{a}{R} \right)^2 + \cdots \right]. \tag{2.29}$$

From Eqs. (2.16) and (2.25), we obtain

$$\begin{aligned} F(q) &= 1 - \frac{1}{3!} \langle r^2 \rangle q^2 + \cdots , \\ |F(q)|^2 &= 1 - \frac{1}{3} \langle r^2 \rangle q^2 + \cdots , \end{aligned} \tag{2.30}$$

for small values of the momentum transfer. These formulae show that the nuclear size can be estimated by studying the form factor for the forward scattering or the behaviour of the form factor in the region of small momentum transfer. Figure 2.5 shows a similar analysis for the radius of proton. It is derived from the analysis of this experimental data that the root-mean-square radius of proton $R_{\mathrm{r.m.s.}}$ is 0.7–0.8 fm.[12,13]

---

[12]The measurement with high accuracy in the extreme forward region, where q^2 is extremely small, is a difficult task and is a challenging subject in these days. In practice, one estimates the radius by analysing the data in the region where the measurement with high accuracy is doable, by assuming, e.g., the exponential or the Gaussian density distribution as shown in Fig. 2.5. Also, it is known that it is necessary to take into account the magnetic effects for the case of nucleons.

[13]As Fig. 2.5 shows, the experimental data for the region of small momentum transfer can be reproduced equally well by assuming the charge distribution in a proton to be either exponential or Gaussian type. Historically, the charge distribution of exponential type has been considered to be a good approximation, and correspondingly, the form factor of dipole type has been used. However, it is getting known from recent studies that the experimental data significantly deviate from the form factor of the dipole type, and it is argued that the charge distribution inside a proton is closer to a Gaussian type.

Exercise 2.6 The radius of the charge distribution of a nucleus can be estimated also from the energy spectrum of the X-ray of the corresponding muonic-atom, or the energy difference between the corresponding energy levels of mirror nuclei.[14] Study the principle of each method.

2.1.3 *Mass Distribution*

All the methods mentioned above are the method to study the distribution of protons, in other words, the *charge distribution* in nuclei, and give no information on the distribution of neutrons. One has to use other methods in order to study the distribution of nucleons including neutrons, which is often called the *mass distribution*, inside a nucleus. The scattering of protons or neutrons by a nucleus, where the strong interaction is involved, and nucleus-nucleus collisions are the examples. Similarly to the case of electron scattering, high incident energies are appropriate. Here we consider two methods.

2.1.3.1 High-Energy Neutron Scattering: Fraunhofer Scattering

Let us first consider the high-energy neutron scattering. Since the interaction between the neutron and the target nucleus is strong, neutrons are absorbed or the target nucleus is excited when the neutrons reach the area of the target nucleus. The associated cross section, i.e., the sum of the cross sections of inelastic scattering and of absorption, is intuitively expected to be given by $\sigma^{(\mathrm{inel})} \sim \pi R^2$ by using the nuclear radius R. On the other hand, in the scattering of relatively high energy by a short range interaction, the elastic scattering corresponds to the shadow scattering in quantum mechanics and the magnitude of its cross section is also expected to be the same, i.e., $\sigma^{(\mathrm{el})} \sim \pi R^2$. In consequence, the total cross section is given by $\sigma^{(\mathrm{total})} \sim 2\pi R^2$. One can therefore obtain the information on nuclear size by measuring the cross section of high-energy neutron scattering.

In order to make the arguments more precise, let us consider the partial wave analysis of neutron scattering by a nucleus, and postulate that the S-matrix of the elastic scattering of each partial wave ℓ is given by

$$S^{(\mathrm{el})} = \begin{cases} 0 & \text{for } \ell \le kR \\ 1 & \text{for } \ell > kR\,, \end{cases} \tag{2.31}$$

[14] The pair nuclei such as ${}^{15}_{7}\mathrm{N}_8$ and ${}^{15}_{8}\mathrm{O}_7$ which have interchanged numbers of protons and neutrons are called *mirror nuclei* to each other. Their energy levels are nearly equal, but shifted.

by emphasizing the above mentioned absorption effects and the characteristic of high-energy scattering that the phase shift is small. By inserting the extreme postulate Eq. (2.31) called the sharp-cut-off-model into the partial wave expansion formulae, Eqs. (A.6), (A.8) and (A.10), for the total cross sections of elastic scattering $\sigma^{(\text{el})}$ and of inelastic scattering $\sigma^{(\text{inel})}$ and for the total cross section $\sigma^{(\text{total})} = \sigma^{(\text{el})} + \sigma^{(\text{inel})}$, or into their integral representations transformed to the form of Poisson sum formula (see Eq. (A.23)), we can confirm that $\sigma^{(\text{el})} \sim \pi R^2$, $\sigma^{(\text{inel})} \sim \pi R^2$, $\sigma^{(\text{total})} \sim 2\pi R^2$.

Experimentally, the total scattering cross section $\sigma^{(\text{total})}$ can be determined relatively easily by the *attenuation method*. In this method, the information on the cross section is obtained by comparing the counts, which are observed with detectors placed in the forward direction, in the measurements with and without scatterer. If we represent the former and the latter as I and I_0, respectively, then it holds that

$$\frac{I}{I_0} = \exp\left(-\sigma^{(\text{total})} n t\right), \tag{2.32}$$

where n is the number density of the target nucleus inside the scatterer, and t the thickness of the scatterer.

Incidentally, the scattering where the S-matrix is given by Eq. (2.31) corresponds to the Fraunhofer diffraction in classical optics. A semi-classical scattering theory leads to the following concise analytic expression for the associated differential cross section of the elastic scattering,

$$\frac{d\sigma}{d\Omega} = R^2 \frac{1}{\theta \sin\theta} |J_1(kR\theta)|^2 . \tag{2.33}$$

The property of the Bessel function $J_1(x)$ suggests that the elastic scattering has large intensities at forward angles. This is nothing but the shadow scattering mentioned at the beginning of this section.

Another important thing is that an oscillatory diffraction pattern is observed. The interval between two successive angles, where the cross section takes local maxima, is given by $\Delta\theta_D \sim \pi/kR$ from the property of the Bessel function. Hence one can deduce a value for the nuclear size from the angular spacing of the diffraction pattern. The validity of the Fraunhofer diffraction method described in this section can be assessed by comparing thus obtained nuclear radius with the nuclear radius obtained from the measurement of the total cross section.

One can also use a high-energy proton–nucleus scattering to study the mass distribution. However, the simple model given by Eq. (2.31) cannot be applied in that case because of the long range Coulomb interaction. In fact, the cross section diverges in the Coulomb scattering. A special consideration is therefore required to handle the scattering by Coulomb interaction.

2.1.3.2 Analysis of High-Energy Nucleon–Nucleus and Nucleus–Nucleus Reaction Cross Sections by Glauber Theory

The high-energy potential scattering can be well described by the *eikonal approximation*, where the scattering phase shift is determined based on the assumption that the reaction takes place along a straight line trajectory. If one applies this idea to the nucleon–nucleus and nucleus–nucleus scatterings, and takes the folding model, where the potential in the eikonal approximation is approximated by folding interactions between either incident nucleon or incident nucleus and constituent nucleons in the target nucleus, then one obtains the following expressions

$$\sigma_R = 2\pi \int_0^\infty b\mathrm{d}b\,[1 - T(b)]\,, \tag{2.34}$$

$$T(b) = \exp(-\bar{\sigma}_{NN} O_v(b))\,, \tag{2.35}$$

$$O_v(b) = \int_{-\infty}^{\infty} \mathrm{d}z \int \mathrm{d}\mathbf{r}\rho_T(\mathbf{r})\rho_P(\mathbf{r} - \mathbf{R})\,, \tag{2.36}$$

which give the total reaction cross section σ_R in terms of the total reaction cross section of the nucleon–nucleon scattering (see Appendix A.4.2). The $\bar{\sigma}_{NN}$ in the *transparency function* $T(b)$ is the average of the total cross section in the proton–neutron and proton–proton scatterings. If one denotes the collision energy in the nucleon–nucleon scattering by $E_{\mathrm{lab}}^{(NN)}$, that in the nucleus–nucleus scattering by $E_{\mathrm{lab}}^{(nn)}$, and the mass number of the projectile nucleus by A_P, then one should use the energy given by the relationship

$$E_{\mathrm{lab}}^{(NN)} = E_{\mathrm{lab}}^{(nn)}/A_P\,, \tag{2.37}$$

(the reasoning is given in Appendix A.4.2). The $O_v(b)$ is called the *overlap function*. It is given by integrating the overlap of the densities of the incident nucleus (or incident nucleon) and of the target nucleus for the impact parameter b over the collision direction (named z-direction) which is perpendicular to b. The $\mathbf{R} = (\mathbf{b}, z)$ is the relative coordinate between the projectile and target nuclei. Since the overlap integral clearly depends on the nuclear size and the distribution of nucleons inside nuclei, one can extract information on those quantities from the analysis of the reaction cross section based on Eq. (2.34).

Equations (2.34)–(2.37) are called the *optical limit formulae* of the Glauber theory, whose derivation is given in Appendix A.4.2. It is necessary to extend them either by taking into account higher order terms in the so-called cumulant expansion method or by using the *diffractive eikonal approximation* which goes beyond the optical limit approximation in order to correctly estimate the size and the reaction cross section of the nuclei in the vicinity of the drip line, whose studies are recently extensively going on, by analyzing their scattering data in terms of the Glauber theory (see, e.g., [8–10]).

2.2 Number Density and Fermi Momentum of Nucleons

2.2.1 *Number Density of Nucleons*

Let us denote the number density of nucleons inside a nucleus by ρ. Since ρ is defined as the mass number A divided by the volume V, we have

$$\rho \equiv \frac{A}{V} = \frac{3}{4\pi r_0^3} \tag{2.38}$$

from Eq. (2.27) for spherical nuclei. As Eq. (2.28) shows, the radius parameter r_0 is almost constant independently of the mass number. Equation (2.38) therefore shows that the number density of nucleons inside a nucleus is almost constant independently of individual nucleus. This property is called the *density saturation*, and is one of the experimental evidences of the liquid-like behaviour of nuclei leading to the *liquid-drop model of nucleus*.

2.2.2 *Fermi Momentum: Fermi-Gas Model, Thomas–Fermi Approximation*

One of the simple pictures for the nucleus as a many-body system of nucleons, which are fermions, is the so-called *Fermi-gas model* or *Thomas–Fermi approximation*. In that method, each nucleon is considered to be moving independently from each other in a mean field of volume V. However, the *Pauli exclusion principle* that each quantum state can be occupied by at most only one fermion is respected. Consequently, the nucleons become to distribute from the state of minimum momentum to that of the highest momentum called the Fermi momentum. Since there exists one quantum state for each volume $(2\pi\hbar)^3$ in the phase space, it holds

$$A = \frac{V}{(2\pi\hbar)^3} \int_0^{\hbar k_\mathrm{F}} p^2 \mathrm{d}p \mathrm{d}\Omega_p \times 2 \times 2 \tag{2.39}$$

if we represent the Fermi momentum as $p_\mathrm{F} = \hbar k_\mathrm{F}$. The last factor 2×2 is the statistical weight in the spin and isospin spaces. Hence we obtain

$$k_\mathrm{F} = \left(\frac{3\pi^2}{2}\right)^{1/3} \rho^{1/3} = \left(\frac{9\pi}{8}\right)^{1/3} \frac{1}{r_0} \tag{2.40}$$

by using Eq. (2.27). Similarly to the density, the Fermi momentum also does not depend on the mass number. The Fermi wave number is proportional to the $1/3$ power of the density as one can expect from the consideration of their dimensions.

Table 2.2 Nucleon number density, average nucleon distance, Fermi wave number, velocity, Fermi energy

r_0 (fm)	$\rho(\text{fm}^{-3})$	a (fm)	k_F (fm^{-1})	v_F/c	E_F (MeV)
1.10	0.179	1.77	1.38	0.29	39.5
1.15	0.157	1.85	1.32	0.28	36.1
1.20	0.138	1.94	1.27	0.27	33.1

Table 2.2 shows the number density of nucleons, average distance between nucleons a, Fermi wave number k_F, the speed of nucleon at the Fermi surface in units of the speed of light in vacuum v_F/c and the Fermi energy E_F for three representative values of r_0. The table shows that the speed of nucleons in the nucleus is at most 30% of the speed of light in vacuum. It is the reason why we adopt the non-relativistic theory to describe nuclei in most part of this book.[15]

Exercise 2.7 Show that the mass density of nuclei ρ_mass is about 2.9×10^{14} g/cm^3 by assuming that $\rho = 0.17\,\text{fm}^{-3}$.

Sidebar 1: Why don't stars collapse? It looks puzzling that stars do not collapse despite the fact that the attractive gravitational forces are acting between the constituent particles. This is because the collapse is prevented by the *degeneracy pressure* of electrons and of other constituent particles in the star. As stated in the main text, all the fermions cannot occupy the smallest momentum state due to the Pauli exclusion principle. Instead, they distribute up to the Fermi surface with high momentum. The resulting pressure of the quantum mechanical origin, i.e., due to the Pauli exclusion principle, is the degeneracy pressure. The Fermi momentum increases and hence the degeneracy pressure increases if the volume of a star decreases by starting collapse. It is why stars do not collapse so easily. On the contrary, if the density of fermions such as electrons inside a star decreases for some reason, then the star starts to collapse because of the break-down of the balance between the attraction by the gravitational force and the degeneracy pressure. For example, a supernova explosion is initiated by the electron capture as we will describe in Sect. 2.3.

Exercise 2.8 Show that the degeneracy pressure of electrons is given by

$$P = \left(\frac{3}{\pi}\right)^{2/3} \frac{1}{20m_\text{e}} h^2 \rho^{5/3} \tag{2.41}$$

with the density of electrons ρ and the mass of electron m_e by using the fact that the pressure P is given by $P = -\partial E/\partial V$ when the temperature is zero, and by identifying the energy E with the total kinetic energy of electrons. Note that particles with

[15] One can take the opposite point of view that 30% is a large number. The relativistic descriptions of nuclear structure and nuclear reactions are also actively going on based on that point of view. The relativistic treatment has the advantage that the spin of nucleons is naturally introduced. In this book, we only touch the general framework of the relativistic mean-field theory in Sect. 6.3.

smaller mass work more effectively, i.e., give larger contribution, to the degeneracy pressure if the number densities of different particles are the same as Eq. (2.41) shows.

Sidebar 2: Why don't atoms collapse? A similar question arises for the size of atoms. Let us consider the neutral hydrogen atom as the example. The attractive Coulomb force acts between the proton and the electron in the hydrogen atom. The corresponding potential energy is given by $V(r) = V_C(r) = -e^2/r$ when they are separated by the distance r. It therefore looks energetically most stable at $r = 0$, i.e., when the proton and the electron stick to each other. Nonetheless, the size of the Hydrogen atom is known to be the Bohr radius given by $a_B = 0.5 \times 10^{-8}$ cm, which is larger than the radius of proton $R_p \sim 0.8 \times 10^{-13}$ cm by 5 orders of magnitude. This is also a consequence of quantum mechanics. The world of quantum mechanics, i.e., the microscopic world, is governed by the uncertainty principle $\Delta x_\alpha \Delta p_\alpha \gtrsim \hbar$. On the other hand, the Hamiltonian describing the motion of the electron around the proton is given by $H = \frac{p^2}{2\mu} + V_C(r)$. Therefore, the uncertainty relation leads to approximately $E \gtrsim \frac{\hbar^2}{2\mu}(\frac{1}{r})^2 - \frac{e^2}{r}$ for the corresponding energy. The kinetic energy increases when r decreases and cancels the gain of the Coulomb energy, and even leads to a positive total energy. The radius of the Hydrogen atom, i.e., the Bohr radius a_B, is obtained as the value of r which minimizes the energy E.

Sidebar 3: Atoms should have collapsed if the uncertainty principle or the commutation relation were absent Since the uncertainty principle is derived from the commutation relation $[\hat{x}_\alpha, \hat{p}_\beta] = i\hbar\delta_{\alpha\beta}$, where $\hat{O}$ is the operator corresponding to an observable O, atoms would become point particles if the commutation relation did not govern the nature.

Sidebar 4: Application to Screening Effect As we learnt in Sect. 1.4 concerning the nuclear reactions in the Sun, many nuclear reactions in stars occur by quantum tunneling through the potential barrier called the Coulomb barrier made by the long-range Coulomb interaction. An interesting phenomenon in this connection is the screening effect caused by the surrounding charged particles which are in the plasma state in hot stars. It leads to an enhancement of the tunneling effect. Recently, the screening effect is argued also in connection with laboratory experiments, where low-energy nuclear reactions are caused in matter, e.g., in host metals [11]. It is reported that the reaction rate is enhanced and the screening effect is considered to be one of the origins of the enhancement. The interest is whether the enhancement of the reaction rate by the screening effect due to the presence of the host material is large and can explain the observed enhancement [12].

Let us denote the effective Coulomb potential, i.e., the screened Coulomb potential, around a nucleus with electric charge Ze by $\varphi(r)$. If we represent the change of the density of a charged particle ζ with the electric charge $Z_\zeta e$ in the surrounding induced by this nucleus as $\delta\rho_\zeta(\mathbf{r})$, then $\varphi(\mathbf{r})$ obeys the Poisson equation

$$\Delta\varphi(\mathbf{r}) = -4\pi \left[Ze\delta(\mathbf{r}) + \sum_{\zeta} Z_{\zeta} e \delta\rho_{\zeta}(\mathbf{r}) \right] . \tag{2.42}$$

The nuclear reactions in matter at laboratories are performed at the room temperature. Hence, we can use the Thomas–Fermi approximation for the density of electrons of the host material, and obtain

$$\varphi(\mathbf{r}) = \frac{Ze}{r} \exp(-q_{\mathrm{TF}} r) \tag{2.43}$$

for the solution of Eq. (2.42) within the framework of the linear approximation, where the inverse of the screening length q_{TF} is given by

$$q_{\mathrm{TF}} = \sqrt{\frac{6\pi\rho_0}{\varepsilon_{\mathrm{F}}^{(0)}} e^2} . \tag{2.44}$$

The ρ_0 and $\varepsilon_{\mathrm{F}}^{(0)}$ are the density of free electrons in the metal and the Fermi energy, respectively, before they are polarized.

On the other hand, for nuclear reactions in hot stars, the densities of the charged particles composing the plasma are well represented by the Boltzmann distribution. Consequently, the screened potential is given by the following well-known *Debye–Hückel formula*,

$$\varphi(\mathbf{r}) = \frac{Ze}{r} \exp(-\kappa_{\mathrm{DH}} r) , \tag{2.45}$$

$$\kappa_{\mathrm{DH}} = \sqrt{4\pi\beta e^2 \sum_{\zeta} Z_{\zeta}^2 \rho_{\zeta}^{(0)}} , \tag{2.46}$$

where $\beta = 1/kT$, where k and T are the Boltzmann constant and the temperature of the star, respectively.

Exercise 2.9 Derive Eqs. (2.44) and (2.46).

2.3 Nuclear Masses

We denote the mass of the nucleus with the mass number A and the atomic number Z by $M(A, Z)$. The nuclear mass is one of the basic quantities of each nucleus. At the same time, it is a fundamental quantity which governs various nuclear reactions.

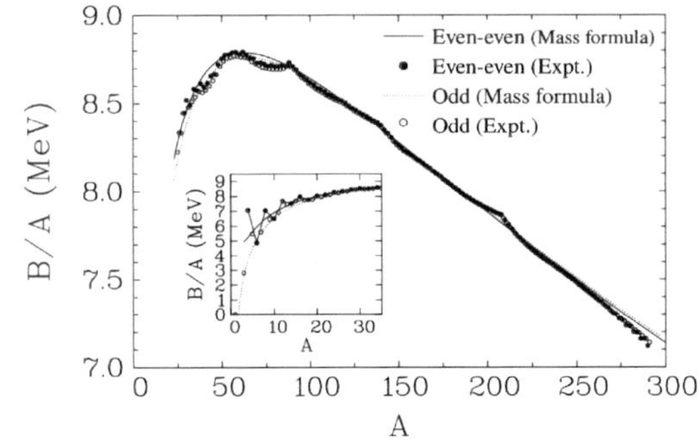

Fig. 2.6 The binding energy per nucleon as a function of the mass number A [13]

2.3.1 *The Binding Energies: Experimental Data and Characteristics*

One usually uses the *binding energy* $B(A, Z)$ for the mass $M(A, Z)$ in various discussions. The $B(A, Z)$ is defined as the energy gain obtained by forming a nucleus instead of being completely apart as individual nucleons. It is therefore given by[16,17]

$$B(A, Z) = ZM_{\text{p}}c^2 + (A - Z)M_{\text{n}}c^2 - M(A, Z)c^2 . \tag{2.48}$$

The filled circles in Fig. 2.6 show the experimental value of the binding energy per nucleon B/A for the most stable even–even nucleus for each mass number, and the open circles that for the most stable nucleus in the case of odd mass number.[18] The inset shows the details for nuclei with small mass number, i.e., for light nuclei. The binding energy has the following characteristics including those shown in these figures.

(1) The binding energy rapidly increases with the mass number in the region of small mass number (see Table 2.3).[19]

---

[16]The *separation energy* is also an important related physical quantity. It is the energy needed to remove a particle such as a nucleon or an alpha particle from the nucleus. For example, the separation energy of neutron is given by $S_n(N, Z) \equiv B(A, Z) - B(A - 1, Z)$ in terms of the binding energies of the relevant nuclei.

[17]The binding energy of a nucleus is usually calculated from the mass of the corresponding neutral atom $M_{\text{atom}}(A, Z)$ as

$$B(A, Z) = ZM_{\text{H}}c^2 + (A - Z)M_{\text{n}}c^2 - M_{\text{atom}}(A, Z)c^2 , \tag{2.47}$$

where M_{H} is the mass of the Hydrogen atom.

[18]Exactly speaking, they have been obtained by using Eq. (2.47).

[19]We name deuteron and triton and denote by d and t in the case of nuclei, while name deuterium and tritium and denote by D and T in the case of neutral atoms.

Table 2.3 Binding energies of some light nuclei

Nuclide	d ($^2_1$H)	t ($^3_1$H)	$^3_2$He	α ($^4_2$He)
Binding energy (MeV)	2.225	8.482	7.718	28.296

Table 2.4 Correlation between the even–odd property of the proton and neutron numbers and the stability of nuclei. The δ represents a systematic variation which appears in the binding energy B in connection with the even–odd property of the proton and neutron numbers. The Δ is called the *odd–even mass parameter*

A	Z	N	Stability	Pairing correlation energy δ (MeV)
Even	Even	Even	Most stable	$\Delta \approx 12/\sqrt{A}$
Odd	Even	Odd	Less stable	0
Odd	Odd	Even	Less stable	0
Even	Odd	Odd	Least stable	$\Delta \approx -12/\sqrt{A}$

(2) The B/A shows a sharp maximum at each light α-nucleus.[20,21]
(3) The binding energy per nucleon B/A is always about 8 MeV except for light nuclei (*saturation property of the binding energy*).
(4) Roughly speaking, the B/A steadily increases with the mass number until the region of Fe whose atomic number is 26,[22] then gradually decreases with A.
(5) The B/A shows local maxima when the number of either protons or neutrons takes special numbers, e.g., $Z = 50$, in the region of large mass number.
(6) There exist systematic dependences on the even–odd property of the number of protons and that of neutrons as shown in Table 2.4.[23,24]

Let us introduce here a few topics associated with the characteristics of the binding energy mentioned from (1) through (6).

Topic 1: The Energy Generation in the Main Sequence Stars The characteristic (1) mentioned above is the reason why the energy is generated by nuclear fusion in stars in the main sequence. The binding energies of light nuclei including the values

[20]Those nuclei such as $^8_4$Be, $^{12}_6$C, $^{16}_8$O whose atomic number and the mass number are the same multiples of those of α particle are called α-nuclei.

[21]The binding energy of $^8_4$Be is less than twice that of α by 0.1 MeV, so that $^8_4$Be is unstable and exists only transiently.

[22]The nucleus which has the largest value of B/A is $^{62}_{28}$Ni (see [14]).

[23]Only 5 nuclei, i.e., d, $^6_3$Li, $^{10}_5$B, $^{14}_7$N, $^{180}_{73}$Ta (isomer), are stable among the so-called odd–odd nuclei, which have odd numbers for both the proton and the neutron numbers. Incidentally, it is $^{209}_{83}$Bi which has the largest atomic number among stable nuclei, although exactly speaking it decays with the half-life of 1.9×10^{19} years. There exist no stable isotopes for Tc and Pm, whose atomic numbers are 43 and 61, respectively.

[24]It is related to the pairing correlation between nucleons, which will be discussed later in detail.

listed in Table 2.3 are the important quantities to determine the energies released by those nuclear reactions and also the energies of the emitted particles such as neutrinos.

The nuclear reactions which power the main sequence stars can be classified into (a) *pp-chain (proton–proton chain)* reactions, which consist of ppI-chain, ppII-chain and ppIII-chain reactions, and (b) *CNO cycle*. Their reactions and the associated energy releases are as follows:

- ppI chain

① $p+p \to {}^2H+e^+ +\nu_e+0.42\,MeV \quad (\nu_{pp})$
② $e^+ +e^- \to 2\gamma+1.02\,MeV$
③ ${}^2H+p \to {}^3He+\gamma+5.49\,MeV$
④ ${}^3He+{}^3He \to {}^4He+2p+12.86\,MeV$

- ppII chain

⑤ ${}^3He+{}^4He \to {}^7Be+\gamma$
⑥ ${}^7Be+e^- \to {}^7Li+\nu_e \quad (\nu_{Be})$
⑦ ${}^7Li+p \to 2\,{}^4He$

- ppIII chain

⑧ ${}^7Be+p \to {}^8B+\gamma$
⑨ ${}^8B \to {}^8Be+e^+ +\nu_e \quad (\nu_B)$
⑩ ${}^8Be \to 2\,{}^4He$

- CNO cycle

⑪ ${}^{12}C+p \to {}^{13}N+\gamma+1.94\,MeV$
⑫ ${}^{13}N \to {}^{13}C+e^+ +\nu_e+1.20\,MeV$
⑬ ${}^{13}C+p \to {}^{14}N+\gamma+7.55\,MeV$
⑭ ${}^{14}N+p \to {}^{15}O+\gamma+7.29\,MeV$
⑮ ${}^{15}O \to {}^{15}N+e^+ +\nu_e+1.74\,MeV$
⑯ ${}^{15}N+p \to {}^{12}C+{}^4He+4.97\,MeV$
⑰ $e^+ +e^- \to 2\gamma+1.02\,MeV$

We introduced the notations ν_{pp}, ν_{Be}, ν_B to distinguish neutrinos of different origins.

By adding the nuclear reactions in ①–④ with appropriate weights, one can see that the energy of 26.72 MeV in total is released by converting four protons into one ^{4}He nucleus in the ppI chain,

$$4p + 2e^- \to {}^4He + 2\nu_e + 26.72\,MeV. \tag{2.49}$$

Also for the CNO cycle, we obtain the same net reaction formula as Eq. (2.49) by adding ⑪–⑯ and subtracting twice of ⑰ on each side. The central temperature of the Sun is about 16 million K, and the standard solar model predicts that about 98.5% of the solar energy is produced by nuclear reactions in the pp chain, while the remaining about 1.5% by the CNO cycle.

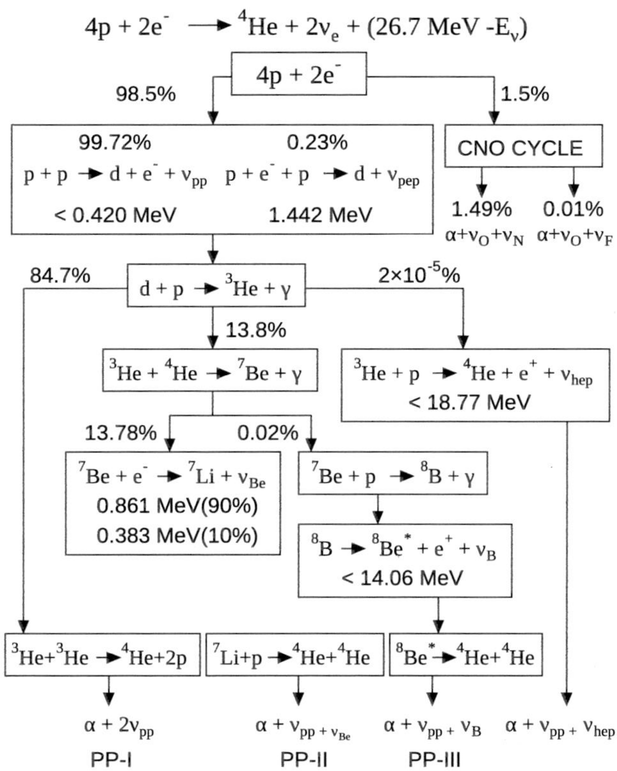

Fig. 2.7 Conceptual representation of the pp chain and CNO cycle. Taken from [15]

Figure 2.7 represents conceptually the pp chain and the CNO cycle, and Fig. 2.8 the spectra of solar neutrinos emitted by various reactions. Note that two-body reactions lead to line spectra, while three-body reactions continuous spectra. Experimentally, the high-energy neutrino ν_B is detected either by water Cherenkov detectors or by using the reaction of ^{37}Cl,

$$\nu_e + {}^{37}\mathrm{Cl} \rightarrow {}^{37}\mathrm{Ar} + \mathrm{e}^- . \tag{2.50}$$

The low-energy neutrino ν_{pp} is studied, on the other hand, through the capture reactions by Ga and deuterium.

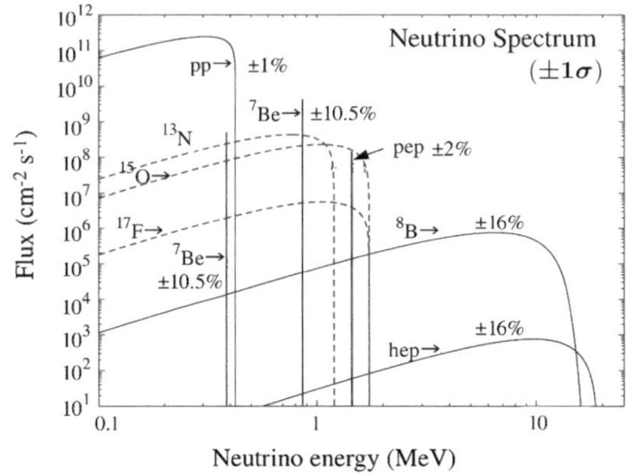

Fig. 2.8 The solar neutrino spectra. Taken from [16]. The *solid lines* are the spectra for the pp chain and the *dashed lines* those for the CNO cycle

The *solar neutrino problem*[25] was a serious problem, which lasted from mid-1960s to the beginning of the twenty-first century. It is now attributed to the neutrino oscillation.

Topic 2: The Energy Problem: *Muon-Catalysed Fusion (μCF)* The binding energies of light nuclei listed in Table 2.3 are important quantities in connection with the so-called energy problem as well. For example, if the nuclear reactions $p+d \rightarrow {}^3He$, $d+t \rightarrow {}^4He+n$ take place, the energy of 5.4 and 17.5 MeV is released per reaction for the two cases, respectively. However, like the $p+p \rightarrow d+e^+ + \nu_e$ reaction in the Sun mentioned in Sect. 1.4, the high Coulomb barrier between the projectile and target nuclei has to be overcome in order for these reactions to occur. The so-called high-temperature fusion reactions attempt to overcome the Coulomb barrier by heating the nuclei to high temperature. In this case, the nuclei of high temperature need to be confined in a small space. To that end, various methods such as the magnetic confinement, where the hot fusion fuel is confined by magnetic fields in a small space in the form of plasma, and the inertial confinement have been developed. However, none of the methods have yet reached the practical usage for the power production. The muon-catalysed fusion is an alternative method, where one tries to induce fusion reactions at low temperature by lowering the Coulomb barrier with the screening effect by μ^- particle.

Figure 2.9 shows the muon-catalysed pd fusion which has been confirmed experimentally for the first time as the evidence of the μCF.[26] Figure 2.10 conceptually shows the muon-catalysed dt fusion which is currently thought to be most promis-

[25]It is the problem that the observed number of solar neutrinos is only about 1/2–1/3 of the theoretical prediction.

[26]In 1956, Alvarez et al. discovered a trajectory of muon which stops in the liquid hydrogen contaminated with deuterium. The flight length in matter is determined by the stopping power. The length 1.7 cm of the trajectory observed by Alvarez et al. exactly matched the flight length of muon with energy of 5.4 MeV which is released in the $p+d \rightarrow {}^3He$ reaction.

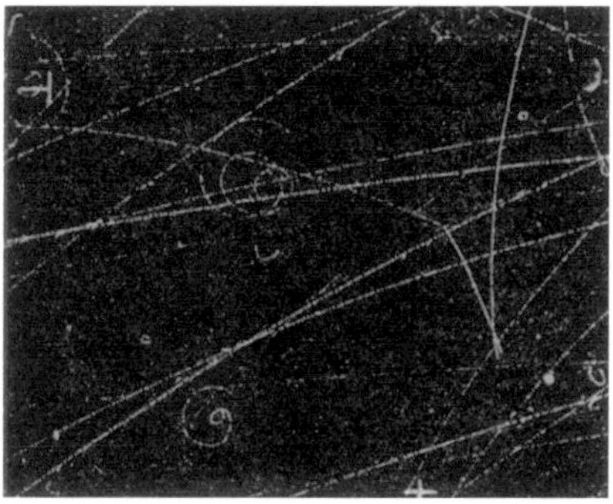

Fig. 2.9 The first experimental evidence of the μCF. Taken from [17]

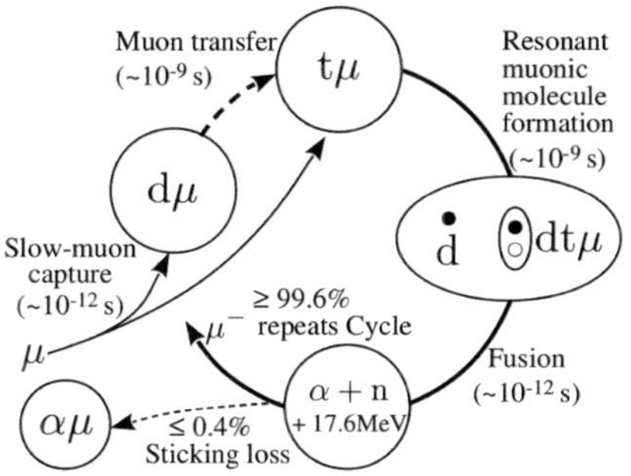

Fig. 2.10 The μCF cycle. $\mathrm{dt}\mu \rightarrow {}^4\mathrm{He}+\mathrm{n}+\mu + 17.5$ MeV reaction. Taken from [18]

ing from the point of view of the practical usage for the power production. In the latter, the liquid target of the mixture of about 50% each of deuterium and tritium is injected by μ^- particle produced by nuclear reactions. Thus produced dμ and tμ atoms collide to form a dtμ molecule. Eventually the nuclear reaction

$$\mathrm{d} + \mathrm{t} \rightarrow \alpha + \mathrm{n} \tag{2.51}$$

takes place. The μ^- particle which stimulated tunneling effect through screening the Coulomb barrier finishes the role of a catalyst by forming a muonic atom by sticking to the α particle produced in the nuclear reaction with a certain probability. However, with a much higher probability it leaves the nucleus by using a part of the energy released in the reaction, and repeats the cycle as a catalyst. Whether the muon-catalysed fusion is useful for the practical usage for the energy generation or not depends on the number of cycle per muon, whose mean life is 2.2 μs, and the reactivation probability by collision of the muon lost by forming a muonic atom such as $\alpha\mu$. The dtμ reaction is more promising than the pdμ reaction, because the dtμ molecule is produced resonantly by the so-called *Vesman mechanism*, and also because the sticking probability is smaller (see [18] for details).

Finally, we wish to refer to the nuclear reaction in matter [11] as a closely related research. It is the study, for example, to see whether the rates of the d+d→t+p, ^{3}He+n reactions are significantly enhanced over those in the free space if a metal target such as Pd and PdO is continuously bombarded with deuterons.

Topic 3: The Alpha Particle Model and Cluster Structures *Symmetry* plays important roles in physics. The characteristic (2) of the binding energy mentioned in Sect. 2.3.1 originates from the property of nuclear force that favours states with good spatial symmetry. Historically, based on this property, the model called *the α-particle model* has been proposed which views the light α-nuclei as aggregates of α particles and describes their ground states and low-lying excited states by special spatial configurations or relative motions of the constituent α particles or by the rotations of the whole nucleus. In this model, for example, the ground state of ^{12}C is thought to be an equilateral triangle configuration of three α particles, while the ground state of ^{16}O a regular tetrahedron of four α particles. In these days, the standard model for nuclear structure is the *shell model* which views the nucleus as an ensemble of nucleons which are moving independently from each other in a common mean field of a single center in the zeroth order approximation (see Chap. 5). On the contrary, the α-particle model is a *molecular point of view* or a *molecular structure model* which assumes the existence of multi-centers or of density localization. This idea has been extended to variational calculations and currently to the *cluster model* (the study of cluster structure) by using the mean-field theory or the *molecular dynamics calculations*, and intensive studies are going on from that points of view. The molecular structures are expected to appear at the energies near the threshold energy for each molecular structure, and are characterized by having a large decay width to each corresponding decay channel as an indicator. Figure 2.11 shows[27] the excitation energy of various molecular structures in light nuclei expected from this idea. It is called the Ikeda diagram (see [20, 21] for details of the study of cluster structure).

Figure 2.12 shows the energy-level diagram of ^{12}C in the region of low excitation energies [22] in order to help a deeper understanding. The second 0^+ state denoted by 0_2^+ at the excitation energy 7.65 MeV, which is close to the threshold energy for the α-decay 7.37 MeV, is a state for which the 3α-particle model works well.[28] The ground state locates somewhat far from the threshold energy for the α-decay. Consequently, the density localization becomes less significant. Nevertheless, the picture that the state consists of 3α-clusters which form an equilateral triangle configuration is still valid to some extent. One can think that it is an experimental evidence that there appears a 3^- state at a low excitation energy, i.e., at 9.64 MeV, which is expected from the $D_{3\mathrm{h}}$ symmetry of the point group associated with the equilateral triangle configuration. The existence of a rotational band, $0_{\mathrm{g.s.}}$, 2_1^+ (4.44 MeV), 4_1^+ (14.08 MeV $\approx$ 4.44 MeV $\times$ 3.33) also supports this point of view. However, if there

---

[27]The threshold energies for α-decay, and those for emitting several α particles, and also the threshold energy of the decay of ^{24}Mg into two ^{12}C nuclei, taken from [19].

[28]The synthesis of ^{12}C in stars proceeds as $^8\mathrm{Be}+\alpha \to {}^{12}\mathrm{C}(0_2^+)$, $^{12}\mathrm{C}(0_2^+) \to {}^{12}\mathrm{C}(0_{\mathrm{g.s.}}^+)$ In that sense, the 0_2^+ state of ^{12}C plays a crucial role in the synthesis of heavy elements, and is named *Hoyle-state* after F. Hoyle who predicted the existence of this state.

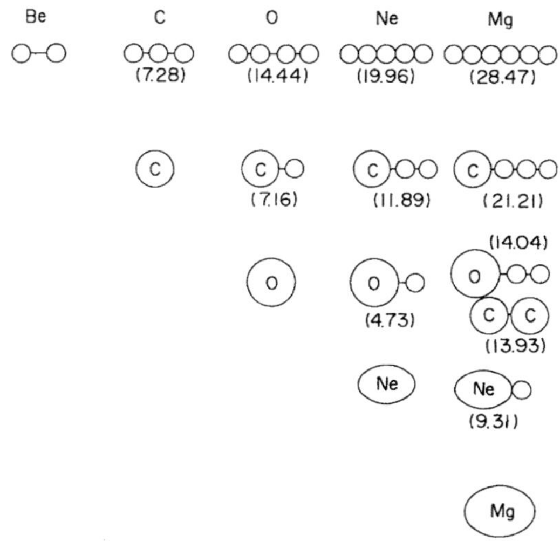

Fig. 2.11 The Ikeda diagram: Threshold energies corresponding to various cluster structures [19]

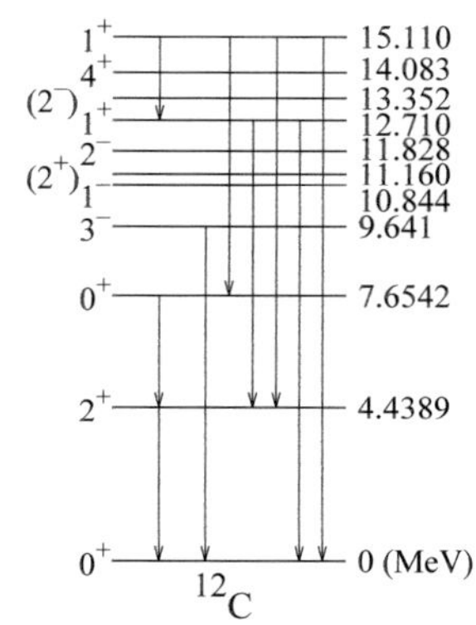

Fig. 2.12 The low-lying energy levels of ^{12}C

appear states only up to 4^+ member, the rotational property of the spectrum can be explained also by a usual shell model without assuming the density localization (*Elliott's $SU(3)$ model*). A more accurate description of the ground state and the ground-state rotational band is provided by a *hybrid model*, which hybridizes the α-cluster model and the shell model in order to take into account the broken spatial symmetry due to the spin–orbit interaction [23].[29] Incidentally, the cluster structure

[29]In the case of ^{12}C, the spin–orbit interaction reduces the extent of deformation and increases the excitation energy of the 2_1^+ state. From the point of view of shell model, a force with smaller spin–orbit splitting leads to a high degeneracy and to a larger deformation due to the Jahn–Teller effect, while a force with a strong spin–orbit force leads to a large splitting between the $p_{3/2}$ and

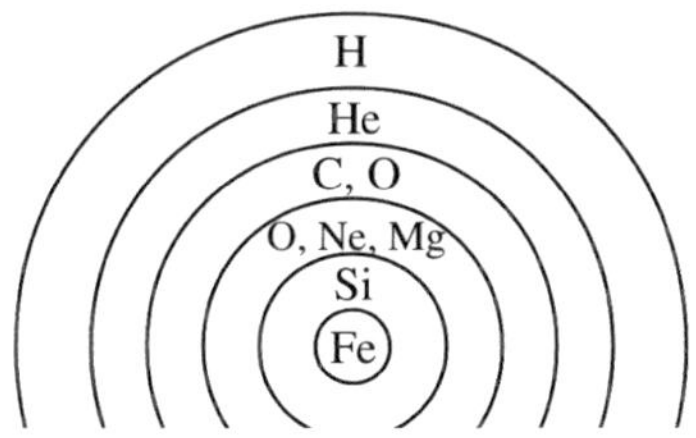

Fig. 2.13 The onion layer structure of a presupernova star

and the deformation of nuclei, which is described in detail in Chap. 7 can be thought to be a kind of *Jahn–Teller effect*.[30]

Exercise 2.10 Let us assume that three α particles form an equilateral triangle, i.e., the D_{3h} state of the point group. Show that in this case the component of the angular momentum along the axis perpendicular to the plane made by the three α particles, which is called the K quantum number, takes only multiples of 3.

Topic 4: The Onion Layer Structure of a Massive Star and the Iron Core In massive stars whose mass M exceeds 12 times the solar mass, i.e., $M > 12M_{\odot}$, there exists a core consisting of Fe and Ni at the center and successive layers of gradually smaller atomic number distribute towards the surface (*onion layer structure*). Figure 2.13 shows the inner structure of a presupernova star. The core part consists of Fe and Ni, because the binding energy per particle becomes the largest in the Fe region as remarked in Sect. 2.3.1 as the characteristic (4) of the binding energy per nucleon B/A.[31,32] For the same reason, nuclei with small atomic number and mass

(Footnote 29 continued)
$p_{1/2}$ states. Consequently, the degeneracy disappears and the spherical shape is predicted to appear. Inversely, if one considers from the point of view of the cluster model, too large deformation or clusterization is predicted and consequently the excitation energy of the 2_1^+ state becomes too low compared with the experimental data if one assumes the cluster state which extends the pure state of the maximum spatial symmetry of the shell model denoted by $(0s)^4(0p)^8[444]^{11}L$ configuration by introducing a finite distance between three alpha clusters, but ignores the spin–orbit interaction, and the experimental data can be well reproduced by admixing states with lower spatial symmetry by the spin–orbit interaction.

[30]In the case when the electronic states degenerate in energy in a polyatomic molecule where the atoms have a highly symmetric geometrical configuration such as a regular polyhedron, it becomes energetically more stable if the degeneracy is removed by assuming a configuration of atoms with lower symmetry. This is known as the Jahn–Teller effect. Similar situations occur in many physical systems including metal microclusters.

[31]A statement often appears in the literature that the central part of massive stars consists of $^{56}_{26}$Fe due to the fact that the nucleus which has the largest binding energy per nucleon is $^{56}_{26}$Fe. However, in reality, the nucleus which has the maximum binding energy per nucleon is $^{62}_{28}$Ni. It is understood that the reason why the major component of the core is $^{56}_{26}$Fe despite this fact is because $^{62}_{28}$Ni is hard to be synthesized by nuclear reactions in stars, while $^{56}_{26}$Fe can be easily synthesized as the last nucleus in the Si-burning reaction.

[32]The temperature of the central part has to be sufficiently high in order for the core of Fe and Ni to be formed in the central part of stars. In the stars whose mass is not so large, the central part

number proceed towards Fe and Ni by fusion reactions. On the contrary, nuclei with large atomic number and mass number proceed to Fe and Ni by fission.

Topic 5: The Supernova Explosion It is known that there exist two types of supernova explosion, i.e., that initiated by carbon detonation and by gravitational collapse.[33] In the latter case, and in massive stars whose core part is made of Fe and Ni, the gravitational collapse is triggered when Fe in the core captures surrounding electrons by *electron capture reaction*,

$$^{56}_{26}\mathrm{Fe} + \mathrm{e}^- \rightarrow {}^{56}_{25}\mathrm{Mn} + \nu_\mathrm{e} , \tag{2.52}$$

and consequently when the degeneracy pressure of electrons decreases. Also, if the central temperature exceeds 5×10^9 K, an endothermic photo-disintegration reaction of Fe,

$$^{56}_{26}\mathrm{Fe} + \gamma \rightarrow 13^4\mathrm{He} + 4\mathrm{n} - 125\,\mathrm{MeV}, \tag{2.53}$$

takes place in addition, and at higher temperatures, the photo-disintegration reaction of ^{4}He,

$$^4\mathrm{He} + \gamma \rightarrow 2\mathrm{n} + 2\mathrm{p} - 28.3\,\mathrm{MeV}, \tag{2.54}$$

occurs as well. The internal energy which should have increased by contraction does not lead to the increase of the pressure by being used in these endothermic reactions. Thus, the contraction of the star cannot be prevented. Consequently, the gravitational collapse of the star proceeds and eventually is bounced back by the hard core in the nuclear force (see Sect. 3.7) when the density increases so high as nucleons start to overlap. The collapse thus changes into explosion and neutrinos and many other particles are emitted into space. This is a scenario of the supernova explosion. The supernova explosion is an interesting subject to be studied also in connection with the synthesis of elements, e.g., as a site to synthesize heavy elements such as U by *rapid neutron capture reactions* called *r-process*.

Exercise 2.11 Show that the thermal pressure in high temperature ideal gas is given by

$$P_T = \rho T \tag{2.55}$$

in terms of the temperature T and the density ρ. (Hint: Assume that the momentum distribution of the gas is given by the Boltzmann distribution. See Sect. 6.2.5.)

(Footnote 32 continued)
therefore consists of nuclei with smaller mass number than that of Fe, though they have an onion layer structure. For example, the central part of stars whose mass lies between $8M_\odot < M < 12M_\odot$ consists of O, Ne and Mg.

[33]There is an alternative way to classify the supernova explosion into the type II and I, respectively, depending on whether the spectrum contains the hydrogen line or not.

2.3.2 *The Semi-empirical Mass Formula (The Weizsäcker–Bethe Mass Formula)—The Liquid-Drop Model*

The bulk behaviour of the binding energy can be well represented by a simple formula

$$B = b_{\mathrm{vol}} A - b_{\mathrm{surf}} A^{2/3} - \frac{1}{2} b_{\mathrm{sym}} \frac{(N-Z)^2}{A} - b_{\mathrm{C}} \frac{Z^2}{A^{1/3}} + \delta(A)\,, \tag{2.56}$$

which takes into account major physical effects. Equation (2.56) is called the *Weizsäcker–Bethe mass formula*.[34] The first and the second terms on the right-hand side are called the volume and the surface[35] terms, respectively, because the nuclear radius is proportional to $A^{1/3}$. The property that the bulk binding energy is the volume term which is proportional to the number of constituent particles comes from the fact that the nuclear force which stabilizes a nucleus as an isolated bound system is of short range as we learn in the next chapter. The existence of the surface term will be easily understood from the point of view of the liquid-drop model for nuclei. The third term is called the *symmetry energy term* and tends to locate stable nuclei along the diagonal line in the nuclear chart. We will learn its origin later somewhat in detail. The fourth term represents the effect of *Coulomb energy*. As we have already remarked in Sect. 2.3.1 the last term originates from the pairing correlation between nucleons and represents the fact that the binding energy systematically changes according to the even–odd property of the atomic number and of the neutron number. It will be described in detail in Sect. 5.8.

The values of the parameters in the mass formula somewhat vary depending on the experimental data to be reproduced. The solid line in Fig. 2.6 represents the theoretical values given by the mass formula when the parameters b_{vol}, b_{surf}, b_{sym}, b_{C} are determined by the least-square fit to the binding energies of even–even nuclei for which $N, Z > 7$ and the experimental error of the binding energy B is smaller than 200 keV [25]. We fixed the δ for the pairing correlation to be $\delta = 12/\sqrt{A}\,\mathrm{MeV}$. The dotted line represents the similar fit to odd nuclei. In both cases, the fitting was performed to the most stable isobar for each mass number A, i.e., the experimental data shown in the figure. We observe that the mass formula well reproduces the bulk behaviour of the binding energy. Thus obtained values of the parameters are listed in Table 2.5.

Comment: The Reason Why Stable Nuclei Locate along the Diagonal Line of the Nuclear Chart: The Symmetry Energy and Its Origin As stated above, the symmetry energy locates stable nuclei nearly along the diagonal line in the nuclear chart. Let us consider the contribution of the kinetic energy to the binding energy in

---

[34]At present, there exist several refined mass formulae [24] reflecting various studies such as those of unstable nuclei, superheavy elements and nucleosynthesis of elements through r-process. In those refined mass formulae, for example, the effects of surface diffuseness and of the symmetry energy correction to the volume as well as the surface terms are considered.

[35]If we define the *nuclear surface tension* σ by $\sigma \equiv b_{\mathrm{surf}} A^{2/3}/4\pi(r_0 A^{1/3})^2$, then $\sigma \sim 1\,\mathrm{MeV/fm}^2$.

Table 2.5 The values of parameters in the mass formula used to obtain the solid and dotted lines in Fig. 2.6. All the values are in units of MeV

	b_{vol}	b_{surf}	b_{sym}	b_{C}	δ
Even–even (solid line)	16.2	19.0	47.0	0.755	$12/\sqrt{A}$
Odd (dotted line)	16.3	19.2	45.4	0.770	0

order to understand the origin of the symmetry energy. If we use the Thomas–Fermi approximation, we have

$$E_{\mathrm{kin}} = E_{\mathrm{kin}}^{(\mathrm{n})} + E_{\mathrm{kin}}^{(\mathrm{p})} = \frac{3}{5}\varepsilon_{\mathrm{F}}^{(\mathrm{n})} N + \frac{3}{5}\varepsilon_{\mathrm{F}}^{(\mathrm{p})} Z \approx \frac{3}{5}\varepsilon_{\mathrm{F}} A + \frac{1}{3}\varepsilon_{\mathrm{F}} \frac{(N-Z)^2}{A}. \tag{2.57}$$

By comparing the second term on the right-hand side of Eq. (2.57) with the symmetry energy term in the mass formula, we obtain

$$b_{\mathrm{sym}}^{(\mathrm{kin})} = \frac{2}{3}\varepsilon_{\mathrm{F}} \sim 25\,\mathrm{MeV}. \tag{2.58}$$

The comparison of this value with the value of the parameter of the symmetry energy b_{sym} given in Table 2.5 shows that about half of the symmetry energy arises from the kinetic energy due to the Pauli exclusion principle (Pauli effect). The other half originates from the fact that the nuclear force in the isospin singlet term, i.e., that between n and p in the isospin singlet configuration, is stronger than that in the isospin triplet terms, i.e., that in the isospin triplet configurations of the pp, np and nn systems as we learn in the next chapter.

2.3.3 *Applications of the Mass Formula (1): The Stability Line, the Heisenberg Valley*

The mass formula is useful, for example, to predict the masses of unknown nuclei and also to discuss the bulk aspects concerning the stability of nuclei such as the stability with respect to fission. Here and in the next subsection, let us study some of the applications of the mass formula.

2.3.3.1 The Formula for the Stability Line

In the region of small atomic number, the stable nuclei locate along the diagonal line or in the vicinity in the nuclear chart. This is due to the symmetry energy term in the mass formula. However, the contribution of the Coulomb energy becomes non-negligible as the atomic number gets large. Consequently, the location of stable nuclei gradually deviates from the diagonal line and the neutron number of stable nuclei becomes about 1.6 times the proton number ($N \sim 1.6Z$) in the region of large atomic number such as U. This trend can be derived from the mass formula. If we denote the atomic number of the most stable nucleus for a given mass number A_0 by $Z_{\beta s}(A_0)$, then $\frac{\partial}{\partial Z}B(A, Z)|_{A=A_0} = 0$ at $Z = Z_{\beta s}(A_0)$. Hence we obtain

$$Z_{\beta s}(A) = \frac{A/2}{1 + \frac{b_C}{2b_{\mathrm{sym}}}A^{2/3}} \approx \frac{A/2}{1 + 0.00803A^{2/3}} \tag{2.59}$$

from the mass formula. We used the values listed in Table 2.5 in obtaining the third term. If we denote the value of Z in the case when we let $N = 1.6Z$ by $\tilde{Z}$, for example, $Z_{\beta s} = 61$, $\tilde{Z} = 58$ for $A = 150$, and $Z_{\beta s} = 81$, $\tilde{Z} = 80$ for $A = 208$, indicating that the values of $Z_{\beta s}$ and $\tilde{Z}$ are very close to each other. In fact, $^{150}_{62}$Sm and $^{208}_{82}$Pb are stable nuclei.[36] In this way, the ratio between the proton number and the neutron number for heavy nuclei is determined by the balance between the Coulomb energy and the symmetry energy.

2.3.3.2 The Heisenberg Valley

Figure 2.14 is a schematic three-dimensional representation of the nuclear mass M as a function of the neutron number N and the proton number Z. There exist three planes corresponding to the even–even, even–odd or odd–even, and odd–odd nuclei due to the pairing term δ in the mass formula. The line along a cross section represents the variation of the masses for a given isobars. It behaves like a quadratic function because of the symmetry energy. This property is shown in Fig. 2.15 for the nuclei with odd-mass number, and in Fig. 2.16 for even-A nuclei. As these figures

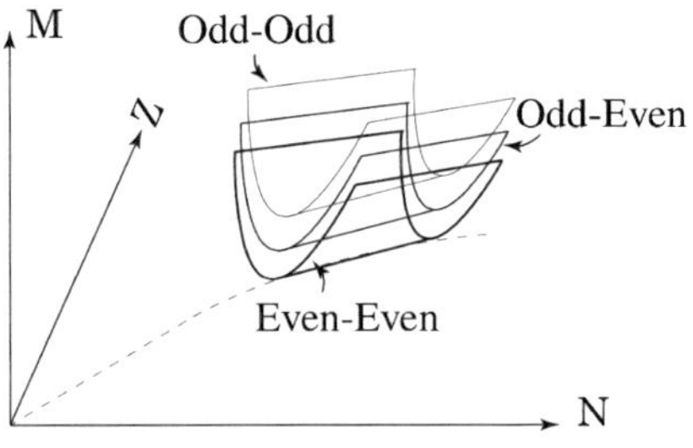

Fig. 2.14 The Z, N dependence of the nuclear mass: The Heisenberg valley

[36] $^{150}_{58}$Ce and $^{208}_{80}$Hg are unstable nuclei. The relationship $N \sim 1.6Z$ should be taken only as a rough relationship that holds in the region of large mass number. To be accurate, it is safer to use Eq. (2.59).

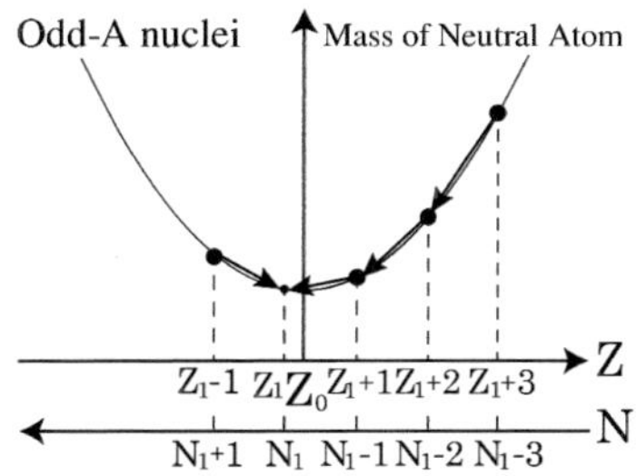

Fig. 2.15 The Z, N dependence of the nuclear masses of the isobars with an odd A

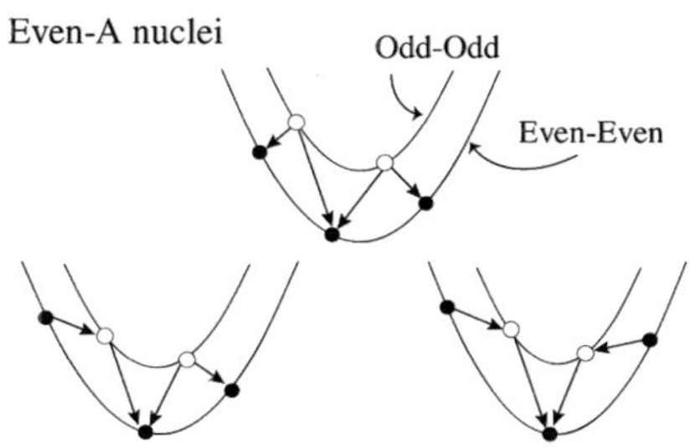

Fig. 2.16 The Z, N dependence of the nuclear masses of the isobars with an even A

show, there exist one stable nucleus in the case of odd A and 1–3 stable nuclei in the case of even A which are stable with respect to β-decay or electron capture.

These stable nuclei form a kind of valley in Fig. 2.14 as the mass number is varied. This valley is called the *Heisenberg valley*. Its projection on the NZ-plane follows almost along Eq. (2.59). The synthesis of light elements up to Fe by nuclear fusion and the synthesis of heavy elements up to Bi by the slow neutron capture reaction, called *s-process*, proceed along the Heisenberg valley.

Incidentally, the conditions for the β^--decay, β^+-decay and electron capture to occur are given by the following equations using the mass $M(A, Z)$ of the neutral atom of atomic number Z and mass number A, the mass of electron m_e and the binding energy Δ_e of the electron to be captured in the electron capture in the atom before it is captured,

$$M(A, Z) > M(A, Z+1) \quad \text{for } \beta^- \text{decay}, \tag{2.60}$$

$$M(A, Z) > M(A, Z-1) + 2m_e \quad \text{for } \beta^+ \text{decay}, \tag{2.61}$$

$$M(A, Z) > M(A, Z-1) + \Delta_e/c^2 \quad \text{for electron capture}. \tag{2.62}$$

Here, we ignored the change of the binding energy of electron associated with the change of atom.

Exercise 2.12 Prove Eqs. (2.60)–(2.62).

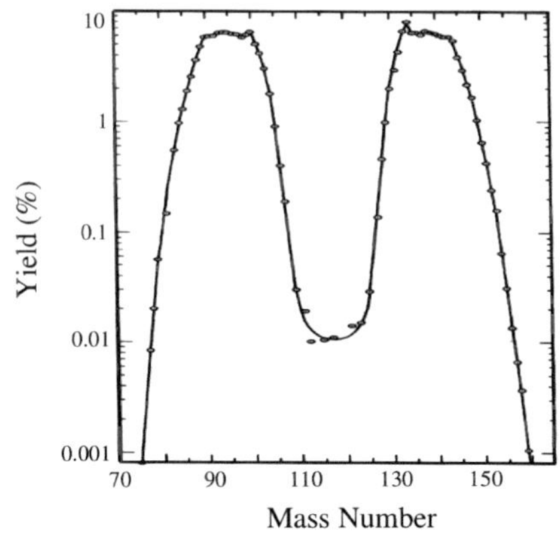

Fig. 2.17 The mass distribution of the fission fragments in the thermal-neutron-induced fission of $^{235}_{92}U$. Taken from [26]

2.3.4 *Applications of the Mass Formula (2): Stability with Respect to Fission*

Fission or nuclear fission is a process where a nucleus with a large atomic number such as U, Th and Pu disintegrates into two fragments, called fission fragments, with nearly equal mass numbers. As an example, Fig. 2.17 shows the mass distribution of the fission fragments in the thermal-neutron-induced fission of ^{235}U.[37] The fission was discovered by Hahn and Strassmann in 1938 [28].[38] Bohr and Wheeler gave a theoretical interpretation soon after the discovery of fission. The following discussions follow their paper [29].[39]

Figure 2.18 schematically represents a process, where a nucleus viewed as a liquid-drop disintegrates into two fission fragments of equal mass, i.e., of a *symmetric fission*.[40] The configuration d in the figure is called *scission point*.

---

[37]In the spontaneous fission or in the induced fission by a small energy, hence with low excitation energy, fission with a striking mass-asymmetry occurs as shown in Fig. 2.17 because of the shell effect which we study in Chap. 5. Also, a few neutrons are emitted simultaneously in each spontaneous fission and in each thermal neutron-induced fission. For example, the average numbers of neutrons emitted in thermal neutron-induced fission of ^{235}U and $^{239}_{94}Pu$ are 2.4 and 2.9, respectively [27].

[38]Hahn and Strassmann, who were studying the decay of U by irradiation with neutrons, discovered fission by identifying the fission fragments, which had been believed to be $_{88}Ra$, $_{89}Ac$, $_{90}Th$, as $_{56}Ba$, $_{57}La$, $_{58}Ce$. Lise Meitner made also a significant contribution to the series of studies as a collaborator of Hahn.

[39]The basic data and theoretical descriptions of fission are given in detail in [27].

[40]As we learn in Chap. 7 many nuclei which undergo spontaneous fission such as actinides to which U belongs are deformed even in their ground states due to the shell effect, i.e., a quantum effect, and the long axis is about 30% longer than the short axis. Also, as shown in Fig. 2.17, most of the fission is an asymmetric fission, where the masses of the two fission fragments are significantly different. Following the original picture of Bohr–Wheeler which is based on the liquid-drop model,

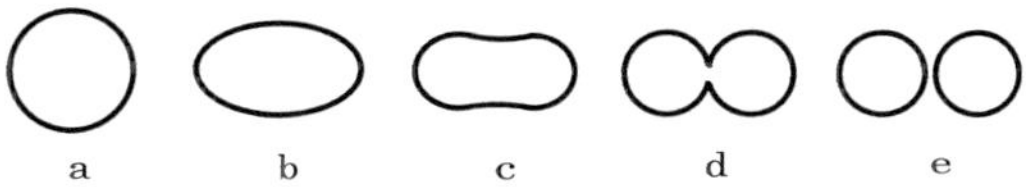

Fig. 2.18 A schematic representation of a symmetric fission

2.3.4.1 The Q-Value of a Spontaneous Fission

Let us first evaluate the Q-value of a spontaneous fission by assuming that a nucleus with a large mass number disintegrates into two fission fragments with equal mass.[41] We have

$$Q \equiv 2B(A/2, Z/2) - B(A, Z) \tag{2.63}$$

$$= A^{2/3} b_C (1 - 2^{-2/3}) \left[\frac{Z^2}{A} - \frac{b_{\text{surf}}}{b_C} \frac{(2^{1/3} - 1)}{(1 - 2^{-2/3})} \right] \tag{2.64}$$

$$\approx A^{2/3} b_C (1 - 2^{-2/3})(Z^2/A - 17.7) \tag{2.65}$$

from the mass formula. Hence, if $Z^2/A > 17.7$, i.e., if $Z > 35$ when we assume $A \sim 2Z$, the nucleus eventually undergoes fission even if we leave it alone. This is called *spontaneous fission*.[42] For example, if one assumes the fission ${}^{236}_{92}\text{U} \rightarrow 2 \times {}^{118}_{46}\text{Pd}$, then the Q-value is about 182 MeV.[43]

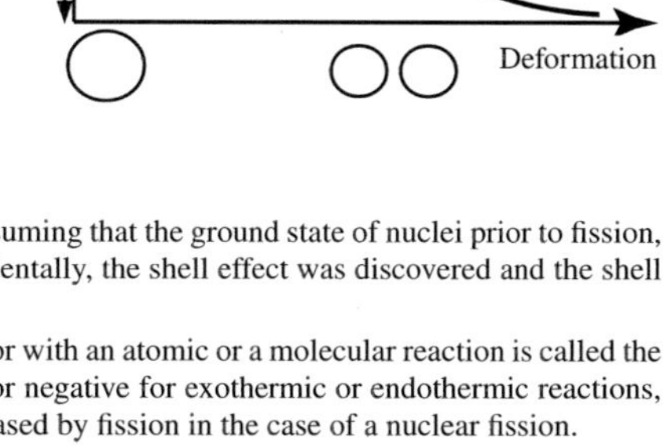

Fig. 2.19 An illustration of the fission barrier by the liquid-drop model

(Footnote 40 continued)
we develop in this book our discussions by assuming that the ground state of nuclei prior to fission, which is a metastable state, is spherical. Incidentally, the shell effect was discovered and the shell model has been established much later.

[41]The heat associated with a nuclear reaction or with an atomic or a molecular reaction is called the Q-value of the reaction. It becomes positive or negative for exothermic or endothermic reactions, respectively. It corresponds to the energy released by fission in the case of a nuclear fission.

[42]Fission is stimulated if the nucleus is irradiated by a photon, or neutron or by other particles. Such reaction is called *induced fission* and is distinguished from the spontaneous fission. It is represented as, for example, (γ,f) and (n,f) for photofission and neutron-induced fission, respectively.

[43]About 200 MeV is a measure of the Q-value for the fission of actinides such as U.

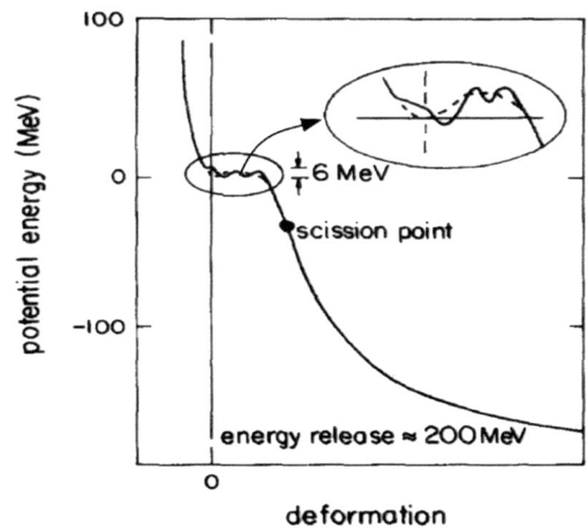

Fig. 2.20 Illustration of the fission barrier by the liquid-drop model with shell effect [27]

The fact that the Q-value is positive does not necessarily mean that fission takes place instantaneously. Figure 2.19 schematically shows the potential barrier governing fission. How to take fission coordinates is itself an important question. Here, we take the nuclear deformation as the fission coordinate corresponding to the conceptual picture for fission shown in Fig. 2.18. This potential barrier has to be overcome by a quantum tunneling in order for fission to occur.[44] Consequently, the lifetime for fission is extremely long. For example, the half-life is $T_{1/2}^{\mathrm{sf}} \sim 10^{17}$ years in the case of ^{238}U. On the other hand, heavy nuclei decay also by emitting α particle. As we learn in Sect. 8.1, the α-decay also occurs by a quantum tunneling. In that case, the abscissa of the potential barrier which corresponds to Fig. 2.19 is the relative coordinate between the α particle and the remaining nucleus in the α-decay, called the daughter nucleus.[45] In the case of ^{238}U, the half-life of α-decay is $T_{1/2}^{\alpha} \sim 10^{9}$ years. As is clear in the WKB formula for a quantum tunneling Eqs. (8.6), (8.7) or a revised formula in the uniform approximation Eq. (2.77), the tunneling probability dramatically decreases following the exponential law as the mass parameter μ of the decaying particle or of the decaying degree of freedom gets large. The μ is nearly equal to the mass of the α particle in the case of α-decay, while the reduced mass of the fission fragments gives a rough measure of μ in the case of fission if the relative coordinate between fission fragments is used as the fission coordinate. Obviously, the

[44] In reality, fission is a quantum tunneling in a multidimensional space with many degrees of freedom.

[45] In the case of α-decay, the potential barrier to be transmitted by a quantum tunneling is located outside the sum of the radii of the α particle and the daughter nucleus. In other words, if we consider the α-decay as a kind of fission process, then the potential barrier appears outside the scission point. That is the reason why the coordinate for the α-decay can be taken to be the coordinate of the relative motion.

latter is much larger. In addition, roughly speaking, the barrier height is proportional to the product of the charges of the fission fragments. One can therefore expect that the barrier height to be transmitted is much higher for fission. It is why the half-life for fission is much longer than that for α-decay. Consequently, the decay of heavy elements occurs mainly by α-decay.

Figure 2.20 shows schematic potential surfaces for fission by Nix. The solid and the dotted lines are the potential surfaces with and without shell correction, which will be discussed later, respectively. It should be remarked that the shell correction introduces a dip in the potential surface at the barrier region.

2.3.4.2 Potential Barrier with Respect to Symmetric Spontaneous Fission and the Fissility Parameter

(1) Fission Coordinate: Deformation Parameter Let us next study how the potential surface with respect to fission shown conceptually in Fig. 2.19 depends on the atomic number and the mass number. Since the isotopes of ^{235,238}U exist in nature with finite but long lifetimes, one can expect that the potential surface for fission has a potential barrier as depicted in Fig. 2.19 up to at least $Z = 92$. On the other hand, there exist no stable nuclei with an extremely large atomic number. One can therefore also expect that the potential barrier ultimately disappears for large enough atomic number. How large is the atomic number at the border?

As stated before, it is not clear at all how to choose the coordinates or parameters for fission. It will be natural to use the distance between two fission fragments as the fission coordinate once they are more apart than the scission point in the late stage of fission. However, the deformation parameters introduced in the following will be more natural to describe the initial stage of fission.

Let us consider fission as a process, where the deformation of a nucleus grows with axial symmetry. The radius $R(\theta)$ of an axial-symmetrically deformed nucleus can be expanded as

$$R(\theta) = R_0 \left[1 + \alpha_0 + \sum_{\lambda=2,4,\ldots} \alpha_\lambda P_\lambda(\cos\theta) \right] \tag{2.66}$$

by using the Legendre functions $P_\lambda(\cos\theta)$, which form a complete set. Here, α_λ are called the deformation parameters. Among them, α_2, α_4 are named quadrupole and hexadecapole deformation parameters, respectively.[46] α_0 is given as a function of $\alpha_{\lambda\neq0}$ from the condition of volume conservation by considering nucleus as an incompressible liquid. Let us truncate the expansion in Eq. (2.66) by the $\lambda = 2$ term in order to understand the physical meaning of the deformation parameter. If we let the radius for $\theta = 0$ and $\theta = \pi/2$ to be a and b, respectively,

---

[46] We do not consider $\lambda = 1$, because it describes the translational motion of the nucleus as a whole. There holds the relationship $\alpha_\lambda = \sqrt{(2\lambda+1)/4\pi}\,\alpha_{20}$ between the α_λ in this section and the deformation parameters $\alpha_{\lambda\mu}$, which will be introduced in Chap. 7.

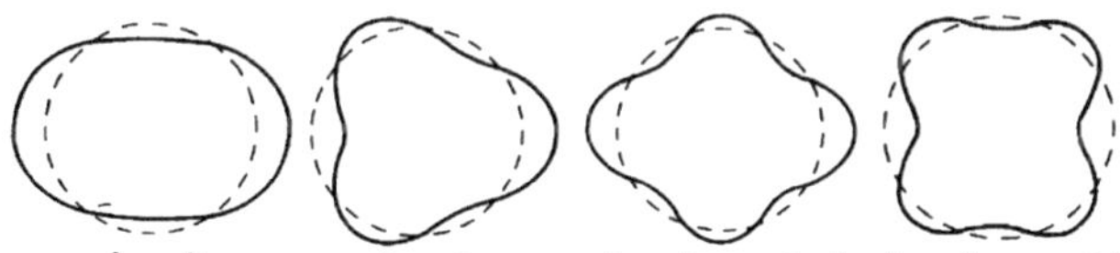

Fig. 2.21 Schematic picture for quadrupole, octupole and hexadecapole deformations

we obtain $a \equiv R(0) = R_0(1+\alpha_2)$, $b \equiv R(\pi/2) = R_0(1-\alpha_2/2) \sim R_0/\sqrt{1+\alpha_2}$ by using $P_2(\cos\theta) = (3\cos^2\theta - 1)/2$. One can see that one obtains a nuclear shape which is elongated (shrunk) along the symmetry axis for positive (negative) α_2. Also, as the expressions for a and b show, the volume is conserved for small deformation in accord with the property of a nucleus as an incompressible liquid. In Eq. (2.66), we assumed the symmetric fission, i.e., the reflection symmetry with respect to the plane which equally divides the symmetry axis. Correspondingly, the deformation parameters of odd order such as the octupole deformation $\lambda = 3$ have been omitted. In Fig. 2.21, we show the nuclear shape corresponding to $\lambda = 2, 3$ and 4.

Exercise 2.13 Confirm that the relation shown in Fig. 2.21 between the nuclear shape and the sign of $\alpha_4 = \sqrt{9/4\pi}\alpha_{40}$ in the presence of hexadecapole deformation holds by using the explicit functional form of the Legendre polynomial $P_4(z)$, $P_4(z) = (35z^4 - 30z^2 + 3)/8$.

(2) Stability with Respect to Fission: Fissility Parameter Both the surface energy E_{surf} and the Coulomb energy E_{C} change when a nucleus changes its shape. Hence it is necessary to study how they depend on the deformation parameters in order to evaluate the potential barrier with respect to fission. The following formulae can be obtained from the mass formula:

$$E_{\text{surf}} = b_{\text{surf}} A^{2/3}\left[1 + \frac{2}{5}\alpha_2^2 + \frac{5}{7}\alpha_3^2 + \sum_{n\geq 4}\frac{(n-1)(n+2)}{2(2n+1)}\alpha_n^2\right], \tag{2.67}$$

$$E_{\text{C}} = b_{\text{C}}\frac{Z^2}{A^{1/3}}\left[1 - \frac{1}{5}\alpha_2^2 - \frac{10}{49}\alpha_3^2 - \sum_{n\geq 4}\frac{5(n-1)}{(2n+1)^2}\alpha_n^2\right]. \tag{2.68}$$

If we consider within the framework of the quadrupole deformation and keep terms only up to the second order of the deformation parameter, the change of the energy by deformation is given by

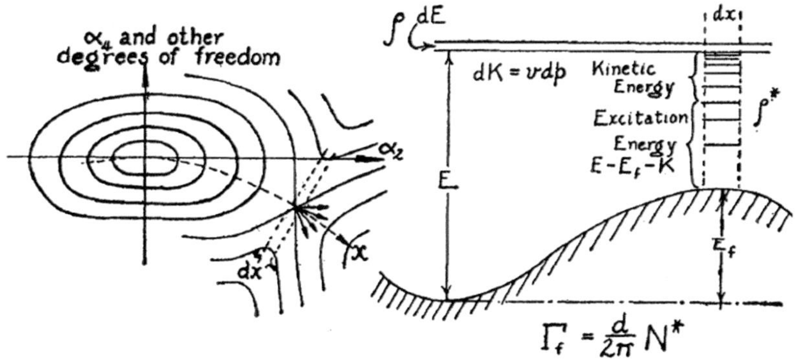

Fig. 2.22 A schematic illustration of the multi-dimensional potential surface for fission and the potential barrier. After [29]

$$\Delta E = \frac{1}{5} b_C \alpha_2^2 A^{2/3} \left(\frac{Z^2}{A}\right)_{cr} (1 - \chi), \tag{2.69}$$

$$\left(\frac{Z^2}{A}\right)_{cr} = 2 \times \frac{b_{surf}}{b_C} \sim 50.3, \tag{2.70}$$

$$\chi \equiv \left(\frac{Z^2}{A}\right) \Big/ \left(\frac{Z^2}{A}\right)_{cr}. \tag{2.71}$$

We see that there appears a potential barrier for the nuclei with $\chi < 1$ and they become metastable with respect to fission. On the contrary, nuclei with a large atomic number for which $\chi > 1$ have no potential barrier and immediately disintegrate by fission.[47] The χ is called the *fissility parameter*.

(3) The Height of the Fission Barrier In reality, the spontaneous fission is a quantum tunneling in a multidimensional space specified either by various deformation parameters or by appropriately chosen coordinates. Figure 2.22, which has been quoted from a paper by Bohr and Wheeler, schematically illustrates fission. The lines in the figure on the left side represent equipotential surfaces. In a simplified picture, one considers that a spherical metastable state decays along a path going through the *saddle point*. Let us define the height of the fission barrier as the height of the saddle point measured from the potential minimum corresponding to the metastable state for fission, and denote it by E_f. The value of E_f can be approximately written down as a function of χ if the fissility parameter χ is small or close to 1. An important point in doing so is that the position of the saddle point moves from a configuration

[47]The critical nucleus is $(A_{cr}, Z_{cr}) \sim (340, 131)$ if we use $N \sim 1.6Z$. However, in more refined estimates which go beyond the quadrupole deformation we used here the potential barrier by the liquid-drop model almost disappears when the atomic number exceeds the critical value $Z_{cr} \sim 100$ [30].

with small deformation near a in Fig. 2.18 to a configuration near the scission point d as the fissility parameter gets smaller starting from 1. If we take a simplification that the scission point coincides with the saddle point when $\chi = 0$, E_f can be estimated as

$$E_f = 2b_{\rm surf}\left(\frac{A}{2}\right)^{2/3} - b_{\rm surf}A^{2/3} + 2b_{\rm C}\left(\frac{Z}{2}\right)^2 \Big/ \left(\frac{A}{2}\right)^{1/3} + \frac{5}{3}b_{\rm C}\left(\frac{Z}{2}\right)^2 \Big/ 2\left(\frac{A}{2}\right)^{1/3} - b_{\rm C}Z^2/A^{1/3}\,. \tag{2.72}$$

By taking the ratio to the surface energy, we have

$$\begin{aligned}\frac{E_f}{E_{\rm surf}} &\equiv f(\chi)\\ &= (2^{1/3}-1) + \left(2^{1/3} + \frac{5}{3}\frac{1}{2\times 2^{2/3}} - 2\right)\chi\\ &\approx 0.260 - 0.215\chi\,.\end{aligned} \tag{2.73}$$

On the contrary, the saddle point appears in the region of small deformation when χ is close to 1. Hence it makes sense to express the energy surface in the expansion with respect to deformation parameters. It is, of course, needed to include terms beyond the second order in order to determine the position and the height of the saddle point. By keeping up to the fourth order with respect to α_2 and second order concerning α_4, we have

$$\begin{aligned}\Delta E_{{\rm surf}+C} = b_{\rm surf}A^{2/3}&\left(\frac{2}{5}\alpha_2^2 + \frac{116}{105}\alpha_2^3 + \frac{101}{35}\alpha_2^4 + \frac{2}{35}\alpha_2^2\alpha_4 + \alpha_4^2\right)\\ - b_{\rm C}\frac{Z^2}{A^{1/3}}&\left(\frac{1}{5}\alpha_2^2 + \frac{64}{105}\alpha_2^3 + \frac{58}{35}\alpha_2^4 + \frac{8}{35}\alpha_2^2\alpha_4 + \frac{5}{27}\alpha_4^2\right)\,.\end{aligned} \tag{2.74}$$

If we determine α_4 for a given α_2 by a variational principle, we obtain

$$\alpha_4 = -\frac{243}{595}\alpha_2^2\,, \tag{2.75}$$

since there exists a coupling term between α_2 and α_4. In consequence, when χ is close to 1, the height of the potential barrier is given by

$$\frac{E_f}{E_{\rm surf}} \equiv f(\chi) = \frac{98}{135}(1-\chi)^3 - \frac{11368}{34425}(1-\chi)^4 + \cdots\,, \tag{2.76}$$

in units of the surface energy.

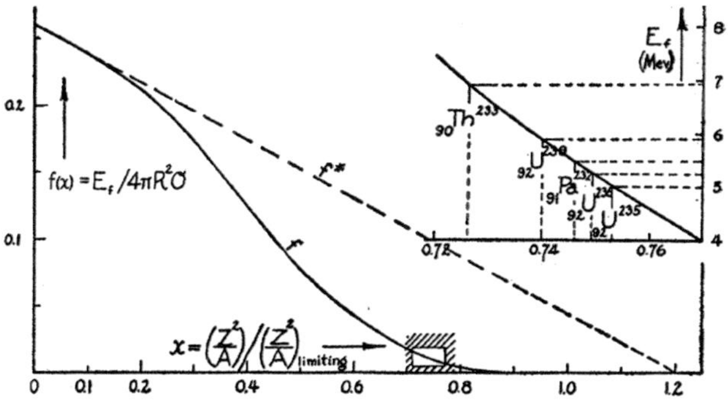

Fig. 2.23 The fissility parameter dependence of the height of the fission barrier plotted as the ratio to the surface energy. After [29]

Figure 2.23 shows the behaviour of $f(\chi)$ after the paper by Bohr and Wheeler [29]. A line which interpolates Eqs. (2.73) and (2.76) is shown as $f(\chi)$. The $f^*(\chi)$ represents $f(\chi)$ given by Eq. (2.73). The value of E_f which is expected when the surface energy for the shaded region around $\chi \sim 0.75$ is set to 530 MeV is shown in the inset on the right corner, since the surface energy does not so change in that region.[48]

The experimental value of E_f can be deduced, for example, from the excitation function of (n,f), (t,pf), (t,df), (d,pf), (γ,f) reactions which are induced by bombarding a target nucleus with either light nuclei or photons.[49] As an example, Fig. 2.24 shows the excitation functions of the fission probabilities for the ^{239}U(n,f) and for the ^{239}U(d,pf) reactions. They are denoted by the solid line and the dots, respectively. Both fission excitation functions exhibit a characteristic threshold rise followed by a flat plateau. A reasonable estimate of the barrier height can be obtained by identifying it to the energy at which the fission probability becomes 1/2 of its plateau value. It is because the tunneling probability or the barrier transmission probability becomes $1/2$ when the decay energy E coincides with the height of the fission barrier, as can be seen from the following formula for the barrier transmission probability in the uniform approximation:

$$P_{\mathrm{UA}} = \frac{1}{1 + \exp(2\int_{r_<}^{r_>} \mathrm{d}r \sqrt{2\mu(V(r) - E)}/\hbar)}, \tag{2.77}$$

which is valid from below to above the barrier as long as the fission barrier is a quadratic function or can be well approximated by a quadratic function. In Eq. (2.77),

[48]Bohr and Wheeler estimated the value of E_f for the other nuclei by assuming $(Z^2/A)_{\mathrm{cr}}$ to be 47.8 from the experimental data that E_f is about 6 MeV for ^{239}U.

[49]The scattering cross section plotted as a function of the collision energy is called the excitation function.

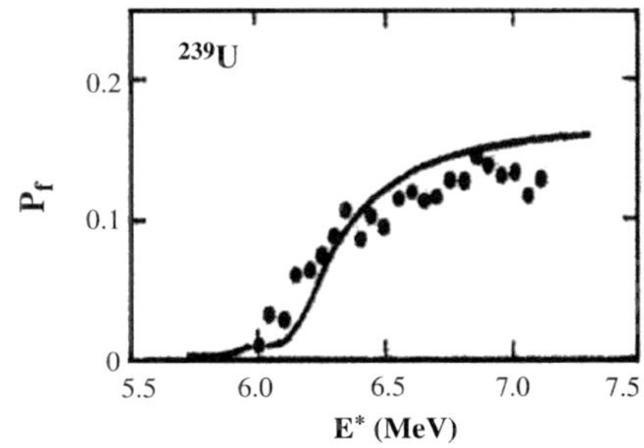

Fig. 2.24 The excitation functions of the induced-fission probabilities of ^{239}U [27]. The *solid line* is for (n,f) and the *dots* are for (d,pf). The abscissa is the excitation energy

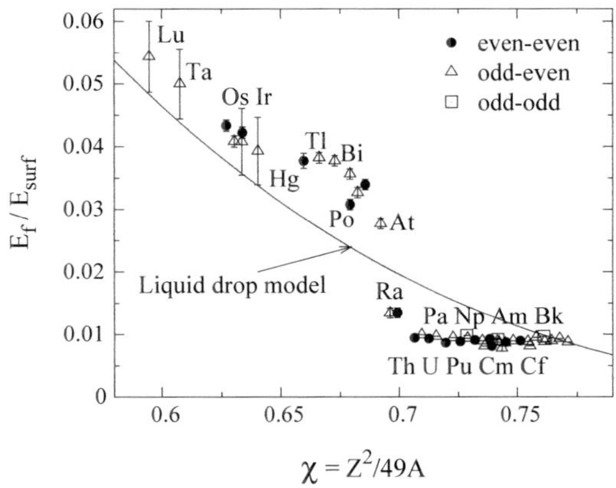

Fig. 2.25 Fission barriers. The data are taken from the compilations in [27]

$r_<$ and $r_>$ are the classical turning points on both sides of the potential barrier: $V(r_<) = V(r_>) = E$, and μ is the reduced mass.[50] The two turning points coincide when the decay energy E matches the height of the fission barrier, so that the resulting barrier transmission probability becomes $1/2$.

Figure 2.25 shows the experimental values of E_f (MeV) in units of the surface energy, which has been assumed to be given by $E_{\rm surf} = 17A^{2/3}$ MeV. The solid line has been calculated by keeping only the first term on the right-hand side of Eq. (2.76). Also, we presented in Table 2.6 the experimental values of $E_f^{\rm exp.}$ together with the predictions by Bohr and Wheeler $E_f^{\rm BW}$ and the binding energy of a neutron $B_{\rm n}^{\rm exp.}$ in the compound nucleus.

What is important in connection with the reactor, which we learn in Sect. 2.3.5, is the sign of $E_f - B_{\rm n}$. If the sign is negative, fission is easily induced even if the energy of the incident neutron is 0. On the other hand, if the sign is positive, an appropriate

[50]In classical mechanics, a particle is reflected at the position where the total energy of the particle becomes identical with the potential energy. Such positions are therefore called *classical turning points* or simply *turning points*.

Table 2.6 The threshold energy for fission E_f and the binding energy of neutron B_n, in units of MeV. The former and the latter are taken from [27, 31], respectively

Nucleus (compound nucleus)	$E_f^{\text{exp.}}$	$B_n^{\text{exp.}}$	$E_f^{\text{exp.}} - B_n^{\text{exp.}}$	E_f^{BW}
^{233}Th	6.4	4.8	1.6	6.9
^{232}Pa	6.3	5.6	0.7	5.5
^{235}U	5.8	5.3	0.5	5.0
^{236}U	5.9	6.5	−0.6	5.25
^{239}U	6.2	4.8	1.4	5.95
^{240}Pu	5.9	6.5	−0.6	

amount of energy has to be supplied in order to induce fission with a large probability. As we learn later, it is the reason why fission can be induced for ^{235}U by a thermal neutron with an extremely small energy, while it is necessary to inject a neutron with the energy of the order of MeV in order to induce fission of ^{238}U. The difference between the two arises from the difference between neutron binding energies in the corresponding compound nuclei, ^{236}U and ^{239}U. Because of the pairing correlation appearing in the mass formula, the B_n for the former is larger by Δ than the value estimated in the simple liquid-drop model without the correction term δ. On the other hand, it gets smaller by Δ in the latter. In fact, the difference between the two $B_n(^{236}\text{U}) - B_n(^{239}\text{U}) = 1.7\,\text{MeV}$ is close to $2\Delta \approx 2 \times \frac{12}{\sqrt{238}} \sim 1.6\,\text{MeV}$.[51]

(4) Spontaneous Fission Half-Lives Figure 2.26 shows the spontaneous fission half-lives $t_{1/2}$ for several actinides (U, Pu, Cm, Cf, Fm) as a function of the fissility parameter χ (filled circles are for even–even nuclei, open circles are for even–odd nuclei. Exactly speaking, the abscissa is slightly different from the χ defined in this book, because a symmetry energy correction is added to the surface energy term in the mass formula). Note that the experimental data lie by and large along a straight line in the semi-logarithmic graph. The detailed structures come from the deviation of the ground state masses from the prediction of the liquid-drop model because of the shell effect. Indeed, if we denote the deviation of the ground state mass from that of the smooth liquid-drop model by δm (MeV), then the values of $\log_{10} t_{1/2}\,(\text{yr}) + 5\delta m$ beautifully align as a function of χ [32] as shown in Fig. 2.27.

The feature shown in Figs. 2.26 and 2.27 can be understood as follows from the point of view of quantum tunneling. Let us denote the curvature of the potential surface for fission in the vicinity of the potential pocket which yields the metastable state by $\omega_0 \equiv (\frac{\partial^2 V}{\partial r^2}/\mu_f)^{1/2}$, where r is an appropriately chosen fission coordinate and μ_f is the corresponding mass parameter. Let us then approximate the potential surface in the vicinity of the potential barrier by a quadratic function with curvature Ω_{fb}. Furthermore, assuming that fission occurs with energy much lower than the

[51] The reactor must have become fairly different from the present one if there existed no pairing correlation between nucleons.

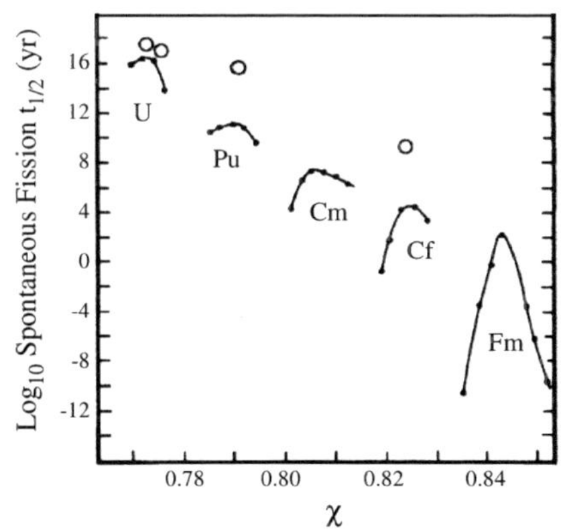

Fig. 2.26 Spontaneous fission half-lives of even–even (*filled circles*) and even–odd (*open circles*) nuclides as a function of the fissility parameter [27]

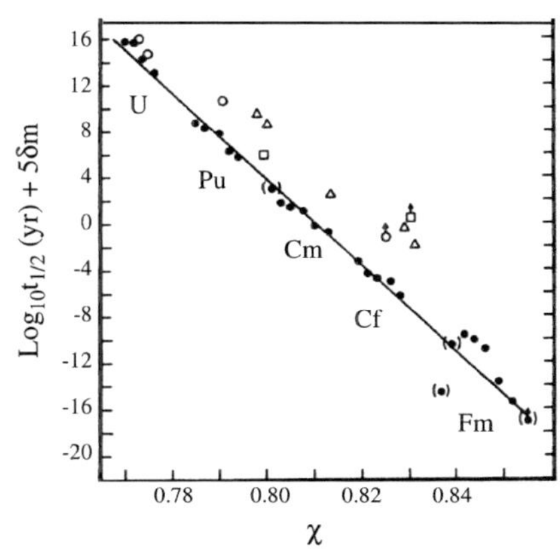

Fig. 2.27 Spontaneous fission half-lives corrected with shell correction [27]

height of the fission barrier, let us represent the tunneling probability in the standard WKB formula by ignoring 1 in the denominator of the formula in the uniform approximation (2.77). The decay width for fission is then given by

$$\Gamma_f = 5 \times \frac{\hbar\omega_0}{2\pi} P_{\text{fis}} \tag{2.78}$$

$$\sim 5 \times \frac{\hbar\omega_0}{2\pi} \exp\left[-\frac{2\pi}{\hbar\Omega_{\text{fb}}}(E_f - E)\right]. \tag{2.79}$$

In Eq. (2.78), following Bohr and Wheeler, we introduced the statistical factor 5 to represent the degeneracy of the quadrupole oscillation which induces fission. Although E_f is given by Eq. (2.76), it will be well represented as $E_f = a - b\chi$ in the linear approximation, since Figs. 2.26 and 2.27 are intended for a narrow region of χ. Finally, we obtain

$$\log_{10} t_{1/2} \sim A - B\frac{2\pi}{\hbar\Omega_{\rm fb}}\chi \tag{2.80}$$

using two constants A and B. One can thus understand why a linear relationship holds between the fission half-life and the fissility parameter in the semi-logarithmic plot.

Exercise 2.14 Estimate the curvature of the fission barrier $\hbar\Omega_{\rm fb}$ for actinides from Fig. 2.26 or 2.27.[52]

2.3.5 *Application to Nuclear Power Generation*

The nuclear power generation in operation today exploits the energy release, called *nuclear energy*, due to fission. In this subsection, we learn the principles of reactor and several key words.

2.3.5.1 Nuclear Reactions in Reactor, Principles of Reactor, Chain Reaction, Neutron Moderator, Neutron Absorber

Figure 2.28 illustrates the nuclear reactions induced by injecting a neutron into a mass of uranium which is a mixture of ^{235}U and ^{238}U. ^{235}U disintegrates into two large nuclei with about half the atomic and mass numbers of ^{235}U (see Fig. 2.17), and two to three neutrons on average.

The phenomenon shown in Fig. 2.28 can be understood as follows. As already stated concerning the Q-value for fission, the nuclear energy of around 200 MeV is

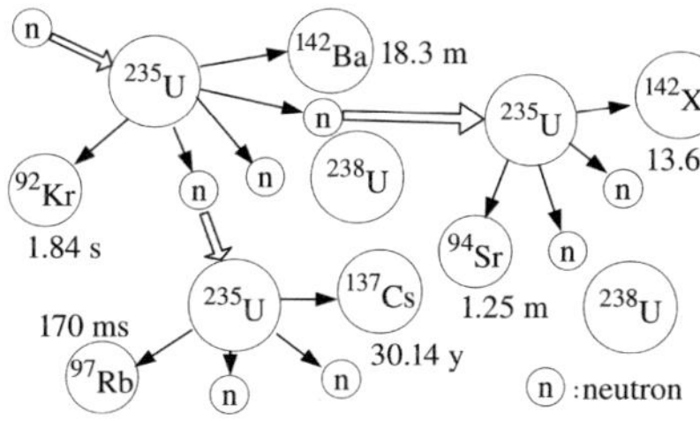

Fig. 2.28 Illustration of nuclear reactions in reactor (*chain reaction*)

[52]The information on both the potential surface and the effective mass is needed in order to theoretically evaluate the fission width. Nowadays, the potential surface can be obtained with relatively high accuracy using, for example, the macroscopic–microscopic method, which refines the liquid-drop model by adding the shell correction and will be described in Sect. 7.6.1. On the contrary, it is difficult to estimate the mass parameter with high reliability. One can hope to obtain important phenomenological information on the effective mass for fission by using the results of this exercise.

released per fission. Though most of the released energy is used as the kinetic energy of the fission fragments, a part of it becomes their excitation energies. Roughly speaking, the fission fragment loses its excitation energy by emitting one neutron if the excitation energy amounts to 8 MeV, which is an average binding energy of nucleon. The resulting nucleus is usually a neutron-rich unstable nucleus, so that it continues to lose its energy by repeating β-decay. Since the β-decay proceeds in obedience to selection rules [33], the resulting daughter nucleus is not necessarily in the ground state, but sometimes is above the threshold energy for neutron emission. In that case, the state decays by emitting a neutron. Since the β-decay usually takes place in the time scale of the order of seconds to minutes, those neutrons emitted after the β-decay are emitted with a time delay compared with the neutrons emitted immediately after fission. Those neutrons are therefore called *delayed neutrons*. In comparison, the neutrons emitted immediately after fission are called *prompt neutrons*. In consequence, two large fragments and two to three neutrons are emitted by the decay of U.[53] The average number of neutrons emitted in the thermal-neutron-induced fission of ^{235}U is 2.47 as stated before, although it fluctuates due to the existence of various decay processes as shown in Fig. 2.17.

The emitted neutrons are absorbed by the surrounding ^{235}U, and successively induce fission leading to a *chain reaction*.[54] It is the principle of reactor to utilize the nuclear energy by appropriately removing the emitted neutrons with an absorber called *control rod* in order to keep the number of thermal neutrons in the space constant, and thus by safely sustaining the chain reaction. A minimum condition is needed for the amount of U fuel in order to sustain the chain reaction, which is called the *critical mass*.

2.3.5.2 The Neutron Absorption Cross Section, Enriched Uranium, and Neutron Moderator

The cross section of the neutron-induced fission of U strongly depends on the isotopes and the energy of the incident neutron. Figure 2.29 shows the neutron-induced fission cross section of ^{235}U and ^{238}U (cited from [33], see also [34]). The figure shows that ^{238}U does not cause fission unless the energy of neutron becomes as high as 1 MeV. This is the reason why no fission of ^{238}U is seen in Fig. 2.28. On the other hand, the energy of thermal neutron $1/40$ eV is sufficient to induce fission of ^{235}U. The difference between the threshold energies for the induced fission of ^{235}U and ^{238}U originates from the pairing correlation between nucleons as stated before.

---

[53] Most of the neutrons are the prompt neutrons which are emitted at times shorter than 4×10^{-14} s. Though the ratio of the delayed neutron is small, it plays an important role in controlling nuclear reactors [6, 33, 34].

[54] It is Fermi and his collaborators who succeeded in the chain reaction of the induced fission of U by a slow neutron for the first time, and the experiment was performed at the University of Chicago on December 2, 1942.

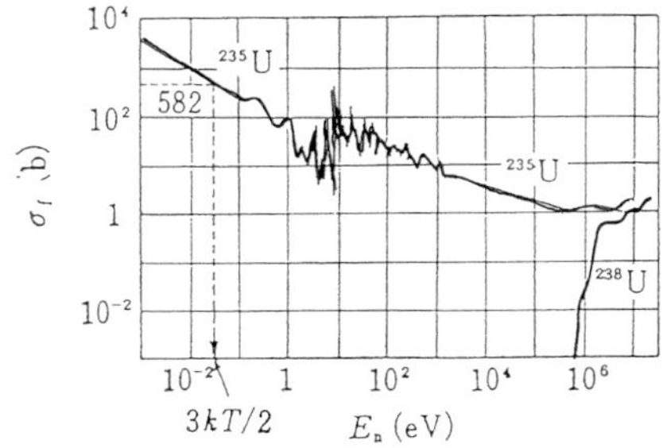

Fig. 2.29 Neutron induced fission cross section [33]

Because of this difference, the *enriched uranium*, which has the composition of ^{235}U with a larger percentage than that of the naturally occurring uranium 0.720%, is used as the fuel for reactors.[55]

Figure 2.29 shows that the induced fission cross section of ^{235}U decreases nearly proportionally to the inverse of the velocity of the incident neutron v as the energy of neutron increases (*1/v-law*). We learn the reason in Sect. 9.3. On the other hand, the energy spectrum of the neutrons emitted in the thermal neutron induced fission of ^{235}U can be well represented as

$$N(E) \propto \sqrt{E}\exp(-E/E_0) \sim \sqrt{E}\exp(-E/1.29\,\text{MeV}) \tag{2.81}$$

using the laboratory energy E, and distributes with its maximum at near $E = E_0/2 \sim 0.65$ MeV. Figure 2.30 shows the laboratory neutron kinetic energy spectrum for thermal neutron-induced fission of ^{235}U. The average energy of the emitted neutrons is $\bar{E} = 1.5E_0 \sim 2$ MeV.[56] Although these fast neutrons can induce fission of ^{238}U, the cross section is much smaller than that of the thermal neutron-induced fission of ^{235}U as Fig. 2.29 shows. Moreover, the induced fission cross section of ^{235}U by high-energy neutrons is small, so that the cross section becomes effectively small. The neutron moderator is therefore introduced in reactors so as to reduce the kinetic energy of neutrons to that of the thermal neutrons.[57]

Exercise 2.15 Summarize why enriched uranium is used for nuclear reactors.

Sidebar: Natural Nuclear Fission Reactor A number of uranium ores, whose isotope ratio of ^{235}U is significantly smaller than the standard value 0.720% in the

[55] In commercial reactors, the enrichment of ^{235}U is typically about 3–4%.

[56] The center of mass energy spectrum of neutrons emitted from a nucleus in the excited state with temperature T is expected to be given by $\varepsilon_n \exp(-\varepsilon_n/T)$, and is well described by the theory called the *evaporation theory*. The dashed line in Fig. 2.30 represents the energy spectrum expected from the evaporation theory transformed to that in the laboratory frame by taking into account the finite velocity of the fission fragment.

[57] The reactor which uses high energy neutrons, instead of neutrons which have been cooled down to thermal neutrons due to the elastic scattering in the neutron moderator, is called *fast reactor* or *fast neutron reactor*.

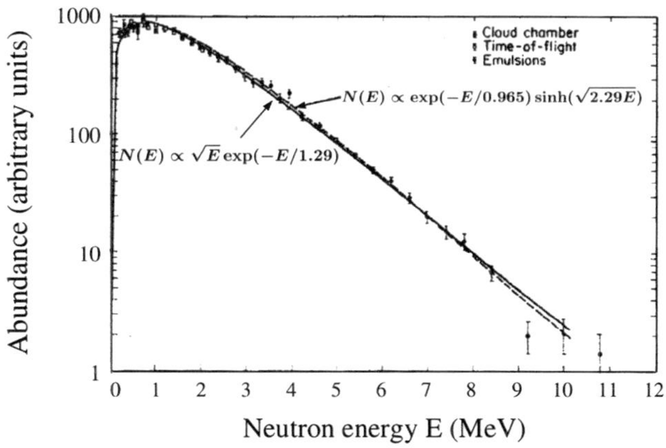

Fig. 2.30 Kinetic energy spectrum of neutrons (Laboratory system) [27]

naturally occurring uranium, have been discovered in the uranium mines at Oklo in Gabon (Gabonese Republic), Africa. This suggests that there has been once natural nuclear fission reactors in the uranium deposits at Oklo.

2.3.5.3 Nuclear Waste Problem: Transmutation

Many of the fission products are radioactive nuclei with finite lifetimes as written in Fig. 2.28. In particular, the relatively long half-lives of ${}^{137}_{55}$Cs and ${}^{90}_{38}$Sr are 30.14 and 28.78 years, respectively, which are of the same order as the human life span. In addition, transuranium elements such as ${}^{237}_{93}$Np, whose half-life exceeds 200 million years, are produced. In the nuclear power industry this is a serious nuisance. These radioactivities cause a biological hazard in discarded nuclear waste for a long time. It is of critical importance to handle these long-lived nuclear wastes, and the studies of nuclear transmutation are going on.

2.3.6 Fission Isomers

At the end of this section, let us learn about *fission isomers*. Figure 2.31 shows the excitation function of the neutron-induced fission of ^{230}Th, i.e., ${}^{230}_{90}$Th(n,f). Unlike Fig. 2.24, there appears a clear resonance structure with the width of 40 keV at the incident energy of neutron of about 720 keV. Figure 2.32 compares the excitation functions of the ${}^{230}_{90}$Th(n,f) and ${}^{230}_{90}$Th(d,pf) reactions, which have been obtained by

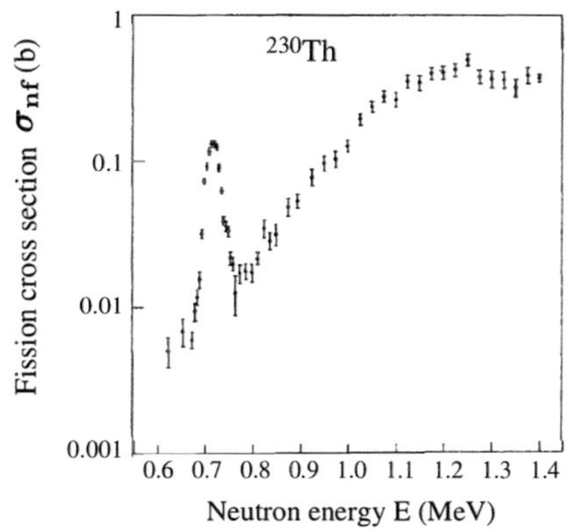

Fig. 2.31 The excitation function of the neutron induced fission of ^{230}Th, showing a resonance. Taken from [35]

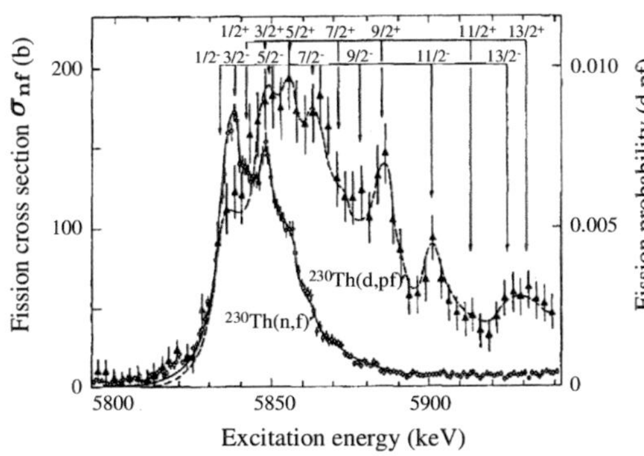

Fig. 2.32 Detailed structure of the excitation function of two induced fissions of ^{230}Th. Taken from [36]

later experiments with high resolution. The abscissa of Fig. 2.32 is the excitation energy of $^{231}_{90}$Th.

The excitation function of fission will be a monotonic function if the potential surface for fission has only a single barrier as predicted by the simple liquid-drop model and as has been shown in Fig. 2.19. In reality, the potential surface has multiple-barriers depending on the nucleus due to the shell correction energy representing the shell effect. The situation is shown in Fig. 2.33. It shows how the understanding for the structure of fission barrier has evolved with time.[58]

The structure seen in Fig. 2.31 was originally interpreted as the meta-stable β-vibrational level produced in the second potential minimum. However, using the fine

[58]One of the powerful methods to obtain the potential surface with multiple-potential barriers is the macroscopic–microscopic method which modifies the liquid-drop model by adding shell correction (see Sect. 7.6.1). In the fission of ^{231}Th, the first fission barrier is lowered by axial asymmetry. The axial symmetry is recovered in the second fission barrier. However, the barrier height gets lower by considering mass asymmetry, and furthermore there appears the third potential minimum.

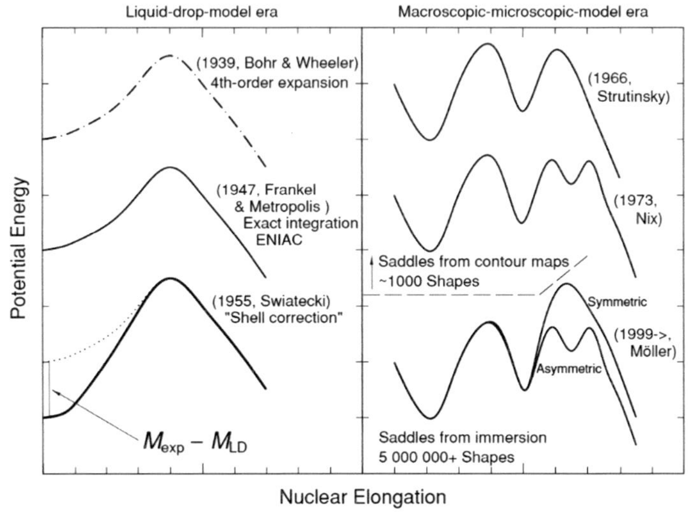

Fig. 2.33 Time evolution of the theory for fission barrier, taken from [37]

structures seen in Fig. 2.32 as the experimental evidence, these structures are now understood to be associated with the rotational motion involving parity doublet, i.e., the positive and negative parity rotational bands, which have almost the same moment of inertia, yielded in the third potential minimum associated with the mass asymmetric fission [36, 38] (see the right-bottom curves in Fig. 2.33). The position, width and spin–parity of each resonance state shown in the figure have been determined by the high-resolution measurement of the excitation function using, for example, the time of flight (TOF), the measurement of the angular distribution of fission fragments, the simultaneous measurements of (n,f) and (d,pf) reactions, and the theoretical analyses by the parity-doublet rotational model. The ratio of the length of the long axis to that of the short axis obtained from the level splitting is 3:1, and matches the deformation at the third potential minimum.[59]

Figure 2.34 shows a few energy levels near the ground state of ^{238}U and metastable states in the second potential well together with their spin and parity. The latter states are called fission isomers.

The fission isomers are a kind of either *superdeformed states*, which we learn in Sect. 7.6.2 and whose ratio of the length of the long axis to that of short axis is about 1.6–1.8, or *hyperdeformed states* which have even larger ratio of the lengths of two axes.[60] As we learn in Chap. 7, a characteristic of deformed states is that there

[59]The ratio of the length of the long axis to that of the short axis is 2:1 for the second potential minimum.

[60]The ratio of the length of the long axis to that of the short axis is about 1.3 for usual deformed nuclei such as the ground state of U. In comparison, the moment of inertia of fission isomers is 2.2

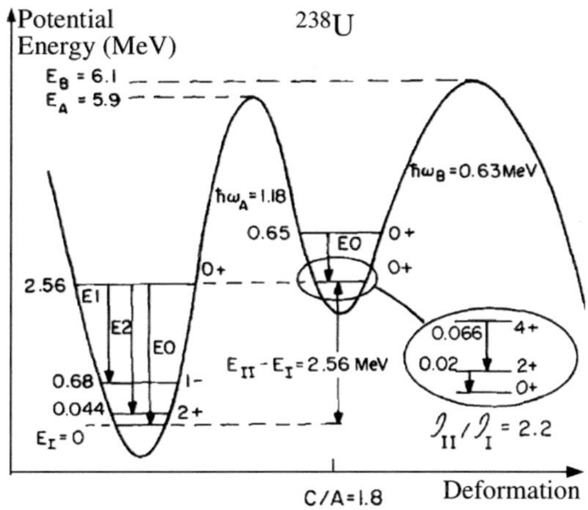

Fig. 2.34 The structure of ^{238}U. Fission isomers. Taken from [39]

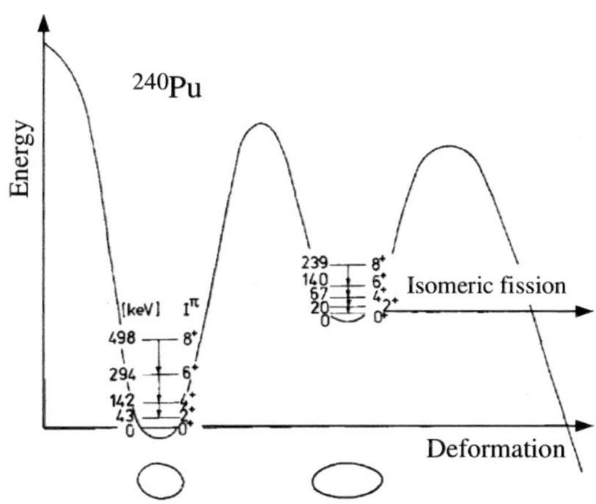

Fig. 2.35 The structure of ^{240}Pu. Fission isomers. Taken from [40]

appear a characteristic group of states called rotational bands. Figure 2.35 compares the rotational bands built on the ground state and on a fission isomer by taking the case of ^{240}Pu as an example. The large difference of the deformation between the two bands can be deduced from the difference of the corresponding level spacing.

(Footnote 60 continued)
times that of the ground state for ^{238}U as shown in Fig. 2.34. ^{8}Be, which has the dumb-bell structure of two touching α particles, has also 2:1 for the ratio of the length of the long axis to that of the short axis, and hence is one of the hyperdeformed states.

References

1. E. Segré, *From X-Rays to Quarks: Modern Physicists and Their Discoveries* (W. H. Freeman and Company, San Francisco, 1980)
2. M. Nogami, *Nuclear Physics*, Japanese edn. (Shoukabou, Tokyo, 1973)
3. T. de Forest Jr., J.D. Walecka, Adv. Phys. **15**, 1 (1966)
4. B. Hahn, D.G. Ravenhall, R. Hofstadter, Phys. Rev. **101**, 1131 (1956)
5. A. Tamii et al., Phys. Rev. Lett. **107**, 062502 (2011)
6. W.N. Cottingham, D.A. Greenwood, *An Introduction to Nuclear Physics*, 2nd edn. (Cambridge University Press, Cambridge, 2001)
7. R. Hofstadter, F. Bumiller, M.R. Yearian, Rev. Mod. Phys. **30**, 482 (1958)
8. N. Takigawa, M. Ueda, M. Kuratani, H. Sagawa, Phys. Lett. B **288**, 244 (1992)
9. J.S. Al-Khalili, J.A. Tostevin, Phys. Rev. Lett. **76**, 3903 (1996)
10. B. Abu-Ibrahim, Y. Suzuki, Phys. Rev. C **61**, 051601 (2000)
11. J. Kasagi, H. Yuki, T. Baba, T. Noda, T. Ohtsuki, A.G. Lipson, J. Phys. Soc. Jpn. **71**, 2881 (2002)
12. Y. Kato, N. Takigawa, Phys. Rev. C **76**, 014615 (2007). And references therein
13. G. Audi, A.H. Wapstra, C. Thibault, Nucl. Phys. A **729**, 337 (2003)
14. R. Shurtleff, E. Derringh, Am. J. Phys. **57**, 552 (1989)
15. T. Arima, Master thesis, Tohoku University
16. W.-M. Yao et al. (Particle Data Group), J. Phys. G 33, 1 (2006)
17. L. Alvarez et al., Phys. Rev. **105**, 1127 (1957)
18. S.E. Jones, Nature **321**, 127 (1986)
19. K. Ikeda, N. Takigawa, H. Horiuchi, Prog. Theor. Phys. Suppl. **E68**, 464 (1968)
20. K. Ikeda, T. Marumori, R. Tamagaki, H. Tanaka, J. Hiura, H. Horiuchi, Y. Suzuki, F. Nemoto, H. Bando, Y. Abe, N. Takigawa, M. Kamimura, K. Takada, Y. Akaishi, S. Nagata, Prog. Theor. Phys. Suppl. **52** (1972); K. Ikeda, R. Tamagaki, S. Saito, H. Horiuchi, A. Tohsaki-Suzuki, M. Kamimura, Prog. Theor. Phys. Suppl. **62** (1977)
21. H. Horiuchi, K. Ikeda, K. Kato, Prog. Theor. Phys. Suppl. **192**, 1 (2012)
22. C.M. Lederer, V. Shirley, *Table of Isotopes*, 7th edn. (Wiley, New York, 1978)
23. N. Takigawa, A. Arima, Nucl. Phys. A **168**, 593 (1971)
24. P. Möller, J.R. Nix, W.D. Myers, W.J. Swiatecki, At Data Nucl. Tables **59**, 185 (1995); P. Möller, J.R. Nix, Nucl. Phys. A **536**, 20 (1992); H. Koura, M. Uno, T. Tachibana, M. Yamada, Nucl. Phys. A **674**, 47 (2000); H. Koura, T. Tachibana, M. Uno, M. Yamada. Prog. Theor. Phys. **113**, 305 (2005)
25. S. Yusa, private communication
26. K.F. Flynn, L.E. Glendenin, ANL-7749 (1970); cited in [27]
27. R. Vandenbosch, J.R. Huizenga, *Nuclear Fission* (Academic Press, New York, 1973)
28. O. Hahn, F. Strassmann, Naturwiss. **27**, 11 (1939)
29. N. Bohr, J.A. Wheeler, Phys. Rev. **56**, 426 (1939)
30. F. Ivanyuk, K. Pomorski, Phys. Rev. C **79**, 054327 (2009); Int. J. Mod. Phys. E **19**, 514 (2010)
31. A.H. Wapstra, G. Audi, Nucl. Phys. A **432**, 1 (1985)
32. W.J. Swiatecki, Phys. Rev. **100**, 937 (1955)
33. K. Yagi, *Nuclear Physics*, Japanese edn. (Asakura, Tokyo, 1971)
34. W.S.C. Williams, *Nuclear and Particle Physics* (Clarendon press, Oxford, 1991)
35. G.D. James, J.E. Lynn, L.G. Earwaker, Nucl. Phys. A **189**, 225 (1972)
36. J. Blons, Nucl. Phys. A **502**, 121c (1989)
37. P. Möller, A.J. Sierk, T. Ichikawa, A. Iwamoto, R. Bengtsson, H. Uhrenholt, S. Åberg, Phys. Rev. C **79**, 064304 (2009)
38. P. Möller, J.R. Nix, in *Proceedings of the Third IAEA Symposium on the Physics and Chemistry of Fission, Rochester, 1973*, Vol. I (International Atomic Energy Agency, Vienna, 1974), p. 103
39. R. Vandenbosch, Nucl. Phys. A **502**, 1c (1989)
40. V. Metag, D. Habs, H.J. Specht, Phys. Rep. **65**, 1 (1980); The original paper is H.J. Specht, J. Weber, E. Konecny, D. Heunemann. Phys. Lett. B **41**, 43 (1972)

Chapter 3
The Nuclear Force and Two-Body Systems

Abstract The strong interaction between nucleons which is called the nuclear force plays the central role in nuclear physics. Compared with the Coulomb interaction between two charged particles, which plays the exclusive role in condensed matter physics, the nuclear force is fairly complicated reflecting the fact that nucleons are composite particles and have spin and isospin degrees of freedom. There are several different approaches to uncover nuclear force. The one is to consider the most general possibility under the constraints from the consideration of symmetry or transformation invariance. The other is a phenomenological approach. The meson theory is also an attractive historical approach, which is often combined with phenomenological approaches. Recently, approaches based on QCD are also introduced. A difficulty of nuclear force is that the effective force inside nucleus is fairly different from the bare interaction between two nucleons in free space. In this chapter we overview the basic properties and the current understanding of the nuclear force.

3.1 The Fundamentals of Nuclear Force

3.1.1 The Range of Forces—A Simple Estimate by the Uncertainty Principle

The four forces in nature, i.e., the gravitational force, the electromagnetic force, the weak interaction and the strong interaction, can be thought to originate from exchanging gauge particles, i.e., graviton, photon, weak bosons ($W^{\pm}$ and Z^0 particles) and mesons (or gluons at the more fundamental level of the quantum chromodynamics (QCD)), which mediate each force, between two particles. Following this idea, one can relate the range of each force to the mass of the corresponding gauge particles using the uncertainty principle.

If we denote the mass of a gauge particle by m, then the uncertainty of the energy associated with the propagation of the gauge particle is $\Delta E \sim mc^2$. Hence, the corresponding uncertainty of time is $\Delta t \sim \hbar/\Delta E \sim \hbar/mc^2$. Assuming that the gauge particle moves with the light velocity during that time interval, the range of the force d can be estimated to be

N. Takigawa and K. Washiyama, *Fundamentals of Nuclear Physics*,
DOI 10.1007/978-4-431-55378-6_3

$$d \sim c \times \Delta t \sim \frac{\hbar}{mc} , \tag{3.1}$$

and coincides with the *Compton wavelength* of the gauge particle. As is well known, the electromagnetic force reaches infinite distance, because the mass of photon is zero. On the other hand, the range of the weak interaction is very short, since the masses of $W^{\pm}$ and Z^0 particles are large. If we assume that the strong interaction is mediated by exchanging mesons as we learn later in the meson theory, the range of the force which originates from the exchange of the lightest meson, i.e., π-mesons, can be estimated to be about 1.4 fm by using that the rest energy of the π-mesons is about 140 MeV. Incidentally, although the graviton has not yet been discovered, its mass is expected to be zero, because the gravitational force has an infinite range like the electromagnetic force.

3.1.2 The Radial Dependence

Though we made a rough estimate of the range of forces using the uncertainty principle in the previous section, the strength of the nuclear force acting between two particles varies as a function of the distance between them similarly to that the strength of the Coulomb interaction gets weaker in inverse proportion to the square of the distance r between the two charged particles as their distance increases.

Corresponding to the fact that the strength of the Coulomb force is proportional to $1/r^2$, the Coulomb potential is given by $V_C(r) = Q_1 Q_2 e^2/r$, where $Q_1 e$ and $Q_2 e$ are the electric charges of the two charged particles. This formula can be derived by thinking that one of the charged particles produces an electric field through interaction with the electromagnetic field, and then exerts a force on the other charged particle via the electric field. The equation which determines the electric field in this case is known as the D'Alembert equation.

The same idea can be applied to the strong interaction. If we denote the field of the strong interaction by $\phi(\mathbf{r}, t)$, then the equation called Klein–Gordon equation, which is a generalization of the D'Alembert equation for the electric force, is given by

$$\left[\nabla^2 - \frac{1}{c^2}\frac{\partial^2}{\partial t^2} - \left(\frac{mc}{\hbar}\right)^2\right]\phi(\mathbf{r}, t) = -4\pi\rho(\mathbf{r}, t) , \tag{3.2}$$

where m is the mass of the gauge particle. It is the mass of a meson in the meson theory for the nuclear force. There appears on the left-hand side the Compton wavelength of the gauge particle because of the finiteness of the mass of the gauge particle. The $\rho(\mathbf{r}, t)$ appearing on the right-hand side represents the density at the spatial position $\mathbf{r}$ at time t of the charged particles which create the electric force in the D'Alembert equation for the Coulomb potential. It is the density of hadrons such as nucleons, which are the sources of mesons or gluons, in the case of the strong interaction.

In the static limit to ignore the time variation, the solution of Eq. (3.2) is given by

$$\phi(\mathbf{r}, t) \approx \int \rho(\mathbf{r}', t) \frac{\mathrm{e}^{-mc|\mathbf{r}-\mathbf{r}'|/\hbar}}{|\mathbf{r} - \mathbf{r}'|} \mathrm{d}\mathbf{r}' . \tag{3.3}$$

Exercise 3.1 Derive Eq. (3.3) by the following two methods 1 and 2.

1. The method of Fourier transformation
   a. Show that the Fourier transformation of $\phi(\mathbf{r})$ is given by

$$\tilde{\phi}(\mathbf{k}) = \frac{4\pi}{k^2 + (mc/\hbar)^2} \tilde{\rho}(\mathbf{k}, t) . \tag{3.4}$$

   b. Derive Eq. (3.3) by making the inverse Fourier transformation of $\tilde{\phi}(\mathbf{k})$.
2. The Green function method
   Define the Green function in the static case by

$$\left[\nabla^2 - \left(\frac{mc}{\hbar}\right)^2\right] G_N(\mathbf{r} - \mathbf{r}') = -4\pi\delta(\mathbf{r} - \mathbf{r}') , \tag{3.5}$$

   and then show that it becomes

$$G_N(\mathbf{r} - \mathbf{r}') = \frac{\mathrm{e}^{-mc|\mathbf{r}-\mathbf{r}'|/\hbar}}{|\mathbf{r} - \mathbf{r}'|} . \tag{3.6}$$

Exercise 3.2 Prove the following equation.

$$\nabla^2 \frac{1}{|\mathbf{r} - \mathbf{r}'|} = -4\pi\delta(\mathbf{r} - \mathbf{r}') . \tag{3.7}$$

The nuclear potential $\phi_N(r)$ can be obtained by letting $\rho(\mathbf{r}') = \rho_0\delta(\mathbf{r}')$ in the integral on the right-hand side of Eq. (3.3) in the limit of ignoring the size of the particle such as a nucleon which is the source of the force. Here, we took the position of the hadron which is the source of the force as the origin of the spatial coordinate. We can see that the distance dependence changes from the $1/r$ law for the electric force to $\mathrm{e}^{-\mu r}/r$ with $\mu = mc/\hbar$. This functional form including the exponential damping factor is called the Yukawa type, and the strength of the nuclear force becomes $\mathrm{e}^{-1} \approx 1/2.7$ in the distance of $1/\mu$. This distance corresponds to the range of the force which we obtained in the previous section and coincides with the Compton wavelength of the gauge particle.

3.1.3 *The State Dependence of Nuclear Force*

In the following, we discuss the interaction between two nucleons, called nuclear force, among the strong interactions. Since each nucleon has the spin degree of freedom, and also the degree of freedom in the isospin space, which has been touched in Sect. 1.2, in addition to that in the usual space, the nuclear force is much more complex than the Coulomb force, and causes a great variety of phenomena. In this section, we learn the classification of the states of two-nucleon systems, the exchange operators and the projection operators in preparation for studying more details about the nuclear force.

3.1.3.1 The Classification of the States of Two-Nucleon Systems

The wave function of two-nucleon systems ψ can be expressed as the product of the wave functions in the usual space, the spin space and the isospin space as

$$\psi(x_1, x_2) = \varphi(\mathbf{r})\zeta(s_1, s_2)\eta(t_1, t_2) \tag{3.8}$$

if there is no interaction between the coordinates in the usual and spin spaces such as the spin–orbit interaction which we will learn in Sect. 3.2.2. The s and t are the z-components of the coordinates in the spin and isospin spaces, the lower indices 1 and 2 refer to the first and the second nucleons, $\mathbf{r} \equiv \mathbf{r}_1 - \mathbf{r}_2$ is the coordinate for the relative motion, $\mathbf{r}_1$ and $\mathbf{r}_2$ the spatial coordinates of the first and the second nucleons, respectively. Also, x is used to represent the coordinates in the usual, spin and isospin spaces altogether simultaneously, i.e., $x = (\mathbf{r}, s, t)$.

If one uses the simultaneous eigenstates of the square $\hat{\mathbf{S}}^2$ of the total spin $\hat{\mathbf{S}} \equiv \hat{\mathbf{s}}_1 + \hat{\mathbf{s}}_2$ and its z-component $\hat{S}_z$, $|S, S_z\rangle$, then the wave function in the spin space ζ is one of the following states,

$$|1, 1\rangle_{\mathscr{S}} = \alpha_1\alpha_2\ , \tag{3.9}$$

$$|1, 0\rangle_{\mathscr{S}} = \frac{1}{\sqrt{2}}[\alpha_1\beta_2 + \beta_1\alpha_2]\ , \tag{3.10}$$

$$|1, -1\rangle_{\mathscr{S}} = \beta_1\beta_2\ , \tag{3.11}$$

$$|0, 0\rangle_{\mathscr{S}} = \frac{1}{\sqrt{2}}[\alpha_1\beta_2 - \beta_1\alpha_2]\ . \tag{3.12}$$

We ignored the coordinates on the left-hand side, but used the index $\mathscr{S}$ to explicitly indicate that it is the wave function in the spin space. On the right-hand side, we expressed the coordinates s_1 and s_2 simply with the lower index 1 and 2. The α and β represent the spin-up and spin-down states, $|\frac{1}{2}\frac{1}{2}\rangle$ and $|\frac{1}{2} - \frac{1}{2}\rangle$, respectively. The states with $S = 1$ and $S = 0$ are symmetric and antisymmetric concerning the two nucleons, respectively, and does not change and changes the sign when the two nucleons are interchanged. The former and the latter are called the spin-triplet and the spin-singlet states, respectively.

Similarly, the wave function in the isospin space η can be expressed using the following simultaneous eigenstates of the square of the total isospin and the magnitude of its z-component,

$$|1,1\rangle_{\mathscr{I}} = \mathrm{n_1 n_2}\,, \tag{3.13}$$

$$|1,0\rangle_{\mathscr{I}} = \frac{1}{\sqrt{2}}[\mathrm{n_1 p_2} + \mathrm{p_1 n_2}]\,, \tag{3.14}$$

$$|1,-1\rangle_{\mathscr{I}} = \mathrm{p_1 p_2}\,, \tag{3.15}$$

$$|0,0\rangle_{\mathscr{I}} = \frac{1}{\sqrt{2}}[\mathrm{n_1 p_2} - \mathrm{p_1 n_2}]\,, \tag{3.16}$$

and each state can be classified into either isospin-triplet or isospin-singlet states. The lower index $\mathscr{I}$ means isospin space. As defined in Eq. (1.3), n and p represent neutron (isospin-up) and proton (isospin-down) states, respectively.

On the other hand, the wave function for the relative motion can also be classified into either symmetric or antisymmetric states with respect to the exchange of two nucleons. Since the exchange of two particles corresponds to the parity transformation (or the parity inversion) of the coordinate of the relative motion **r**, the former state is called E-state in the sense of even, while the latter O-state in the sense of odd state. For example, if the wave function $\varphi(\mathbf{r})$ is given by $\varphi(\mathbf{r}) = R_\ell(r)Y_{\ell m}(\theta,\phi)$ by using the spherical harmonics $Y_{\ell m}$, then the states with an even number for the orbital angular momentum ℓ such as the s-wave are E-states, while those with odd-ℓ values such as the p-wave are O-states.

Let us now represent the state for two-nucleon systems as ${}^{2T+1\,2S+1}\mathrm{E}$ or ${}^{2T+1\,2S+1}\mathrm{O}$ when the total spin and the total isospin are S and T. The symbols E and O represent the symmetry property, i.e., the even–odd property, of the wave function of the relative motion. Since nucleons are fermions, the wave function of two-nucleon systems has to change sign for the simultaneous coordinate interchange in the whole space. Namely, $\psi(x_2,x_1) = -\psi(x_1,x_2)$. Hence only the following four states, ${}^{13}\mathrm{E}$, ${}^{31}\mathrm{E}$, ${}^{11}\mathrm{O}$, ${}^{33}\mathrm{O}$, are allowed for two-nucleon systems. For example, ${}^{13}\mathrm{E}$ is called singlet triplet even state. A characteristic of the nuclear force is that it significantly differs in these four states.

3.1.3.2 The Exchange Operators

(1) The Spin-Exchange Operator Various exchanges take place when two nucleons interact. For example, when a proton in the spin-up state collides with another proton in the spin-down state, the orientation of the spin is interchanged between them with a certain probability (spin exchange). This happens because the nuclear force contains terms which are proportional to the inner product $\hat{\boldsymbol{\sigma}}_1 \cdot \hat{\boldsymbol{\sigma}}_2$ of the spins of the two nucleons. We can rewrite the inner product as

$$\hat{\boldsymbol{\sigma}}_1 \cdot \hat{\boldsymbol{\sigma}}_2 = \frac{1}{2}(\hat{\sigma}_{1+}\hat{\sigma}_{2-} + \hat{\sigma}_{1-}\hat{\sigma}_{2+}) + \hat{\sigma}_{1z}\hat{\sigma}_{2z} \tag{3.17}$$

by using the raising and lowering operators of the spin. The spin exchange processes are caused by the first and the second terms on the right-hand side.

It is convenient to introduce the following operator in order to describe the spin exchange effect,

$$P_\sigma = \frac{1}{2}(1 + \hat{\boldsymbol{\sigma}}_1 \cdot \hat{\boldsymbol{\sigma}}_2) \ . \tag{3.18}$$

One can easily show that $P_\sigma|1M\rangle_{\mathscr{S}} = |1M\rangle_{\mathscr{S}}$ $(M = 1, 0, -1)$ and $P_\sigma|00\rangle_{\mathscr{S}} = -|00\rangle_{\mathscr{S}}$ by operating P_σ on the spin-triplet and spin-singlet states. Since the spin-triplet and spin-singlet states are symmetric and antisymmetric regarding the exchange of the spins of two nucleons, this means that P_σ has the same function as the exchange of the spins of the two nucleons. For this reason, P_σ is called the *spin-exchange operator*, alternatively the *Bartlett operator*.

(2) The Charge-Exchange Operator The exchange of isospin can also occur in two-nucleon systems. For example, a proton and a neutron are interchanged when they collide (charge exchange). In order to describe this phenomenon, it is convenient to introduce the following *charge-exchange operator*, which is also called the *isospin-exchange operator* or the *Heisenberg operator*, by using the isospin operator τ of each nucleon,

$$P_\tau = \frac{1}{2}(1 + \hat{\boldsymbol{\tau}}_1 \cdot \hat{\boldsymbol{\tau}}_2) \ . \tag{3.19}$$

(3) The space-Exchange Operator (Majorana Operator) One can also consider the operator which interchanges the coordinates of two nucleons in the usual space. It is denoted by $P_{\rm M}$,

$$P_{\rm M}\varphi(\mathbf{r}) = \varphi(-\mathbf{r}) \ . \tag{3.20}$$

Since the wave function for two-nucleon systems has to be antisymmetric in whole space,

$$P_{\rm M} P_\tau P_\sigma \psi = -\psi \ . \tag{3.21}$$

Hence,

$$P_{\rm M} P_\tau P_\sigma = -1 \ . \tag{3.22}$$

One can therefore write

$$P_{\rm M} = -P_\tau P_\sigma \ . \tag{3.23}$$

3.1.3.3 The Projection Operators

As we see later, the nuclear force strongly depends on the states of two nucleons through the $\hat{\boldsymbol{\sigma}}_1 \cdot \hat{\boldsymbol{\sigma}}_2$ and $\hat{\boldsymbol{\tau}}_1 \cdot \hat{\boldsymbol{\tau}}_2$ terms in the nuclear force.

One often uses the projection operators defined by

$$\Pi_{st} \equiv \frac{1}{4}(3 + \hat{\boldsymbol{\sigma}}_1 \cdot \hat{\boldsymbol{\sigma}}_2) \ , \quad \Pi_{ss} \equiv \frac{1}{4}(1 - \hat{\boldsymbol{\sigma}}_1 \cdot \hat{\boldsymbol{\sigma}}_2) \ , \tag{3.24}$$

$$\Pi_{it} \equiv \frac{1}{4}(3 + \hat{\boldsymbol{\tau}}_1 \cdot \hat{\boldsymbol{\tau}}_2) \ , \quad \Pi_{is} \equiv \frac{1}{4}(1 - \hat{\boldsymbol{\tau}}_1 \cdot \hat{\boldsymbol{\tau}}_2) \ , \tag{3.25}$$

in order to explicitly show the state dependence of the nuclear force. The indices st, ss, it and is represent to project on the spin-triplet states, spin-singlet state, isospin-triplet states and isospin-singlet state, respectively.

Exercise 3.3 Show that Π_{st} satisfies the following conditions which are required for the projection operator on the spin-triplet states.

$$\Pi_{st}^2 = \Pi_{st} \ , \quad \Pi_{st}\Pi_{ss} = 0 \ , \quad \Pi_{st}|1M\rangle_{\mathscr{S}} = |1M\rangle_{\mathscr{S}} \ , \quad \Pi_{st}|00\rangle_{\mathscr{S}} = 0 \ . \tag{3.26}$$

3.2 The General Structure of Nuclear Force

One of the methods to construct nuclear force or nuclear potential is to take into account various conservation laws suggested by fundamental experiments and to consider the most general structure of the nuclear potential under those constraints. Each conservation law is related to the invariance or symmetry under an associated transformation. For nuclear force, one can first list up that

- the potential is a scalar quantity,
- it is invariant under space reflection, i.e., Parity invariant,
- it is time reversal invariant

as the basic properties. Furthermore, the proton–proton and proton–neutron scatterings at low energies suggest that the nuclear force is the same for the three states of isospin triplets to a high accuracy. This is called the *charge independence* and implies that the nuclear force is a function of the inner product of the isospins of two nucleons $\hat{\tau}_1 \cdot \hat{\tau}_2$. On the other hand, the way of thinking that the force between two neutrons and that between two protons are the same is called the *charge symmetry*. The charge symmetry and the charge independence are suggested also from the facts that the energy levels of mirror nuclei, e.g., ${}^{15}_{7}\mathrm{N}_8$ and ${}^{15}_{8}\mathrm{O}_7$, very much resemble except for the energy shift due to the Coulomb energy, and that there appear corresponding energy levels among isobars such as ${}^{14}_{8}\mathrm{O}_6$, ${}^{14}_{7}\mathrm{N}_7$ and ${}^{14}_{6}\mathrm{C}_8$.[1]

In this section, we give the general form of the nuclear potential based on the considerations of the invariance or the symmetry.

3.2.1 *Static Potentials (Velocity-Independent Potentials)*

Let us first consider the potentials, called static potentials, which do not depend on the velocity of the relative motion between the nucleons.

[1] A group of energy levels which have the same J^π but differ in the z-component of the isospin form an isobaric multiplet. For example, there exist many isobaric triplets for ${}^{14}_{8}\mathrm{O}_6$, ${}^{14}_{7}\mathrm{N}_7$, and ${}^{14}_{6}\mathrm{C}_8$.

3.2.1.1 Central Potentials

The static potentials can be classified into the central potentials, which depend only on the distance r between the nucleons, and the noncentral potentials, which depend on the angle of the relative coordinate $\mathbf{r}$. The most general central static potential can be written as

$$V_C = V_0(r) + (\boldsymbol{\sigma}_1 \cdot \boldsymbol{\sigma}_2)V_\sigma(r) + (\boldsymbol{\tau}_1 \cdot \boldsymbol{\tau}_2)V_\tau(r) + (\boldsymbol{\sigma}_1 \cdot \boldsymbol{\sigma}_2)(\boldsymbol{\tau}_1 \cdot \boldsymbol{\tau}_2)V_{\sigma\tau}(r) \tag{3.27}$$

by considering the symmetry and invariance conditions mentioned at the beginning of this section. The strength and the functional form of $V_0(r)$, $V_\sigma(r)$, $V_\tau(r)$ and $V_{\sigma\tau}(r)$ cannot be determined by the consideration of symmetry alone. Equation (3.27) can be rewritten as

$$V_C = V_{\mathrm{W}}(r) + V_{\mathrm{M}}(r)P_{\mathrm{M}} + V_{\mathrm{B}}(r)P_\sigma + V_{\mathrm{H}}(r)P_\tau \ , \tag{3.28}$$

where the indices W, M, B and H stand for Wigner, Majorana, Bartlett, and Heisenberg, respectively, or

$$V_C = V_{st}(r)\tilde{\Pi}_{st} + V_{ts}(r)\tilde{\Pi}_{ts} + V_{tt}(r)\tilde{\Pi}_{tt} + V_{ss}(r)\tilde{\Pi}_{ss} \tag{3.29}$$

by using the exchange and projection operators. Here, $\tilde{\Pi}_{st}$ is the projection operator on the isospin-singlet, spin-triplet states and is given by $\tilde{\Pi}_{st} \equiv \Pi_{is} \times \Pi_{st}$. $\tilde{\Pi}_{ts}$, $\tilde{\Pi}_{tt}$, $\tilde{\Pi}_{ss}$ are defined in a similar way. The radial functions in the different representations are transformed as

$$\begin{aligned} V_{\mathrm{W}} &= V_0 - V_\sigma - V_\tau + V_{\sigma\tau} \ , \\ V_{\mathrm{M}} &= -4V_{\sigma\tau} \ , \\ V_{\mathrm{B}} &= 2V_\sigma - 2V_{\sigma\tau} \ , \\ V_{\mathrm{H}} &= 2V_\tau - 2V_{\sigma\tau} \ , \end{aligned} \tag{3.30}$$

and

$$\begin{aligned} V_{ts} &= V_{\mathrm{W}} + V_{\mathrm{M}} - V_{\mathrm{B}} + V_{\mathrm{H}} = V_0 - 3V_\sigma + V_\tau - 3V_{\sigma\tau} \ , \\ V_{st} &= V_{\mathrm{W}} + V_{\mathrm{M}} + V_{\mathrm{B}} - V_{\mathrm{H}} = V_0 + V_\sigma - 3V_\tau - 3V_{\sigma\tau} \ , \\ V_{ss} &= V_{\mathrm{W}} - V_{\mathrm{M}} - V_{\mathrm{B}} - V_{\mathrm{H}} = V_0 - 3V_\sigma - 3V_\tau + 9V_{\sigma\tau} \ , \\ V_{tt} &= V_{\mathrm{W}} - V_{\mathrm{M}} + V_{\mathrm{B}} + V_{\mathrm{H}} = V_0 + V_\sigma + V_\tau + V_{\sigma\tau} \ . \end{aligned} \tag{3.31}$$

Exercise 3.4 Show the relationships (3.30) and (3.31).

3.2.1.2 Noncentral Potentials

One can make an operator which satisfies the invariant conditions for nuclear force in the following way,

$$S_{12} = \frac{3(\mathbf{r}\cdot\boldsymbol{\sigma}_1)(\mathbf{r}\cdot\boldsymbol{\sigma}_2)}{r^2} - \boldsymbol{\sigma}_1\cdot\boldsymbol{\sigma}_2\ , \tag{3.32}$$

by using the relative coordinate $\mathbf{r}$ between the nucleons and the spin of each nucleon. Equation (3.32) can be represented in the form,

$$S_{12} = (24\pi)^{1/2}[[\sigma_1^{(1)} \times \sigma_2^{(1)}]^{(\lambda=2)} \times Y_2(\hat{\mathbf{r}})]_0^{(0)}\ , \tag{3.33}$$

by using the notation of the tensor product. The S_{12} is therefore called the *tensor operator*. Using the total spin of two-nucleon system,

$$\mathbf{S} = \frac{1}{2}(\mathbf{s}_1 + \mathbf{s}_2)\ , \tag{3.34}$$

S_{12} can be rewritten also as

$$S_{12} = 2\left[3\frac{(\hat{\mathbf{S}}\cdot\mathbf{r})^2}{r^2} - \hat{\mathbf{S}}^2\right]. \tag{3.35}$$

Equation (3.35) shows that S_{12} has a finite value only when it is operated on the spin-triplet states. Hence one can represent the noncentral potential as

$$V_{\mathrm{T}} = \left[V_T^{\mathrm{even}}(r)\Pi(T=0, L_{\mathrm{even}}) + V_T^{\mathrm{odd}}(r)\Pi(T=1, L_{\mathrm{odd}})\right] S_{12}\ , \tag{3.36}$$

where $\Pi(T=0, L_{\mathrm{even}})$ and $\Pi(T=1, L_{\mathrm{odd}})$ are the projection operators on the isospin 0, even parity states and on the isospin 1, odd parity states, respectively. This force is called the *tensor force*. As the form of Eq. (3.32) shows, the tensor force resembles the force acting between two magnetic dipoles. We learn later that the tensor force is responsible for binding the deuteron among various two-nucleon systems, and plays a vital role in the existence of nuclei, e.g., in locating the stable nuclei along the diagonal line in the nuclear chart.

Exercise 3.5 Show Eqs. (3.33) and (3.35).

3.2.2 *Velocity-Dependent Potentials*

A representative of the velocity-dependent potentials is the *spin–orbit interaction*, whose corresponding potential is given by

$$V_{\mathrm{LS}} = \left[V_{\mathrm{LS}}^{\mathrm{even}}(r)\Pi(T=0, L_{\mathrm{even}}) + V_{\mathrm{LS}}^{\mathrm{odd}}(r)\Pi(T=1, L_{\mathrm{odd}})\right]\hat{\mathbf{L}}\cdot\hat{\mathbf{S}} \,. \tag{3.37}$$

Here, $\hat{\mathbf{L}}$ is the angular momentum operator of the relative motion, and is given by

$$\mathbf{L} = \mathbf{r}\times\mathbf{p} = \mathbf{r}\times(\mathbf{p}_1 - \mathbf{p}_2) \,. \tag{3.38}$$

As we learn later, the spin–orbit interaction plays an important role in nuclear physics, e.g., in governing the magic numbers.

3.3 The Properties of Deuteron and the Nuclear Force

3.3.1 *The Effect of Tensor Force: The Wave Function in the Spin–Isospin Space*

Of two-nucleon systems, only deuteron is stable. Neither proton–proton nor neutron–neutron bound state exists. In that sense, the deuteron is the simplest stable nucleus in nature. The detailed analyses of the experiments concerning deuteron are therefore important in obtaining various information on the nuclear force together with the analyses of the nucleon–nucleon scatterings, which we learn in the next section.

We show in Table 3.1 the properties of the deuteron in comparison with those of the proton and the neutron.

As we learn in detail in Sect. 3.6 (for example, see Fig. 3.5), the central force is attractive in distant regions for all two-nucleon systems. Moreover, it is stronger in the spin-singlet state, to which proton–proton and neutron–neutron systems belong, than in the spin-triplet states of deuteron. Nevertheless, only deuteron exists as a stable state in nature. This is because even an attractive force cannot yield a stable

Table 3.1 Properties of the proton, the neutron and the deuteron. u: Unified atomic mass unit; Q: Electric quadrupole moment. Taken from [1]. Values of the magnetic and electric quadrupole moments for deuteron are taken from [2, 3], respectively

	Proton	Neutron	Deuteron
Electric charge	e	0	e
Mass (u)	1.00727646688(13)	1.0086654(4)	2.01410219(11)
Mass (MeV)	938.272029(80)	939.550(5)	1875.61282(16)
Binding energy (MeV)			2.22452(20)
Spin	1/2	1/2	1
Mean life	$>10^{31}$ to 10^{33} years	887.5 ± 0.8 s	Stable
Magnetic moment (μ_N)	2.792847351(28)	−1.9130427(5)	0.8574382308(72)
Q ($e\,\mathrm{fm}^2$)	0	0	0.285783(30)

Fig. 3.1 Model experiment of the tensor force. Toys courtesy of Gyo Takeda

state unless it is strong enough in the three dimensional world governed by quantum mechanics, and also because an extra attractive force leading to a stronger attraction as a whole is added to deuteron due to the tensor force, since the spin of the deuteron is 1. In order to demonstrate the mechanism why only deuteron is stable among two-nucleon systems, Fig. 3.1 shows an assembly of toys of magnetic dipoles floating on the surface of oil. The directions of the magnetic dipoles shown by the arrows align due to the interaction between the magnetic dipoles when two toys approach close enough. Deuteron becomes stable for a similar reason. Since the nuclear force is of short range, the relative motion between the two nucleons is desirable to be in the s-wave which has a finite probability at zero distance in order to be stabilized by the nuclear force. On the other hand, the wave function of two-nucleon systems has to be antisymmetric in the usual, spin and isospin spaces as a whole with respect to the exchange of two nucleons. Consequently, only deuteron which is in the isospin-singlet state (spin-triplet states) and can supplement the insufficiency of the central force by the tensor force becomes to exist in nature as a stable system.

Here, let us examine what information on the effects and magnitude of the tensor force can be obtained from the magnetic and electric quadrupole moments of deuteron, although the binding energy is an important quantity in determining the nuclear force [4].

The operator of the magnetic moment of deuteron is given by (see Sect. 4.1.1),

$$\hat{\boldsymbol{\mu}}_{\rm d} = \mu_{\rm p}\hat{\boldsymbol{\sigma}}_{\rm p} + \mu_{\rm n}\hat{\boldsymbol{\sigma}}_{\rm n} + \hat{\mathbf{L}}_{\rm p} \,, \tag{3.39}$$

where $\mathbf{L}_{\rm p}$ is the orbital angular momentum of proton and is given by $\mathbf{L}_{\rm p} = \mathbf{L}/2$ by using the angular momentum of the relative motion between proton and neutron $\mathbf{L}$. All the magnetic moments are given in units of the nuclear magnetic moment μ_N. The $\hat{\boldsymbol{\mu}}_{\rm d}$ can be rewritten as

$$\hat{\boldsymbol{\mu}}_{\rm d} = (\mu_p + \mu_{\rm n})\,\hat{\mathbf{J}} - \left(\mu_p + \mu_{\rm n} - \frac{1}{2}\right)\hat{\mathbf{L}} \tag{3.40}$$

by using the total angular momentum of deuteron $\mathbf{J} \equiv \mathbf{L} + \frac{1}{2}(\boldsymbol{\sigma}_{\rm p} + \boldsymbol{\sigma}_{\rm n})$ and $\mathbf{L}$.[2] As we learn in Sect. 4.3.2, the magnetic moment is defined as the expectation value of the z-component of $\hat{\boldsymbol{\mu}}$ in the case when the z-component of the angular momentum M takes its maximum value $M = J$,

$$\mu = \langle J\, M = J | \hat{\mu}_z | J\, M = J \rangle \ . \tag{3.41}$$

Since the operator of the magnetic moment is a tensor of rank 1, the value of the magnetic moment becomes 0 and disagrees with the experimental data if one assumes that the spin of deuteron is 0. For finite J, the projection theorem (see Appendix A.6.4) gives,

$$\mu = \frac{1}{J+1} \langle J\, M = J | \hat{\boldsymbol{\mu}} \cdot \hat{\mathbf{J}} | J\, M = J \rangle \ , \tag{3.42}$$

The spin of deuteron is 1 as shown in Table 3.1. This is due to the existence of the tensor force, which mixes the $L = 2$ state (D-state) to the $L = 0$ state (S-state) for the relative motion as Eq. (3.33) implies. If we denote the amplitude of the mixing by α_D, then the wave function of deuteron can be formally expressed as

$$|\psi_D\rangle = |^{13}S_1\rangle + \alpha_D |^{13}D_1\rangle \ , \tag{3.43}$$

where the lower index 1 shows that the magnitude of the total angular momentum J is 1. Inserting Eq. (3.43) into the state $|J\, M = J\rangle$ in Eq. (3.42), we obtain

$$\mu = \mu_{\rm n} + \mu_{\rm p} - \frac{3}{2}\left(\mu_{\rm n} + \mu_{\rm p} - \frac{1}{2}\right) P_D \ , \tag{3.44}$$

where $P_D = |\alpha_D|^2$ represents the probability of D-state in deuteron.

The fact that the magnetic moment of deuteron is nearly equal to the sum of the magnetic moment of the proton and that of the neutron in Table 3.1 supports that the spin of deuteron is 1, that is, the spins of the proton and the neutron in the deuteron are aligned. Moreover, the difference between them suggests that the deuteron contains the D-state by about 4 %.

The admixture of the D-state means that the deuteron is deformed by that amount. We learn in Chap. 7 that the quadrupole moment of each nucleus gives a measure of the deformation of the nucleus. The admixture of the D-state suggested by the magnetic moment is consistent with the fact that the quadrupole moment is finite in Table 3.1.

Exercise 3.6 Derive Eq. (3.44).

---

[2] We used that $\boldsymbol{\sigma}_{\rm p} - \boldsymbol{\sigma}_{\rm n}$ becomes 0 when it is operated on the $|S = 1, S_z = 1\rangle$ state.

3.3.2 The Radial Wave Function: Estimate of the Magnitude of the Force Between Proton and Neutron

Let us next examine the radial wave function between proton and neutron. For simplicity, we assume that the orbital angular momentum is zero. If we ignore the difference between the neutron and proton masses and represent the radial wave function as $R_0(r) = u(r)/r$, then the Schrödinger equation for $u(r)$ is given by

$$\frac{\mathrm{d}^2 u}{\mathrm{d}r^2} + \frac{M_N}{\hbar^2}\left[E - V(r)\right] u(r) = 0 , \tag{3.45}$$

where M_N is the nucleon mass. If we further assume that the potential is a square well potential of depth V_0 and range a, then the equation can be analytically solved and its solution is given by

$$u(r) = \begin{cases} A \sin Kr & \text{for } r < a \\ B\mathrm{e}^{-\gamma r} & \text{for } r \geq a , \end{cases} \tag{3.46}$$

$$K = \frac{\sqrt{M_N(V_0 - W)}}{\hbar} , \tag{3.47}$$

$$\gamma = \frac{\sqrt{M_N W}}{\hbar} , \tag{3.48}$$

where $W \sim 2.22\,\mathrm{MeV}$ is the binding energy. From the continuity condition at $r = a$, we obtain

$$K \cot Ka = -\gamma . \tag{3.49}$$

Exercise 3.7 The assessment of the depth of the potential.

1. If we assume $V_0 \gg W$, then $Ka \approx \pi/2$ from Eq. (3.49). Using this, derive

$$V_0 a^2 \approx \frac{\pi^2 \hbar^2}{4M_N} . \tag{3.50}$$

2. Show that $V_0 \sim 50\,\mathrm{MeV}$ by assuming that the range of the force nearly equals the Compton wavelength of pions: $a \approx \hbar/m_\pi c \sim 1.45\,\mathrm{fm}$.

The wave function

$$u = C\left[\mathrm{e}^{-\gamma r} - \mathrm{e}^{-\alpha r}\right] \tag{3.51}$$

is also often used for the wave function given by Eq. (3.46), and is called the wave function of Hulthén-type [5]. The normalization constant is given by

$$C \approx \sqrt{2\gamma}\left(1 + \frac{3\gamma}{2\alpha}\right) \tag{3.52}$$

if one assumes $\alpha \gg \gamma$. The wave function of the Hulthén-type has a desirable property that it satisfies the boundary condition that it is zero at the origin.

3.4 Nucleon–Nucleon Scattering

The neutron–proton and proton–proton scattering data provide important information on the nuclear force. The experimental data are given as the differential cross sections and scattering cross sections. More detailed information on the spin dependence can be obtained if one uses polarized projectile or polarized target [6]. Here, we discuss the properties of nuclear force obtained from the scattering data of unpolarized beam and target nucleus.

3.4.1 Low-Energy Scattering: Effective Range Theory

Let us consider the scattering of a neutron by a proton in order to avoid special considerations needed to handle the long range Coulomb interaction. When both neutron and proton are not polarized and their spins randomly orient all the directions with equal probability, the state of two nucleons takes the spin-singlet state and the spin-triplet states with the statistical weight 1:3. The experimentally obtained cross section is therefore given by

$$\sigma = \frac{1}{4}\left(\sigma^{(0)} + 3\sigma^{(1)}\right) \tag{3.53}$$

if we denote the cross sections in the spin-singlet state and in the spin-triplet states by $\sigma^{(0)}$ and $\sigma^{(1)}$, respectively.

Exercise 3.8 Show Eq. (3.53).

Exercise 3.9 Explain experimental methods to determine the cross section such as the direct measurement and the measurement of the absorption cross section.

Let us study how the information on the nuclear force can be obtained from the analysis of the scattering cross section. In general, the differential cross section is given by

$$\frac{\mathrm{d}\sigma}{\mathrm{d}\Omega} = |f(\theta)|^2 \tag{3.54}$$

by using the scattering amplitude $f(\theta)$, where θ is the scattering angle. As we explain in Appendix A.1, one of the standard methods to analyze the experimental data of scattering is to make the partial wave expansion of the wave function and perform the *phase shift analysis*. In this method, the scattering amplitude $f(\theta)$ is given by

$$f(\theta) = \frac{1}{2ik} \sum_{\ell=0}^{\infty} (2\ell + 1)(e^{2i\delta_\ell} - 1) P_\ell(\cos\theta) \tag{3.55}$$

by using the phase shift δ_ℓ for each partial wave ℓ and the Legendre function $P_\ell(\cos\theta)$. k is the incident wave number.

In the case of scattering by a short range force, only the s-wave contributes to the scattering if the incident energy of the scattering is sufficiently low. Namely, the phase shift can be set equal to zero to a good approximation for the partial waves $\ell \geq 1$. This can be understood by noting that the angular momentum is given as the product of the incident momentum and the impact parameter. If the incident energy is low, the impact parameter becomes large for the partial waves $\ell \geq 1$, hence becomes outside of the range of the nuclear force. More important thing is that the s-wave phase shift can be represented as

$$k \cot\delta = -\frac{1}{a} + \frac{1}{2}k^2 r_e \tag{3.56}$$

by using two parameters irrespective of the details of the force (*effective range theory* [7]). In other words, one can determine, at best, only two parameters from the data of low-energy scattering. The parameters a and r_e are called the *scattering length* and the *effective range*, respectively.

Exercise 3.10 Discuss the range of the incident energy for which only the s-wave contributes to the scattering by assuming that the range of nuclear force is about 2 fm.

We obtain

$$\sigma = \frac{4\pi a^2}{(1 - r_e a k^2/2)^2 + (ak)^2} \tag{3.57}$$

by inserting the result of Eq. (3.56) into the relationship between the cross section and the phase shift. We can obtain the information on the scattering length and the effective range by experimentally studying in detail the change of the cross section at low energies as a function of the incident energy. An alternative way to determine the scattering length with high precision is to use the cross sections of the scattering of a thermal neutron by a proton, and by a parahydrogen molecule. Since the energy of the thermal neutron[3] is as low as 1/40 eV, the cross section is given by

$$\sigma_{\text{thermal-neutron}} = \pi \left[3(a^{(1)})^2 + (a^{(0)})^2\right] \tag{3.58}$$

by taking the limit of $k \to 0$. On the other hand, the cross section of the scattering of the thermal neutron by a parahydrogen molecule[4] is given by [8, 9]

$$\sigma_{\text{parahydrogen}} = \pi(3a^{(1)} + a^{(0)})^2 . \tag{3.59}$$

[3]Neutrons which are in thermal equilibrium after multiple scattering inside matters of the normal temperature.

[4]Hydrogen molecule consisting of two protons, where two proton spins aligned antiparallel.

Table 3.2 Nuclear force parameters determined from low-energy scattering and the properties of deuteron, taken from [11]

Scattering system (state indices)	a (fm)		r_e (fm)	
	Expt.	Argonne v_{18}	Expt.	Argonne v_{18}
p+p ($T=1, S=0$)	-7.8063 ± 0.0026	-7.8064	2.794 ± 0.014	2.788
n+p ($T=1, S=0$)	-23.749 ± 0.008	-23.732	2.81 ± 0.05	2.697
n+n ($T=1, S=0$)	-18.5 ± 0.4	-18.487	2.80 ± 0.11	2.840
n+p ($T=0, S=1$)	5.424 ± 0.003	5.419	1.760 ± 0.005	1.753

One can determine $a^{(1)}$ and $a^{(0)}$ by using the observed values of these cross sections.

Exercise 3.11 Prove Eq. (3.59).

The following relation between the binding energy of deuteron, scattering length and the effective range,

$$\frac{1}{a^{(1)}} \approx \alpha - \frac{1}{2} r_e^{(1)} \alpha^2 , \tag{3.60}$$

can be used in order to further determine the effective range for the spin-triplet states. Here, $\alpha = [(M_N/\hbar^2)\varepsilon_D]^{1/2}$, ε_D being the binding energy of deuteron.

Table 3.2 shows the scattering length and the effective range obtained from the analyses of the low-energy scattering and deuteron [10–14]. It can be shown on the way of derivation of the effective range theory, i.e., of Eq. (3.56) that there exist bound states if the sign of the scattering length a is positive, while there exists no bound state if it is negative. Hence the results shown in Table 3.2 show that only the isospin 0, spin-triplet states, i.e., deuteron, is stable, i.e., a bound state, of two-nucleon systems. In this way, the nuclear force has a strong state dependence. Also, it should be remarked that the scattering length is nearly equal in all the proton–proton, neutron–neutron and neutron–proton scatterings which belong to the isospin-triplet states ($T = 1$) (charge independence of the nuclear force).

The effective range theory tells that only two parameters on the nuclear force can be determined from low-energy scattering data. However, it is convenient to have an explicit expression of the functional form and the strength of the potential in order to perform various calculations. Here we mention that there exist also such attempts.

Exercise 3.12 Assume that the nucleon–nucleon potential is given by the sum of the central (V_C) and the tensor (V_T) forces of the Yukawa-type, which have equal weight of Wigner and Majorana terms,

$$V(r) = V_C(r) + V_T(r) , \tag{3.61}$$

$$V_C(r) = V_0 \frac{\mathrm{e}^{-\mu r}}{\mu r}(w + m P_{\mathrm{M}}) = V_0 \frac{\mathrm{e}^{-r/a_C}}{r/a_C}(w + m P_{\mathrm{M}}) , \tag{3.62}$$

$$V_T(r) = V_{0T} \frac{e^{-r/a_T}}{r/a_T} S_{12}(w + mP_M) , \tag{3.63}$$

$$w = 1 - m = 0.5 . \tag{3.64}$$

Show that the parameters should be taken as

$$V_0 = -48.1\,\text{MeV}, \quad V_{0T} = 23.1\,\text{MeV}, \quad a_C = 1.17\,\text{fm}, \quad a_T = 1.74\,\text{fm}, \tag{3.65}$$

in order to reproduce the experimental values of the effective range and the scattering length in the spin-singlet and spin-triplet states.

3.4.2 High-Energy Scattering: Exchange Force

Equation (3.30) shows that the spin and isospin dependence of the nuclear force leads to exchange forces. The existence of those exchange forces manifests itself in the differential cross section of nucleon–nucleon scattering at high energies. Figure 3.2 shows the angular distribution of neutrons scattered by protons in the neutron energy range 85–105 MeV. It is remarkable that a large amount of scattering takes place to backward angles as well as to forward angles.

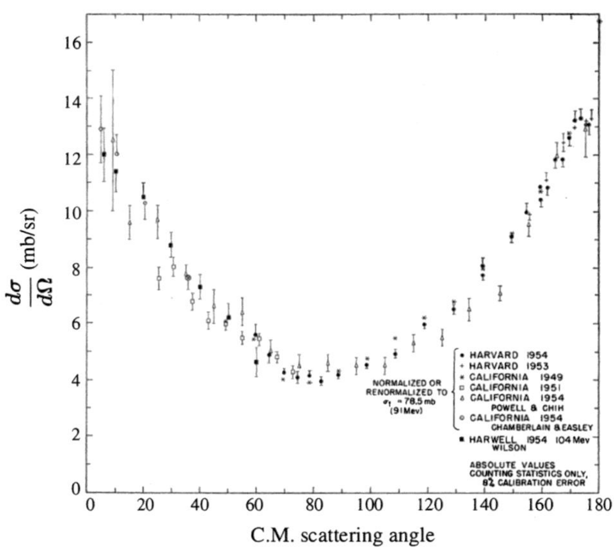

Fig. 3.2 The differential cross section of neutron–proton scattering, taken from [15]

Let us now assume that the nuclear force contains only the Wigner force. Since one can use the Born approximation for high-energy scattering, the scattering amplitude is given by

$$f_{\rm W}(\theta) = -\frac{M_N}{\hbar^2 q}\int_0^\infty V_{\rm W}(r) r \sin qr dr \ , \tag{3.66}$$

with

$$q = 2k\sin(\theta/2) \ , \tag{3.67}$$

to the first order. M_N is the nucleon mass and q the wave number corresponding to the momentum transfer. If we assume the Gauss-type for the form factor of the nuclear force,

$$V_{\rm W}(r) = -V_{W0}\exp\left(-\frac{r^2}{R_{\rm W}^2}\right) \ , \tag{3.68}$$

then we have

$$f_{\rm W}(\theta) = \frac{M_N}{\hbar^2 q} V_{W0}\frac{\sqrt{\pi}}{4}R_{\rm W}^3 \exp\left[-\frac{(qR_{\rm W})^2}{4}\right] . \tag{3.69}$$

Equation (3.69) shows that the cross section should monotonically decrease towards backward angles if the nuclear force consists of the Wigner force alone. This contradicts the experimental data shown in Fig. 3.2.

In order to resolve this contradiction, let us assume that the nuclear force contains a Majorana term

$$V_{\rm M}(r)P_{\rm M} \ , \tag{3.70}$$

and examine its effect in the first-order Born approximation. Since the scattering amplitude is linearly proportional to the nuclear force in the first-order Born approximation, it gets an additional term,

$$f_{\rm M}(\theta) = -\frac{M_N}{\hbar^2 q'}\int_0^\infty V_{\rm M}(r) r \sin q' r dr \tag{3.71}$$

with

$$\mathbf{q}' = \mathbf{k}_i + \mathbf{k}_f \ , \tag{3.72}$$

$$q' = 2k_i \sin[(\pi - \theta)/2] \ , \tag{3.73}$$

due to the Majorana force. If we assume the Gaussian-type for $V_{\rm M}(r)$ in the same way as that for the Wigner force, the differential cross section increases with the scattering angle. One can thus explain the large cross section at backward angles appearing in the experimental data.[5]

[5] The Serber force, which is also called the Serber exchange force, is one of such forces. It consists of the Wigner and the Majorana terms of the same sign with equal weight and is assumed to be proportional to $1 + P_{\rm M}$ (see Table 3.4).

3.4.3 High-Energy Scattering: Repulsive Core

Another characteristic of the nuclear force is that it becomes strong repulsion in the close distance where two nucleons strongly overlap. This repulsive force in the short distance is often called the *repulsive core*. As can be seen from the phase shift analysis, the sign of the potential and the sign of the phase shift δ have a one to one correspondence if the sign of $V(r)$ is fixed independently of the distance r, i.e., δ is positive for attractive potentials and negative for repulsive potentials.

Figure 3.3 shows the behaviour of the phase shift for several partial waves as a function of the incident energy. Now, let us focus on the $^{31}S_0$ channel ($T = 1, S = 0, L = 0$ channel). The figure indicates that the potential for this channel behaves effectively as an attractive potential in the scattering at low energies, while as a repulsive potential in the scattering at high energies. If we consider the uncertainty relation between the distance and the momentum, this implies that the nuclear potential behaves as a repulsive potential at short distances, while as an attractive potential at long distances. Figure 3.4 represents this property in a simple way by combining a square potential barrier and a square well potential.

In the following, we show that this simple model can well explain the energy dependence of the phase shift, and that it gives an estimate of the extension of the repulsive core. We first note that the phase shift for the partial wave of the angular momentum ℓ is given by

$$\delta_\ell(k) = -\frac{Mk}{\hbar^2}\int_0^\infty V(r)j_\ell^2(kr)r^2\mathrm{d}r\ . \tag{3.74}$$

This equation confirms the following relation between the attractive or repulsive property of the potential and the sign of the phase shift

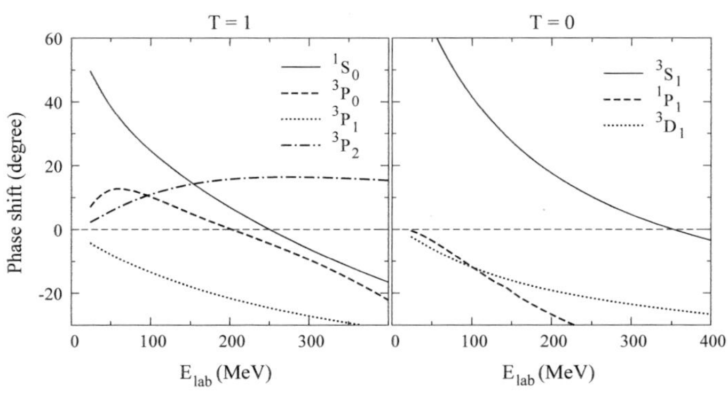

Fig. 3.3 The phase shifts for nucleon–nucleon scattering in several channels. Made from the analysis in [16]

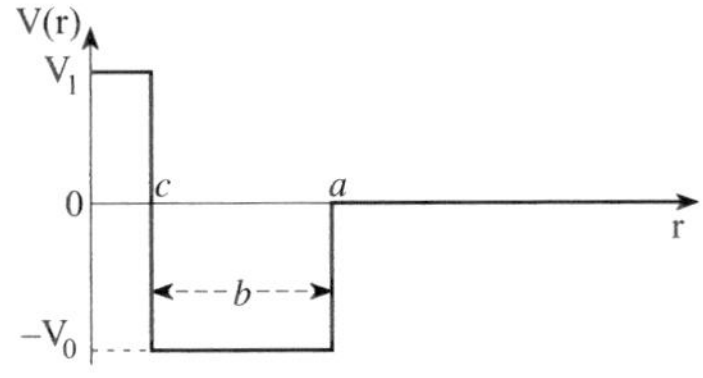

Fig. 3.4 A simple model for the nuclear force (repulsive core plus short range attraction)

$$\delta_\ell > 0 \quad \text{for } V(r) < 0\,, \tag{3.75}$$

$$\delta_\ell < 0 \quad \text{for } V(r) > 0\,, \tag{3.76}$$

which we mentioned before for the case when the sign of the potential is fixed.

The s-wave phase shift is given by

$$\delta_0 = -\frac{M}{2\hbar^2 k}\left[(cV_1 - bV_0) - c(V_1 + V_0)\frac{\sin(2kc)}{2kc} + aV_0\frac{\sin(2ka)}{2ka}\right]\,, \tag{3.77}$$

when the potential is given by Fig. 3.4. Here b is defined as

$$b = a - c\,. \tag{3.78}$$

Let us assume now that

$$V_1 > \frac{b}{c}V_0 \quad \text{and} \quad a \gg c, \tag{3.79}$$

and study the sign of the phase shift in the scattering at low and high energies. In the scattering at low energies, where the conditions $2kc \ll 1,\, 2ka \gg 1$ hold, we obtain

$$\delta_0(k) \approx \frac{MaV_0}{2\hbar^2 k} > 0\,, \tag{3.80}$$

by ignoring the last term of Eq. (3.77) and approximating as $\sin(2kc)/2kc \approx 1$. Equation (3.80) shows that the phase shift is positive. On the other hand, in the scattering at high energies, where the conditions $2kc \gg 1,\, 2ka \gg 1$ hold, we obtain

$$\delta_0(k) \approx -\frac{M}{2\hbar^2 k}(cV_1 - bV_0) < 0\,, \tag{3.81}$$

by ignoring the second and third terms in Eq. (3.77). Thus the phase shift becomes negative. The radius of the repulsive core c is given by $2\bar{k}c \approx 1$ if we denote the wave number which corresponds to the critical energy $E^{\rm cr}_{\rm c.m.}$ where the phase shift becomes zero by $\bar{k}$. $E^{\rm cr}_{\rm c.m.} \sim 125\,$MeV from Fig. 3.3, so that one can estimate $c \approx 0.3\,$fm, though more accurate studies give $c = 0.4$–$0.5\,$fm.

3.4.4 Spin Polarization Experiments

Since 1990s, experiments with polarized incident nucleons and/or target nucleus became popular. Those experiments are very useful to obtain information on the spin-dependent forces.

Suppose that the force between two nucleons is given by

$$V = f(r)\,(V_0 + V_\sigma \boldsymbol{\sigma}_1 \cdot \boldsymbol{\sigma}_2)\ . \tag{3.82}$$

If we define the spin raising and lowering operators by

$$\sigma_\pm \equiv \sigma_x \pm \mathrm{i}\sigma_y \tag{3.83}$$

as is often done in the calculations concerning angular momentum, then the inner product of the spins can be rewritten as

$$\boldsymbol{\sigma}_1 \cdot \boldsymbol{\sigma}_2 = \frac{1}{2}\,(\sigma_{1+}\sigma_{2-} + \sigma_{1-}\sigma_{2+}) + \sigma_{10}\sigma_{20}\ . \tag{3.84}$$

The spin-flip, i.e., the exchange of direction of spins between the incident and target nucleons, takes place through the first and the second terms on the right-hand side of Eq. (3.84). Hence one can expect that detailed information on the spin-dependent terms in the nuclear force can be obtained from the experiments related to the spin polarization.

Exercise 3.13 Study methods to polarize the incident particle or the target nucleus.

Exercise 3.14 Consider the proton–neutron scattering by the following potential,

$$V = (1 + \kappa \hat{\boldsymbol{\sigma}}_1 \cdot \hat{\boldsymbol{\sigma}}_2) f(r)\ , \tag{3.85}$$

and assume that the proton and the neutron are in the spin-up and spin-down states, respectively, at the beginning of the scattering. Obtain the probability of exchanging the direction of spins between the proton and neutron at the end of the scattering as a function of the strength of the spin-dependent term κ in the first-order Born approximation.

3.5 Microscopic Considerations: Meson Theory, QCD

In the meson theory, the force between two nucleons is assumed to work through the exchange of various mesons such as π-mesons, ω-mesons and ρ-mesons. In this section, we discuss the properties of the nuclear force from this point of view.

Let us first derive the force which arises from the exchange of π-mesons. The first step is to properly postulate a Lagrangian density. Since the π-meson field obeys the Klein–Gordon equation, the corresponding Lagrangian is given by

$$\mathscr{L}_{\text{pion}} = \frac{1}{2}\partial_\mu \Pi^a \partial^\mu \Pi^a - \frac{1}{2}m_\pi^2 \Pi^a \Pi^a \,,$$
$$= \frac{1}{2}\left(\frac{\partial^2 \Pi^a}{\partial t^2} - \frac{\partial^2 \Pi^a}{\partial x^2} - \frac{\partial^2 \Pi^a}{\partial y^2} - \frac{\partial^2 \Pi^a}{\partial z^2}\right) - \frac{1}{2}m_\pi^2 \Pi^a \Pi^a \,. \quad (3.86)$$

Here, we used the natural unit ($\hbar = c = 1$). Π is the π-meson field and the upper index a is the index in the isospin space. The $a = 1, 2, 3$ correspond to the x, y, z-components, respectively, and Π^1, Π^2, Π^3 are related to the fields of three charge states π^+, π^0, π^- of π-mesons as

$$\pi^+ = (\Pi^1 + \mathrm{i}\Pi^2)/\sqrt{2}, \quad \pi^- = (\pi^+)^* = (\Pi^1 - \mathrm{i}\Pi^2)/\sqrt{2}, \quad \pi^0 = \Pi^3 \,. \quad (3.87)$$

On the other hand, the Lagrangian density for the interaction between nucleons and pion fields are given by

$$\mathscr{L}_{\text{int}}^{\text{NR}} = g_\pi \varphi^\dagger \sigma_\alpha \tau^a \varphi \nabla_\alpha \Pi^a \,, \quad (3.88)$$

in the non-relativistic approximation by considering that it is a scalar in the combined spaces of the spin and usual space coordinates, it is scalar in the total isospin space including both nucleons and pions, and that π-mesons are pseudoscalar particles. φ is a two-dimensional spinor representing the nucleon field, $\boldsymbol{\sigma}$ and $\boldsymbol{\tau}$ are the spin and isospin operators of nucleons, respectively. Corresponding to Eq. (3.88), the interaction Hamiltonian is given by

$$H = -\int g_\pi \varphi^\dagger \sigma_\alpha \tau^a \varphi \nabla_\alpha \Pi^a \mathrm{d}\mathbf{r} \,. \quad (3.89)$$

It is necessary to determine the meson fields which mediate the force in order to determine the interaction between two nucleons. They are determined by the Euler–Lagrange equation,

$$\frac{\partial}{\partial x^\mu}\left[\frac{\partial \mathscr{L}}{\partial(\partial q_i/\partial x^\mu)}\right] - \frac{\partial \mathscr{L}}{\partial q_i} = 0 \,, \quad (3.90)$$

where q_i represents the meson field Π and the nucleon field φ in a unified way. We obtain the equation,

$$(\nabla^2 - m_\pi^2)\Pi^a(x) = g_\pi \nabla_\alpha \left(\varphi^\dagger \sigma_\alpha \tau^a \varphi\right) \,, \quad (3.91)$$

for the π-meson field by applying the Euler–Lagrange equation to $q = \Pi^a$ when the Lagrangian density is given by Eqs. (3.86) and (3.88). We introduced the static approximation and ignored the term of time derivative of the π-mesons. Note that this equation has the same form as Eq. (3.2). The solution can be found as

$$\Pi^a(\mathbf{r}) = -\frac{g_\pi}{4\pi}\int \mathrm{d}\mathbf{r}' \frac{\mathrm{e}^{-m_\pi|\mathbf{r}-\mathbf{r}'|}}{|\mathbf{r}-\mathbf{r}'|}\nabla'_\alpha \left[\varphi^\dagger(\mathbf{r}')\sigma_\alpha \tau^a \varphi(\mathbf{r}')\right] \quad (3.92)$$

in the same way as that used to derive Eq. (3.3). Inserting this result into Eq. (3.89), we have

$$H = \frac{g_\pi^2}{4\pi} \iint d\mathbf{r} d\mathbf{r}' \varphi^\dagger(\mathbf{r}) \sigma_\alpha \tau^a \varphi(\mathbf{r}) \nabla_\alpha \frac{e^{-m_\pi |\mathbf{r}-\mathbf{r}'|}}{|\mathbf{r}-\mathbf{r}'|} \nabla'_\beta [\varphi^\dagger(\mathbf{r}') \sigma_\beta \tau^a \varphi(\mathbf{r}')] . \quad (3.93)$$

The interaction potential can be obtained as

$$\begin{aligned} V(\mathbf{r}_1, \boldsymbol{\sigma}_1, \boldsymbol{\tau}_1, \mathbf{r}_2, \boldsymbol{\sigma}_2, \boldsymbol{\tau}_2) &= \frac{\delta^2 H}{\delta\rho(\mathbf{r}_1)\delta\rho(\mathbf{r}_2)} \\ &= -\frac{g_\pi^2}{4\pi} (\boldsymbol{\tau}_1 \cdot \boldsymbol{\tau}_2)(\boldsymbol{\sigma}_1 \cdot \nabla_1)(\boldsymbol{\sigma}_2 \cdot \nabla_2) \frac{e^{-m_\pi |\mathbf{r}_1 - \mathbf{r}_2|}}{|\mathbf{r}_1 - \mathbf{r}_2|} \end{aligned} \quad (3.94)$$

through the functional derivative with respect to the density. In deriving Eq. (3.94), we treated the nucleon field in the mean field approximation by identifying $\varphi^\dagger(\mathbf{r})\varphi(\mathbf{r})$ with the mean nucleon density $\rho(\mathbf{r})$ of c-number, and used the rule $\frac{\delta\rho(\mathbf{r})}{\delta\rho(\mathbf{r}_1)} = \delta(\mathbf{r} - \mathbf{r}_1)$ for the functional derivative. By performing the differentiation, Eq. (3.94) can be rewritten as

$$\begin{aligned} & V^{\mathrm{OPEP}}(\mathbf{r} = \mathbf{r}_1 - \mathbf{r}_2, \sigma_1, \tau_1, \sigma_2, \tau_2) \\ &= \frac{g_\pi^2}{3\hbar c} m_\pi c^2 \frac{e^{-\mu r}}{\mu r} (\boldsymbol{\tau}_1 \cdot \boldsymbol{\tau}_2) \left[(\boldsymbol{\sigma}_1 \cdot \boldsymbol{\sigma}_2) + \left(1 + \frac{3}{\mu r} + \frac{3}{(\mu r)^2}\right) S_{12} \right] \\ &\quad - \frac{4\pi}{\mu^3} \frac{g_\pi^2}{3\hbar c} m_\pi c^2 (\boldsymbol{\tau}_1 \cdot \boldsymbol{\tau}_2)(\boldsymbol{\sigma}_1 \cdot \boldsymbol{\sigma}_2) \delta(\mathbf{r}) , \end{aligned} \quad (3.95)$$

where $\mu = m_\pi c/\hbar$. The index OPEP of V^{OPEP} means one-pion exchange potential. Note that there appear spin- and isospin-dependent terms as well as a term which corresponds to the tensor force.

Exercise 3.15 In the relativistic theory, the Lagrangian density for the interaction between the nucleon field and the π-meson field is given by

$$\mathscr{L}_{\mathrm{int}}^{\mathrm{R}} = -g_\pi \bar{\psi} \gamma_5 \gamma_\mu \tau^a \partial^\mu \Pi^a \psi . \quad (3.96)$$

Show that Eq. (3.88) can be obtained from Eq. (3.96) by introducing non-relativistic approximations.

Equation (3.95) shows that the range of the nuclear force due to the exchange of one pion is given by the Compton wavelength of π-meson and that the radial dependence, i.e., the form factor, is given by the Yukawa function $e^{-\mu r}/\mu r$. These results agree with the conclusions obtained in Sects. 3.1.1 and 3.1.2 based on the uncertainty principle and by making reference to the electromagnetic field. In general, the range of the force due to the exchange of a meson of mass m is of the order of the corresponding Compton wavelength $\hbar/mc$. Hence the contribution of the exchange of heavier mesons such as ρ-meson, whose mass (more precisely the rest energy) is 775.5 MeV, ω-meson, whose mass is 782.7 MeV, and σ-particles, whose exchange corresponds to the exchange of two π-mesons, becomes non-negligible as the distance between two nucleons decreases.

The meson theory has difficulty in discussing the nuclear force in the short range region where the repulsive core appears. Another approach, called the quark cluster model, has been developed instead. In this approach, the nucleon is thought to consist of three quarks and the properties of nuclear force are studied microscopically by taking into account the effects of antisymmetrization among quarks [17]. We wish to mention that the study of nuclear force based on the *chiral perturbation theory* (χPT) as an approximate theory (effective field theory) for QCD [18], and more recently, the study of nuclear force using the Monte Carlo calculation of QCD defined on the lattice (*Lattice QCD* calculation) [19] are also going on as alternative approaches.

3.6 Phenomenological Potential with High Accuracy: Realistic Potential

In parallel with the theoretical derivation of the nuclear force from microscopic points of view, attempts to determine semi-phenomenological nuclear forces which carry quantitative accuracy as well as high applicability have also been developed by combining the phenomenology and meson theory. Historically, the Hamada–Johnston potential [20] and the Reid *hard-core* and *soft-core* potentials [21],[6] and later, as refined potentials, the Bonn potential [12], the Argonne potential [11], the Paris potential [13], and the Nijmegen potential [14] are known as the representatives of such semi-phenomenological potentials.

These potentials have common features that they have a repulsive core and are described by OPEP in the distant region and contain many parameters. The parameters are determined so as to reproduce the experimental data of the nucleon–nucleon scattering at laboratory energies below 350 MeV, and the properties of deuteron. In that sense, they are often called realistic potentials. Each potential differs, e.g., in the strength of the tensor force, the treatment of the medium and short distances, non-locality, and off-shell effects. Table 3.3 summarizes the key issues of determination as well as the properties of these realistic phenomenological potentials. As Table 3.3 shows, the refined realistic potentials, i.e., Paris, Argonne v_{18}, Bonn and Nijmegen potentials, take into account the charge independence breaking (CIB)[7] or the charge

[6] We name the potentials whose height of the repulsive core is infinite and finite the hard- and the soft-core potentials, respectively.

[7] One of the main reasons of the CIB is the difference of the masses of three π-mesons.

Table 3.3 Properties of several realistic phenomenological potentials: Core: property of the repulsive core. Inputs: information used to determine the potential parameters; ere: expansion coefficients in the effective range theory, d: properties of deuteron, t: properties of triton, pp: proton–proton scattering data, np: neutron–proton scattering data, nn: neutron–neutron scattering data. E_{max}: Maximum energy of the nucleon–nucleon scattering used to determine the potential, in units of MeV. Components: C: central force. T: tensor force. LS: spin–orbit force. LL: L^2 term. $(LS)^2$: Second-order spin–orbit force. Characteristics: p^2-dep. C: square-of-momentum dependent central force, CIB: charge independence breaking, CD: charge dependence, EM: include detailed electromagnetic potential

Name	Core	Inputs	E_{max}	Components	Characteristics
Hamada–Johnston	Hard	ere, d, pp, np	315	C, T, LL, LS, $(LS)^2$	
Reid hare-core	Hard	ere, d, pp, np	350	C, T, LS	
Reid soft-core	Soft[a]	ere, d, pp, np	350	C, T, LS	
Paris	(Soft)	d, ere, NN	330	C, T, LS, $(LS)^2$	p^2-dep. C
Argonne v_{18}	Soft[b]	ere, d, pp, np, nn	350	C, T, LL, LS, $(LS)^2$	CIB, EM, local
Bonn	(Soft)	d, t, pp, np	350		CD, nonlocal
Nijmegen	Soft	pp, np	350	C, T, LS, $(LS)^2$	CD, local

[a]Yukawa type
[b]Woods–Saxon type

dependence (CD) of the nuclear force[8,9] and are used in accurate calculations of few-body systems and in the study of nuclear structure from microscopic point of view.

Here, we explain the Hamada–Johnston and Reid potentials which have relatively concise expressions and are convenient to see the bulk properties of the nuclear force.

3.6.1 Hamada–Johnston Potential

Figure 3.5 shows the Hamada–Johnston potential. The Reid hard core potential resembles the Hamada–Johnston potential qualitatively.

The Hamada–Johnston potential postulates a repulsive core of infinite height for the short distance inside of $r = r_C = 0.49$ fm, and assumes that the potential is given by

$$V = V_C(\mu r) + V_T(\mu r)S_{12} + V_{LS}(\mu r)\mathbf{L}\cdot\mathbf{S} + V_{LL}(\mu r)L_{12} \quad (3.97)$$

in the region outside of the core. Here, the operator of the spin-orbit interaction of the second order L_{12} is defined as

[8]The details of the treatment and the accuracy depend on the potential. For example, the Nijmegen potential takes into account only a part of the charge symmetry breaking (CSB), while the Argonne potential v_{18} handles the CSB phenomenologically.

[9]The accurate treatment of the CSB of the nuclear force is important in explaining the mass difference between triton and ^{3}He, and more in general, the Nolen–Schiffer (NS) anomaly [22] concerning the energy difference between the corresponding energy levels of mirror nuclei.

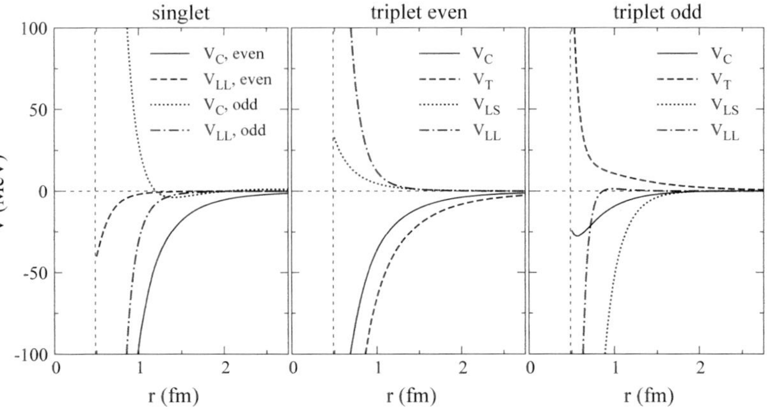

Fig. 3.5 Hamada–Johnston potential [20]. The figure has been made based on Eq. (3.99) through (3.106) and by using the values of parameters given in [23]

$$\begin{aligned} L_{12} &= (\boldsymbol{\sigma}_1 \cdot \boldsymbol{\sigma}_2)\mathbf{L}^2 - \frac{1}{2}\left[(\boldsymbol{\sigma}_1 \cdot \mathbf{L})(\boldsymbol{\sigma}_2 \cdot \mathbf{L}) + (\boldsymbol{\sigma}_2 \cdot \mathbf{L})(\boldsymbol{\sigma}_1 \cdot \mathbf{L})\right] \\ &= (\delta_{LJ} + \boldsymbol{\sigma}_1 \cdot \boldsymbol{\sigma}_2)\mathbf{L}^2 - (\mathbf{L} \cdot \mathbf{S})^2 , \end{aligned} \tag{3.98}$$

and the radial functions are assumed to be given by

$$V_C = V_0(\boldsymbol{\tau}_1 \cdot \boldsymbol{\tau}_2)(\boldsymbol{\sigma}_1 \cdot \boldsymbol{\sigma}_2)Y(x)\left[1 + a_C Y(x) + b_C Y^2(x)\right] , \tag{3.99}$$

$$V_T = V_0(\boldsymbol{\tau}_1 \cdot \boldsymbol{\tau}_2)(\boldsymbol{\sigma}_1 \cdot \boldsymbol{\sigma}_2)Z(x)\left[1 + a_T Y(x) + b_T Y^2(x)\right] , \tag{3.100}$$

$$V_{LS} = g_{LS} V_0 Y^2(x)\left[1 + b_{LS} Y(x)\right] , \tag{3.101}$$

$$V_{LL} = g_{LL} V_0 \frac{Z(x)}{x^2}\left[1 + a_{LL} Y(x) + b_{LL} Y^2(x)\right] , \tag{3.102}$$

with

$$V_0 = \frac{g_\pi^2}{3\hbar c} m_\pi c^2 = 3.65\,\text{MeV}, \tag{3.103}$$

$$x = \mu r = \frac{m_\pi c}{\hbar} r , \tag{3.104}$$

$$Y(x) = \frac{1}{x}\mathrm{e}^{-x} , \tag{3.105}$$

$$Z(x) = \left(1 + \frac{3}{x} + \frac{3}{x^2}\right) Y(x) . \tag{3.106}$$

3.6.2 Reid Potential

In the Reid hard-core potential, the outside region of the repulsive core is assumed to be given by

$$V = V_C(\mu r) + V_T(\mu r)S_{12} + V_{LS}(\mu r)\mathbf{L}\cdot\mathbf{S}\,, \tag{3.107}$$

and the radial dependence is parametrized as

$$\begin{aligned} V_C(x) &= \sum_{n=1}^{\infty} a_n \frac{\mathrm{e}^{-nx}}{x}\,, \\ V_T(x) &= \frac{b_1}{x}\left[\left(\frac{1}{3}+\frac{1}{x}+\frac{1}{x^2}\right)\mathrm{e}^{-x} - \left(\frac{k}{x}+\frac{1}{x^2}\right)\mathrm{e}^{-kx}\right] + \sum_{n=2}^{\infty} b_n \frac{\mathrm{e}^{-nx}}{x}\,, \\ V_{LS}(x) &= \sum_{n=1}^{\infty} c_n \frac{\mathrm{e}^{-nx}}{x}\,. \end{aligned} \tag{3.108}$$

The position of the hard core r_C depends on the total isospin of two nucleons, and is taken as $r_C = 0.42\,\mathrm{fm}$ for the ^1S state and $0.43\,\mathrm{fm}$ for the other states in the case when $T = 1$, and $r_C = 0.55\,\mathrm{fm}$ for the $T = 0$ state. The Reid soft-core potential represents the effect of repulsive core by using a Yukawa function of short-range repulsive type.

The parameters $a_C, \ldots, b_{LL}$ in the Hamada–Johnston potential and a_n, b_n, c_n in the Reid potential are determined phenomenologically through the analyses of the experimental data.

3.7 Summary of the Nuclear Force in the Free Space

Let us here summarize the nuclear force in the free space, i.e., the nuclear force obtained from the analyses of the nucleon–nucleon scattering and of deuteron.

Figure 3.6 shows how various nuclear forces such as the central and tensor forces vary with the distance r between two nucleons. They are shown separately for each state of the two-nucleon systems such as ^{31}E and ^{11}O.[10] The figure shows also the mesons which contribute to the nuclear force in each spatial region.

The main characteristics of the nuclear force appearing in the figure can be summarized as follows.

1. The nuclear force is of short range in contrast with the gravitational force and the electromagnetic force. Note that the unit of the abscissa is 10^{-15}m.

[10] The even–odd property of the spatial motion and the spin multiplicity are shown in the figure. The triplet and singlet in ^{3}E and ^{1}E represent the multiplicity in the spin space.

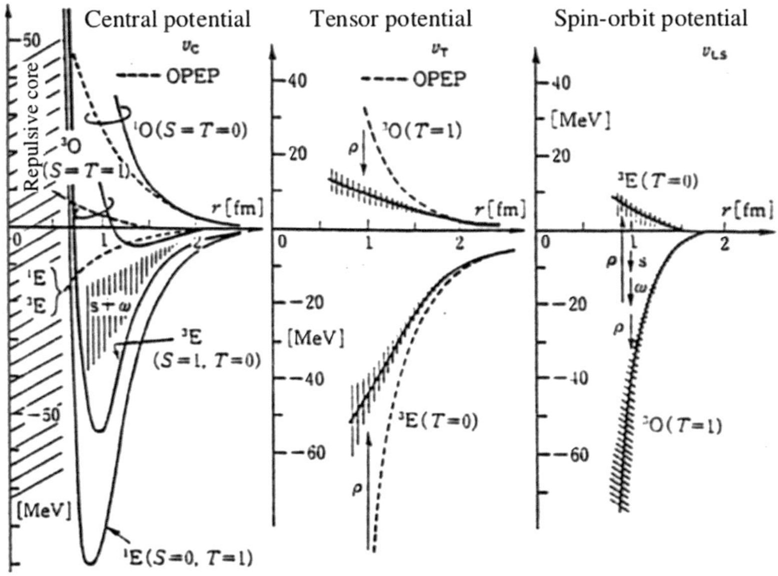

Fig. 3.6 Summary of the nuclear potential, taken from [24]

2. The long distance region is governed by the one-pion exchange potential (OPEP), and is attractive.
3. It turns to a strong repulsion in the short distance, where two nucleons overlap.
4. It strongly depends on the state of two-nucleon systems through, e.g., the tensor force. For example, of the two-nucleon systems, it is only deuteron, whose total isospin is 0 and total spin is 1, for which the attraction is strong enough to yield a bound state.

3.8 Effective Interaction Inside Nucleus

3.8.1 G-Matrix

Further considerations are necessary in order to use for the studies of nuclear structure and nuclear reactions even if the bare nuclear force, i.e., the nuclear force acting between two nucleons in isolation, i.e., in the free space, is known. Although there exist some variations depending on each potential, as we learnt in the previous section, the short range part of the nuclear force is represented either by a repulsive core of infinite strength or a strong repulsive core even if it is not infinite. If we use the

Fig. 3.7 Conceptual representation of the G-matrix. The *wavy line* and the *dotted line* represent G and the bare potential V, respectively

bare nuclear force with a hard core as it is, e.g., in the mean-field theory, which is a standard theory for studying nuclear structure, we get into trouble of the divergence of the matrix elements of the interaction.

One of the methods to circumvent the difficulty associated with the fact that the nuclear force is strong, e.g., the problem of hard core, is the G-matrix theory, in other words the Brueckner theory [25]. In this method, one uses not the bare nuclear force, but the effective nuclear force called G-matrix, which takes into account the effects of multiple scattering and Pauli principle, as the interaction between nucleons in performing the theoretical calculations of nuclear structure and nuclear reactions.[11] Figure 3.7 conceptually represents the G-matrix. In the case of the free space, the effective interaction which takes into account the multiple scattering is called the T-matrix, and is given by

$$T = V + V \frac{1}{E - K_1 - K_2 + i\varepsilon} T , \tag{3.109}$$

in the operator representation when the total energy of the two-nucleon system is E. By repeating the insertion of the whole right-hand side to the T on the right-hand side of Eq. (3.109), one will be able to understand that T is the effective interaction which takes into account multiple scattering. K_1, K_2 are the kinetic energy operators of the nucleons 1 and 2 in the intermediate states.

The G-matrix can be thought to be the extension of the T-matrix to the nuclear force inside nucleus. The important changes from the case of free space are to take into account the Pauli exclusion principle so as to exclude the states which are already occupied by the other nucleons from the intermediate states of nucleons in the multiple scattering, and to use the energy in the medium in calculating the propagator. Consequently, the G-matrix is given by the following equation,

$$G = V + V \frac{Q}{E - H_1^{(0)} - H_2^{(0)}} G , \tag{3.110}$$

where Q is the operator to restrict the intermediate states by following the Pauli exclusion principle. $H_1^{(0)}$ and $H_2^{(0)}$ are the unperturbed Hamiltonian, and correspond, e.g., to the Hamiltonian of the mean field in the shell model. Equation (3.110) is called the *Bethe–Goldstone equation*.

[11]The G-matrix theory is not the standard Rayleigh–Schrödinger perturbation theory, but a perturbation theory which takes into account special higher-order terms, i.e., the ladder scattering.

We give here the matrix representations corresponding to Eqs. (3.109) and (3.110) in order to make the meaning of the word *matrix* clear. In the case of T-matrix, it is given by

$$T^E_{\mathbf{k}'_1\mathbf{k}'_2,\mathbf{k}_1\mathbf{k}_2} = V_{\mathbf{k}'_1\mathbf{k}'_2,\mathbf{k}_1\mathbf{k}_2} + \sum_{\mathbf{p}_1\mathbf{p}_2} V_{\mathbf{k}'_1\mathbf{k}'_2,\mathbf{p}_1\mathbf{p}_2} \frac{1}{E-(\mathbf{p}_1^2/2M)-(\mathbf{p}_2^2/2M)+\mathrm{i}\varepsilon} T^E_{\mathbf{p}_1\mathbf{p}_2,\mathbf{k}_1\mathbf{k}_2} \tag{3.111}$$

by using the wave number vector specifying the plane wave as the state index. $\mathbf{k}_1$, $\mathbf{k}_2$ and $\mathbf{k}'_1$, $\mathbf{k}'_2$ are the wave number vectors of the incoming and outgoing nucleons, respectively. In the case of the G-matrix, using, e.g., the state index of the shell model as the matrix index, we have

$$G^E_{cd,ab} = V_{cd,ab} + \sum_{mn>\varepsilon_F} V_{cd,mn} \frac{1}{E-\varepsilon_m-\varepsilon_n} G^E_{mn,ab} \;. \tag{3.112}$$

Leaving the details of the G-matrix theory to other books [25], here we remark that the effective interaction inside nucleus becomes density dependent through the Fermi energy ε_{F} which appears on the right-hand side of Eq. (3.112) in order to restrict the intermediate states (see Sect. 2.2 as for the relation between the Fermi energy and the density).

3.8.2 Phenomenological Effective Interaction

As we have hitherto studied, it is not easy to determine the nuclear force exactly, and to perform the study of nuclear structure and nuclear reactions from the point of view of many-body problems based on the nuclear force determined from more fundamental points of view such as the meson theory and QCD. The studies using handy phenomenological interactions have therefore also been developed in parallel with the studies based on the practical phenomenological nuclear force with high accuracy described in Sect. 3.6 and those based on the G-matrix theory. Those phenomenological potentials consist of several effective terms such as a strong exchange term (Majorana term) to guarantee the saturation property of density, and/or the term which takes the density dependence into account referring to the G-matrix theory. Here, we mention some examples of those phenomenological effective interactions.[12]

[12] As is clear from the projection operator method [26] by Feshbach. The effective interaction used in a theoretical calculation depends on the size of the model space which is explicitly taken into account in that calculation. The unitary model operator approach (UMOA) is one of such theoretical approaches [27].

Table 3.4 Exchange property of the nuclear force. Each column gives the coefficient of each exchange term

Name of force	Wigner	Majorana	Bartlett	Heisenberg
Serber[a]	0.5	0.5	0.0	0.0
Rosenfeld	−0.13	0.93	0.46	0.26

[a]It well explains the large backward cross sections in the neutron–proton scattering

3.8.2.1 Examples of the Radial Form Factor and Exchange Property

The radial form factor of the nuclear force is often assumed to be either the Yukawa-type or the Gauss-type as described below, or a sum of those terms with different range parameters,

$$V(r) = -V_0 \frac{e^{-\mu r}}{\mu r} \quad \text{(Yukawa potential)}, \tag{3.113}$$

$$V(r) = -V_0 e^{-r^2/r_0^2} \quad \text{(Gauss potential)}. \tag{3.114}$$

As for the exchange property, we show the cases of Serber and Rosenfeld forces in Table 3.4 as representative examples.

3.8.2.2 Effective Interactions Used for Variational Calculations of Light Nuclei

Here, we mention the Volkov interaction [28], which is often used for the study of cluster structure of light nuclei by variational methods, as an example of the effective interaction,[13]

$$V(r) = \left[-V_1 e^{-(r/r_1)^2} + V_2 e^{-(r/r_2)^2}\right](1 - m + m P_M) , \tag{3.115}$$

where $V_1 = V_2 = 60\,\text{MeV}, r_1 = 1.80\,\text{fm}, r_2 = 1.01\,\text{fm}$. The Majorana exchange term is used to guarantee the saturation of nuclei in the Volkov force. Regarding the cluster structure of light nuclei, larger values of m predict more distinct cluster structure.

[13]The Brink–Boeker force [29], the Hasegawa–Nagata force [30], and the Minnesota force [31] have also been often used. These forces are common in assuming the radial factor of Gaussian type so that the calculations of the matrix elements using the wave functions of the harmonic oscillator model can be easily performed. The Hasegawa–Nagata force has a characteristic that it contains the spin–orbit force and the tensor force. Except for the Minnesota force, they determine the parameters so as to reproduce the radii and the relative binding energies of some light nuclei. The Minnesota force determines the parameters so as to reproduce the scattering data. The characteristics of various effective interactions and the way to calculate the matrix elements of the Hamiltonian when the wave function is given by a Slater determinant are described in detail in [32].

3.8.2.3 Effective Interactions Used for Hartree–Fock Calculations

Besides those forces, there exist the Skyrme force and the Gogny force as the effective interactions which are often used for the non-relativistic Hartree–Fock calculations. We describe the former in detail in Chap. 6.

References

1. W.-M. Yao et al. (Particle Data Group), J. Phys. G **33**, 1 (2006)
2. P.J. Mohr, B.N. Taylor, D.B. Newell, Rev. Mod. Phys. **84**, 1527 (2010)
3. M. Pavanello, W.-C. Tung, L. Adamowicz, Phys. Rev. A **81**, 042526 (2012)
4. G.E. Brown, A.D. Jackson, *The Nucleon-Nucleon Interaction* (North-Holland, Amsterdam, 1976)
5. L. Hulthén, K.V. Laurikainen, Rev. Mod. Phys. **23**, 1 (1951)
6. K. Kubo, K. Katori, *Spin and Polarization*, Japanese edn. (Baifukan, Tokyo, 1994)
7. L.I. Schiff, *Quantum Mechanics* (McGraw-Hill Kogakusha, Tokyo, 1968)
8. M. Nogami, *Nuclear Physics*, Japanese edn. (Shoukabou, Tokyo, 1973)
9. N.A. Jelly, *Fundamentals of Nuclear Physics* (Cambridge University Press, Cambridge, 1990)
10. R. Machleidt, Nucl. Phys. A **689**, 11c (2001); L. Jäde, H.V. von Geramb, Phys. Rev. C **57**, 496 (1998); K. Amos et al. Adv. Nucl. Phys. **25**, 276 (2002)
11. R.B. Wiringa, V.G.J. Stoks, R. Schiavilla, Phys. Rev. C **51**, 38 (1995); R.B. Wiringa, R.A. Smith, T.L. Ainsworth. Phys. Rev. C **29**, 1207 (1984)
12. R. Machleidt, K. Holinde, Ch. Elster, Phys. Rep. **149**, 1 (1987); J. Haidenbauer, K. Holinde, Phys. Rev. C **40**, 2465 (1989); R. Machleidt. Adv. Nucl. Phys. **19**, 189 (1989)
13. M. Lacombe, B. Loiseau, J.M. Richard, R. Vinh Mau, J. Côté, P. Pirès, R. de Tourreil, **21**, 861 (1980)
14. M.M. Nagels, T.A. Rijken, J.J. de Swart, Phys. Rev. D **17**, 768 (1978); P.M.M. Maessen, T.A. Rijken, J.J. de Swart, Phys. Rev. C **40**, 2226 (1989); V.G.J. Stoks, R.A.M. Klomp, C.P.F. Terheggen, J.J. de Swart. Phys. Rev. C **49**, 2950 (1994)
15. R.H. Stahl, N.F. Ramsey, Phys. Rev. **96**, 1310 (1954)
16. R.A. Arndt, M.H. MacGregor, Phys. Rev. **141**, 873 (1966)
17. K. Yazaki, Prog. Theor. Phys. Suppl. **91**, 146 (1987)
18. R. Machleidt, D.R. Entem, J. Phys. Conf. Ser. **20**, 77 (2005)
19. N. Ishii, S. Aoki, T. Hatsuda, Phys. Rev. Lett. **99**, 022001 (2007); S. Aoki, T. Hatsuda, N. Ishii, Comput. Sci. Disc. **1**, 015009 (2008); Prog. Theor. Phys. **123**, 89 (2010)
20. T. Hamada, I.D. Johnston, Nucl. Phys. A **34**, 382 (1962)
21. P.V. Reid, Ann. Phys. **50**, 411 (1968)
22. J.A. Nolen, J.P. Schiffer, Ann. Rev. Nucl. Sci. **19**, 471 (1969)
23. Aage Bohr, Ben R. Mottelson, *Nuclear Structure*, vol. I (Benjamin, New York, 1969)
24. R. Tamagaki, *Frontiers of Physics*, vol. 15 (Kyouritsu, Tokyo, 1986). Japanese edition
25. G.E. Brown, *Unified Theory of Nuclear Models and Forces* (North-Holland, Amsterdam, 1967)
26. G.R. Satchler, *Direct Nuclear Reactions* (Clarendon Press, Oxford, 1983)
27. S. Fujii, R. Okamoto, K. Suzuki, Phys. Rev. Lett. **113**, 182501 (2009); K. Suzuki, R. Okamoto, BUTSURI **42**, 263 (1987); K. Suzuki, S.Y. Lee, Prog. Theor. Phys. **64**, 2091 (1980)
28. A.B. Volkov, Phys. Lett. **12**, 118 (1964); Nucl. Phys. **74**, 33 (1965)
29. D.M. Brink, E. Boeker, Nucl. Phys. **91**, 1 (1967)
30. A. Hasegawa, S. Nagata, Prog. Theor. Phys. **45**, 786 (1971)
31. D.R. Thompson, Y.C. Tang, Phys. Rev. **159**, 806 (1967)
32. D.M. Brink, in *Proceedings of the International School of Physics "Enrico Fermi", Course XXXVI*, Varenna, 1965, ed. by C. Bloch (Academic Press, New York, 1966), p.247

Chapter 4
Interaction with Electromagnetic Field: Electromagnetic Moments

Abstract As mentioned in Chap. 1, by measuring the magnetic moment of proton, Stern showed that the proton is not an ideal point particle which can be described by the Dirac equation. Also, we learnt in Chap. 3 that the magnetic moment of deuteron tells that deuteron is in the spin-triplet and isospin-singlet states, and that one can get information on the magnitude of the tensor force through the admixture of the D-state in deuteron. In this way, the electromagnetic properties of nuclei provide valuable information on the structure of nucleons and nuclei. Also, the interaction of nuclei with the radiation field leads to the emission and absorption of a γ-ray, governs the lifetime of each energy level, and provides information on the nuclear structure as well as nuclear collective motions through the strength of the electromagnetic transitions between nuclear levels. Furthermore, the measurement of the angular correlation between the cascade γ-rays enables us to determine the spin of each energy level. The electromagnetic transitions play an important role also in the synthesis of elements through, e.g., the radiative neutron capture (see Cottingham and Greenwood, An Introduction to Nuclear Physics, 2nd edn. (Cambridge University Press, Cambridge, 2001); Thompson and Nunes, Nuclear Reactions for Astrophysics: Principles, Calculation and Applications of Low-Energy Reactions (Cambridge University Press, Cambridge, 2009) [1, 2]). In this chapter we learn the electromagnetic moments such as the magnetic dipole moment and the quadrupole moment which provides important information on the nuclear shape. The electromagnetic transitions by emitting γ-rays will be discussed in Sect. 8.3 after we learn about nuclear structure in Chaps. 5 and 7.

4.1 Hamiltonian of the Electromagnetic Interaction and Electromagnetic Multipole Moments

The Hamiltonian of the interaction between a nucleus and the electromagnetic field is given by

$$H_{\rm em} = \int \rho_C(\mathbf{r})\varphi(\mathbf{r})d\mathbf{r} - \frac{1}{c}\int \mathbf{j}(\mathbf{r}) \cdot \mathbf{A}(\mathbf{r})d\mathbf{r} \ , \qquad (4.1)$$

N. Takigawa and K. Washiyama, *Fundamentals of Nuclear Physics*,
DOI 10.1007/978-4-431-55378-6_4

where $\rho_C(\mathbf{r})$ and $\mathbf{j}(\mathbf{r})$ are the electric charge density and the electric current density of the nucleus, and $\varphi(\mathbf{r})$ and $\mathbf{A}(\mathbf{r})$ are the scalar and vector potentials to represent the electromagnetic field. We obtain

$$H_{\text{em}} = Q\varphi(0) - \mathbf{P}\cdot\mathbf{E}(0) - \boldsymbol{\mu}\cdot\mathbf{H}(0) - \frac{1}{6}\sum_{ij} Q_{ij}\left.\frac{\partial E_j}{\partial x_i}\right|_{\mathbf{r}=0} + \cdots , \tag{4.2}$$

$$Q = \int \rho_C(\mathbf{r})d\mathbf{r} , \tag{4.3}$$

$$\mathbf{P} = \int \mathbf{r}\rho_C(\mathbf{r})d\mathbf{r} , \tag{4.4}$$

$$\boldsymbol{\mu} = \frac{1}{2c}\int \mathbf{r}\times\mathbf{j}(\mathbf{r})d\mathbf{r} , \tag{4.5}$$

$$Q_{ij} = \int \rho_C(\mathbf{r})(3x_ix_j - \delta_{ij}r^2)d\mathbf{r} , \tag{4.6}$$

$$\mathbf{E} = -\nabla\varphi , \tag{4.7}$$

$$\mathbf{H} = \nabla\times\mathbf{A} , \tag{4.8}$$

by expanding $\varphi(\mathbf{r})$ and $\mathbf{A}(\mathbf{r})$ around the center of the nucleus. The indices i, j run from one to three using the notation $\mathbf{r} = (x_1, x_2, x_3)$. δ_{ij} is the Kronecker delta. We assumed that the magnetic field is constant over the nucleus, and used $\mathbf{A} = \frac{1}{2}\mathbf{H}\times\mathbf{r}$, which is valid in that case. Q is the total charge and $\mathbf{P}$, $\boldsymbol{\mu}$, Q_{ij} are called electric dipole moment, magnetic dipole moment and electric quadrupole moment, respectively.

Exercise 4.1 Show that $\mathbf{A} = \frac{1}{2}\mathbf{H}\times\mathbf{r}$ satisfies $\mathbf{H} = \nabla\times\mathbf{A}$ if the magnetic field is uniform.

4.1.1 Operators for the Dipole and Quadrupole Moments

The electric charge density ρ_C and the electric current density $\mathbf{j}$ in Eqs. (4.4)–(4.6) must be considered as the operators $\hat{\rho}_C$ and $\hat{\mathbf{j}}$ in order to compare the experimental values of these multipole moments and the related transition probabilities with their theoretical values. As we describe in Chaps. 5 and 7 and in Sect. 8.3, there exist various ways such as the microscopic methods, e.g., the shell model, which uses the nucleonic degrees of freedom, and the collective model which uses the surface deformation parameters as coordinates to describe nuclear structure and nuclear reactions. Since $\hat{\rho}_C$ and $\hat{\mathbf{j}}$ are needed to be expressed in terms of relevant coordinates, their explicit expressions change depending on the model or the method we take to describe nuclear structure or reactions. Here, we explicitly write down $\hat{\rho}_C$ and $\hat{\mathbf{j}}$ by assuming that we adopt a microscopic model which uses the nucleonic degrees of freedom. If we treat each nucleon as a point particle, and represent the spin and isospin of the k-th nucleon as $\hat{\mathbf{s}}_k$ and $\hat{\mathbf{t}}_k$, then they are given by

$$\hat{\rho}(\mathbf{r}) = e\sum_k \left(\frac{1}{2} - \hat{t}_z(k)\right)\delta(\hat{\mathbf{r}} - \mathbf{r}_k) , \tag{4.9}$$

$$\begin{aligned}\hat{\mathbf{j}}(\mathbf{r}) &= e\sum_k \left(\frac{1}{2} - \hat{t}_z(k)\right)\frac{1}{2}\left[\hat{\mathbf{v}}_k\delta(\hat{\mathbf{r}} - \mathbf{r}_k) + \delta(\hat{\mathbf{r}} - \mathbf{r}_k)\hat{\mathbf{v}}_k\right] \\ &+ \frac{e\hbar}{2M_N}\sum_k g_s(k)\nabla \times \hat{\mathbf{s}}_k\delta(\hat{\mathbf{r}} - \mathbf{r}_k) ,\end{aligned} \tag{4.10}$$

where

$$\hat{\mathbf{v}}_k = \frac{\mathrm{i}}{\hbar}\left[\hat{H}, \hat{\mathbf{r}}_k\right] \sim \frac{1}{m}\hat{\mathbf{p}}_k , \tag{4.11}$$

$$g_s(k) \equiv \frac{1}{2}[g_{\mathrm{n}}(k) + g_{\mathrm{p}}(k)] + \hat{t}_z(k)[g_{\mathrm{n}}(k) - g_{\mathrm{p}}(k)] . \tag{4.12}$$

g_s is called the spin g-factor. We use the values of the anomalous magnetic moments $g_{\mathrm{n}} = 2\mu_{\mathrm{n}}/\mu_N = -3.826$ and $g_{\mathrm{p}} = 2\mu_{\mathrm{p}}/\mu_N = 5.586$ discussed in Chap. 1 for g_{n} and g_{p}. The first and second terms on the right-hand side of Eq. (4.10) are called the convection current and the magnetization current, respectively. The former is the usual electric current due to the orbital motion of protons, while the latter the electric current associated with the magnetic moments of protons and neutrons. Equation (4.10) can be derived from the Dirac equation, which the Fermi particle should obey, by making non-relativistic approximation.

Exercise 4.2 In the Dirac theory, the current density is given by $j^k = c\psi^\dagger\alpha^k\psi$ with $\alpha^k \equiv \begin{pmatrix} 0 & \sigma^k \\ \sigma^k & 0 \end{pmatrix}$. On the other hand, the two-dimensional spinor χ representing the third and fourth components of ψ, i.e., the small components if the velocity is small, can be approximated as $\chi \approx (\boldsymbol{\sigma} \cdot \mathbf{p}/2mc)\varphi$ by using the first and second components φ if the velocity of the particle is sufficiently smaller than the light velocity. Noting these properties, derive Eq. (4.10).

By inserting Eqs. (4.9) and (4.10) into Eqs. (4.3)–(4.6), we obtain

$$\hat{Q} = e\sum_k \left(\frac{1}{2} - \hat{t}_z(k)\right) , \tag{4.13}$$

$$\hat{\mathbf{P}} = e\sum_k \left(\frac{1}{2} - \hat{t}_z(k)\right)\hat{\mathbf{r}}_k , \tag{4.14}$$

$$\hat{\boldsymbol{\mu}} = \frac{1}{2c}\int \mathbf{r} \times \left[\hat{\mathbf{j}}_c(\mathbf{r}) + \hat{\mathbf{j}}_{\mathrm{mag}}(\mathbf{r})\right]d\mathbf{r} \equiv \hat{\boldsymbol{\mu}}_{\mathrm{convection}} + \hat{\boldsymbol{\mu}}_{\mathrm{mag}} , \tag{4.15}$$

$$\hat{\boldsymbol{\mu}}_{\mathrm{convection}} = \mu_N\sum_k \left(\frac{1}{2} - \hat{t}_z(k)\right)\hat{\boldsymbol{\ell}}_k , \tag{4.16}$$

$$\hat{\boldsymbol{\mu}}_{\mathrm{mag}} = \mu_N\sum_k g_s(k)\hat{\mathbf{s}}_k , \tag{4.17}$$

$$\hat{Q}_{ij} = e \sum_k \left(\frac{1}{2} - \hat{t}_z(k) \right) \left[3\hat{x}_i(k)\hat{x}_j(k) - \delta_{ij}\hat{r}^2(k) \right] . \tag{4.18}$$

4.1.2 Various Corrections

To make discussions with high accuracy, one must take into account various corrections such as (1) the effect of the finite size of proton, (2) that the velocity $\hat{\mathbf{v}}_k \equiv \frac{i}{\hbar}[\hat{H}, \hat{\mathbf{r}}_k]$ deviates from $\hat{\mathbf{p}}_k/M_N$ in the presence of velocity dependent forces such as the spin–orbit interaction, (3) the effect of meson exchange current, i.e., that the electromagnetic field operates on the mesons exchanged between the nucleons, (4) the quenching effect, i.e., that the anomalies of the magnetic moment of nucleons get reduced inside nucleus because of the Pauli exclusion principle, and (5) relativistic effects.

4.1.3 Measurement of the Magnetic Moment: Hyperfine Structure

The detailed study of the energy levels of atoms shows that the level structure is more complex than what is expected from the motion of the electron alone. An additional structure called the hyperfine structure appears. Consider a nucleus with an odd mass number. The spin of the nucleus is a half integer. We represent the corresponding nuclear spin operator as $\hat{\mathbf{I}}$. The magnetic moment operator of the nucleus $\hat{\boldsymbol{\mu}}_{\text{nucleus}}$ is then given by $\hat{\boldsymbol{\mu}}_{\text{nucleus}} = g\mu_N \hat{\mathbf{I}}$ by using the g-factor. The energy levels of the whole atom including both electrons and the nucleus split into close-lying energy levels due to the influence of the magnetic field made by the electron on this nuclear magnetic moment. This is why the hyperfine structure arises in the level structure. As an example, Fig. 4.1 shows the low-lying energy levels of a neutral hydrogen atom [3].

As one can expect from the above arguments, the study of hyperfine structure is a powerful method to determine the magnitude of the magnetic moment of the nucleus.[1] In the following, let us study somewhat in detail the principle of the method.

If we denote the magnetic field made by electrons by $\hat{\mathbf{H}}_e$, then the interaction Hamiltonian between the electrons and the nucleus is given by

$$\hat{H}_{\text{hyperfine}} = -\hat{\boldsymbol{\mu}}_{\text{nucleus}} \cdot \hat{\mathbf{H}}_e . \tag{4.19}$$

[1] Stern developed a tenuous molecular beam method in order to realize the interaction-less situation which is often postulated in idealized theoretical arguments. He discovered the anomalous magnetic moment of proton by the experiment which sends the tenuous stream of molecules through an inhomogeneous magnetic field similarly to the Stern–Gerlach experiment which showed experimentally the quantization of spatial orientation of angular momentum.

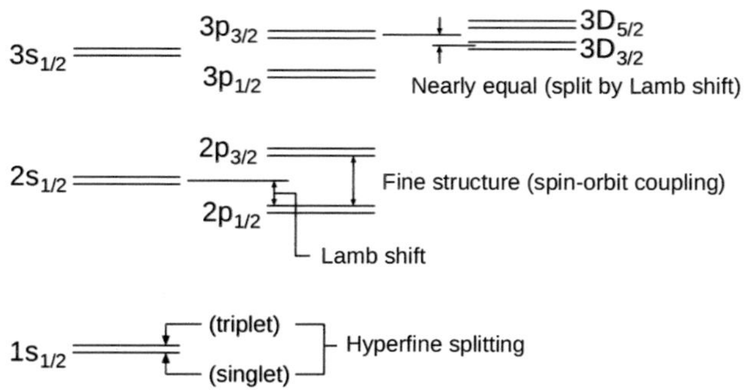

Fig. 4.1 Low-lying energy levels of atomic hydrogen. Taken from [3]

To be specific, let us consider either the hydrogen atom or an alkali atom, where the orbital angular momentum of the outer electron is 0 in the ground state. In that case, $\hat{\mathbf{H}}_e$ is given by [4]

$$\hat{\mathbf{H}}_e = -\frac{16\pi}{3}\mu_B|\psi(0)|^2\hat{\mathbf{s}} , \tag{4.20}$$

if we denote the Bohr magneton by μ_B, the value of the wave function of the outer electron at the origin by $\psi(0)$, and the spin operator of the outer electron by $\hat{\mathbf{s}}$. Hence, we have

$$\hat{H}_{\text{hyperfine}} = \hat{H}_{\text{spin–spin}} = \frac{16\pi}{3}g\mu_N\mu_B|\psi(0)|^2(\hat{\mathbf{s}}\cdot\hat{\mathbf{I}}) . \tag{4.21}$$

Equation (4.21) shows that each energy level splits into two energy levels by the presence of the inner product $(\hat{\mathbf{s}}\cdot\hat{\mathbf{I}})$. In order to emphasize this feature, the index spin–spin has been attached to the Hamiltonian operator in the second term of Eq. (4.21) in the sense of spin–spin interaction.

Equation (4.21) can be used to deduce the g-factor from the measurement of the hyperfine splitting. Thus the magnetic moment can be determined. Note that the probability for electrons to exist at the origin is finite for s-orbits. In general, the operator of the magnetic interaction of the electrons with the nucleus is of the form of $\hat{H}_{JI} = a(\hat{\mathbf{J}}\cdot\hat{\mathbf{I}})$ if the total spin of the electrons is denoted by $\hat{\mathbf{J}}$. Hence each energy level with given J splits into either $2I+1$ levels if $I \le J$ or $2J+1$ levels if $I \ge J$. One can therefore obtain information on J and I by counting the number of energy levels appearing in the hyperfine structure.[2]

[2] The energy spacing between successive energy levels labeled with the value of the total spin $\hat{F} = \hat{J} + \hat{I}$ in the hyperfine structure deviates from the rule expected from the $(\hat{\mathbf{J}}\cdot\hat{\mathbf{I}})$ factor for the magnetic hyperfine splitting considered in this section due to the quadrupole hyperfine splitting if the quadrupole moment of the nucleus is finite. In other words, deviations of the level interval from the interval rule for the magnetic hyperfine structure provide important information on the quadrupole moment or the shape of the nucleus.

Exercise 4.3 Show that the strength of the magnetic field made by the electron at the position of nucleus is 10^5–10^6 Gauss depending on the species of atom by obtaining the magnitude of $|\psi(0)|^2$ as a function of the atomic number Z and the principal quantum number n by using the wave function of the hydrogen atom.

Exercise 4.4 Besides the study of hyperfine structure, there exist various methods to measure the magnetic moment such as the method of nuclear magnetic resonance, molecular beam method, Rabi's molecular beam resonance method (Rabi's refocus method) and the perturbed angular correlation method (PAC). Study the principle and features of each method.

4.2 Electromagnetic Multipole Operators

It is convenient to begin with the generalized definition in order to discuss in a unified way not only the low-order moments introduced in Sect. 4.1.1, but also all the electromagnetic properties of nuclei including higher-order multipole moments as well as the emission and absorption of photons.

Let us first consider the static electric potential $\phi(\mathbf{r})$ made by a nucleus in the far distance from the nucleus. We denote the proton density inside the nucleus by $\rho_{\mathrm{p}}(\mathbf{r})$. Using the formula of the multipole expansion of $1/|\mathbf{r}-\mathbf{r}'|$ for the case of $r > r'$, we have

$$\begin{aligned}\phi(\mathbf{r}) &= e\int \rho_{\mathrm{p}}(\mathbf{r}')\frac{1}{|\mathbf{r}-\mathbf{r}'|}\mathrm{d}\mathbf{r}' \\ &= \frac{4\pi e}{r}\sum_{\ell m}\frac{1}{2\ell+1}\left[\int \mathrm{d}\mathbf{r}'\rho_{\mathrm{p}}(\mathbf{r}')(r')^{\ell}Y_{\ell m}(\Omega_{r'})\right]\frac{1}{r^{\ell}}Y^{*}_{\ell m}(\Omega_r)\,. \end{aligned} \tag{4.22}$$

If we define the μ component of the electric multipole operator by

$$\hat{Q}_{\lambda\mu} = \int \rho_{\mathrm{p}}(\mathbf{r})r^{\lambda}Y_{\lambda\mu}(\Omega_r)\mathrm{d}\mathbf{r}\,, \tag{4.23}$$

then, the static electric potential is given by

$$\phi(\mathbf{r}) = 4\pi e\sum_{\ell m}\frac{\hat{Q}_{\ell m}}{2\ell+1}\frac{Y^{*}_{\ell m}(\Omega_r)}{r^{\ell+1}}\,. \tag{4.24}$$

Similarly we define the μ component of the magnetic multipole operator by

$$\hat{M}_{\lambda\mu} \equiv \int \boldsymbol{\mu}(\mathbf{r})\cdot\nabla(r^{\lambda}Y_{\lambda\mu}(\Omega_r))\mathrm{d}\mathbf{r} \tag{4.25}$$

$$= \frac{1}{c(\lambda+1)}\int [\mathbf{r}\times\mathbf{j}(\mathbf{r})]\cdot\left[\nabla r^{\lambda}Y_{\lambda\mu}(\Omega_r)\right]\mathrm{d}\mathbf{r}\,. \tag{4.26}$$

We used $\mathbf{A}\cdot(\mathbf{B}\times\mathbf{C})=\mathbf{B}\cdot(\mathbf{C}\times\mathbf{A})=\mathbf{C}\cdot(\mathbf{A}\times\mathbf{B})$ and the relationship $\mathbf{j}(\mathbf{r})=c\nabla\times\boldsymbol{\mu}(\mathbf{r})$ in order to transform from Eq. (4.25) to (4.26).

By inserting the results of Eqs. (4.9) and (4.10) into Eqs. (4.23) and (4.26), we obtain the expressions for the electromagnetic multipole operators in terms of nucleonic coordinates as

$$\hat{Q}_{\lambda\mu}=e\sum_k\left(\frac{1}{2}-\hat{t}_z(k)\right)r_k^{\lambda}Y_{\lambda\mu}(\theta_k,\varphi_k)\ , \tag{4.27}$$

$$\hat{M}_{\lambda\mu}=\mu_N\sum_k\left[g_s(k)\hat{\mathbf{s}}_k+\frac{2}{\lambda+1}g_\ell(k)\hat{\boldsymbol{\ell}}_k\right]\cdot\left[\nabla r^{\lambda}Y_{\lambda\mu}(\theta,\varphi)\right]_{\mathbf{r}=\mathbf{r}_k}\ . \tag{4.28}$$

4.3 Properties of the Electromagnetic Multipole Operators

4.3.1 Parity, Tensor Property and Selection Rule

4.3.1.1 Parity

If we denote the operator of the Parity transformation, which inverts all spatial coordinates, by $\hat{\mathscr{P}}$, and represent the coordinate using the polar coordinates (r,θ,φ), the definition of $\hat{\mathscr{P}}$,

$$\hat{\mathscr{P}}(r,\theta,\varphi)\hat{\mathscr{P}}^{-1}\equiv(r,\pi-\theta,\varphi+\pi)\ , \tag{4.29}$$

leads to

$$\hat{\mathscr{P}}Y_{\lambda\mu}(r,\theta,\varphi)\hat{\mathscr{P}}^{-1}=Y_{\lambda\mu}(r,\pi-\theta,\varphi+\pi)=(-1)^{\lambda}Y_{\lambda\mu}(r,\theta,\varphi)\ . \tag{4.30}$$

Equations (4.27) and (4.28) therefore give the following properties of the electric and magnetic multipole operators concerning the parity transformation,

$$\hat{\mathscr{P}}\hat{Q}_{\lambda\mu}\hat{\mathscr{P}}^{-1}=(-1)^{\lambda}\hat{Q}_{\lambda\mu}\ , \tag{4.31}$$

$$\hat{\mathscr{P}}\hat{M}_{\lambda\mu}\hat{\mathscr{P}}^{-1}=(-1)^{\lambda+1}\hat{M}_{\lambda\mu}\ . \tag{4.32}$$

Note that both the spin and the angular momentum operators are invariant under the parity transformation.

4.3.1.2 Tensor Property

Furthermore, it can be derived from Eqs. (4.27) and (4.28) that both $\hat{Q}_{\lambda\mu}$ and $\hat{M}_{\lambda\mu}$ are the μ component of a rank-λ tensor. The former can be easily proved by noting that it is directly proportional to the spherical harmonics $Y_{\lambda\mu}$ and that $Y_{\lambda\mu}$ is the μ

component of a rank-λ tensor. The latter looks complicated at first glance because of the existence of the inner product of vector operators in front of $Y_{\lambda\mu}$. However, the tensor property does not change and is identical with that of $Y_{\lambda\mu}$, since the inner product of two vectors is a scalar, i.e., a rank-0 tensor. Details of tensor algebra are given in Appendix A.6.

4.3.1.3 Selection Rule

It is necessary to calculate the expectation values of multipole operators in order to compare the experimental values of the multipole moments with theoretical values. If we represent the nuclear state as $|\psi_{IM}\rangle$, where I and M are the magnitude of the angular momentum and its z-component in the space-fixed coordinate system, then the expectation values of multipole operators are defined by

$$Q_{\lambda\mu} = \langle\psi_{IM}|\hat{Q}_{\lambda\mu}|\psi_{IM}\rangle \,, \tag{4.33}$$

$$M_{\lambda\mu} = \langle\psi_{IM}|\hat{M}_{\lambda\mu}|\psi_{IM}\rangle \,. \tag{4.34}$$

The following selection rules can be obtained from the parity and tensor properties of the multipole operators mentioned above,

$$Q_{\lambda\mu} = 0 \quad (\text{for odd } \lambda), \tag{4.35}$$

$$M_{\lambda\mu} = 0 \quad (\text{for even } \lambda), \tag{4.36}$$

$$Q_{\lambda\mu},\ M_{\lambda\mu} \neq 0 \quad (\text{only when } \mu = 0 \text{ and } 0 \leq \lambda \leq 2I). \tag{4.37}$$

Equations (4.35) and (4.36) hold when one constructs the wave function $|\psi_{IM}\rangle$ by considering only the strong interaction, because $|\psi_{IM}\rangle$ then has a definite parity.

4.3.2 *Definition of the Electromagnetic Moments*

4.3.2.1 Wigner–Eckart Theorem

If we represent the μ component of a rank-λ tensor as $\hat{T}_{\lambda\mu}$, the following theorem called the Wigner–Eckart theorem holds,[3]

$$\langle\beta I_2 M_2|\hat{T}_{\lambda\mu}|\alpha I_1 M_1\rangle = (2I_2+1)^{-1/2}\langle I_1\lambda M_1\mu|I_2M_2\rangle\langle\beta I_2\|\hat{T}_\lambda\|\alpha I_1\rangle \,. \tag{4.38}$$

Equation (4.38) implies that it is sufficient to know the matrix element for a particular z-component in order to fully determine the matrix elements of a tensor operator. The

[3]There exists an alternative definition which omits the factor $(2I_2+1)^{-1/2}$ on the right-hand side.

matrix elements for the other z-components are automatically determined according to the Clebsch–Gordan coefficients.

4.3.2.2 Magnetic Dipole Moment

Since the operator of the magnetic dipole moment is a tensor of rank-1, we define the magnetic dipole moment which should be compared with the experimental data by

$$\mu \equiv \langle II|\hat{\mu}_z|II\rangle \tag{4.39}$$

$$= \sqrt{\frac{4\pi}{3}}\langle II|\hat{M}_{10}|II\rangle\ , \tag{4.40}$$

following the Wigner–Eckart theorem.[4]

4.3.2.3 The Electric Quadrupole Moment

Similarly we define the electric quadrupole moment by

$$Q \equiv \sqrt{\frac{16\pi}{5}}\langle II|\hat{Q}_{20}|II\rangle\ . \tag{4.41}$$

The tensor component indices of both the wave functions and the operator in Eq. (4.41) refer to the space-fixed coordinate system, i.e., the laboratory system. On the contrary, in the case of a statically deformed nucleus, one defines the intrinsic electric quadrupole moment by

$$Q_0 \equiv \sqrt{\frac{16\pi}{5}}\langle\hat{Q}'_{20}\rangle \tag{4.42}$$

$$= \int \rho_p(\mathbf{r}')(3z'^2 - r'^2)d\mathbf{r}'\ , \tag{4.43}$$

by taking the axes of the coordinate system along the directions of the principal axes, i.e., by taking the body-fixed coordinate frame (or the intrinsic coordinate system). For example, if the nucleus is a spheroid, Q_0 is given by

$$Q_0 = \frac{2}{5}Ze(a^2 - b^2) \tag{4.44}$$

$$= \frac{2}{5}ZeR_0^2 3\alpha_2\left(1 + \frac{1}{4}\alpha_2\right) \tag{4.45}$$

---

[4]Note that the z-component of a vector corresponds to the zeroth component of a tensor of rank-1, i.e., $V_z = \hat{T}_{10}$ (see Appendix A.6.2).

$$\sim ZeR_0^2 \frac{3}{4\pi}\sqrt{\frac{16\pi}{5}}\alpha_{20} , \tag{4.46}$$

by using the length of the semi-axis in the direction of the symmetry axis a and that in the perpendicular direction b, and the deformation α_2 and the radius parameters R_0 introduced in Eq. (2.66). These equations show that one can learn the shape of the nucleus, more precisely the shape of the distribution of protons inside the nucleus, if information on the intrinsic electric quadrupole moment can be obtained. $Q_0 > 0$ and $Q_0 < 0$ mean the shapes prolonged and shrunk in the direction of the symmetry axis and are called prolate and oblate shapes, respectively. In Eq. (4.46), we used the deformation parameter α_{20}, which we will introduce in Chap. 7, and ignored the second order term with respect to α_{20}.

What one measures experimentally is not the value of Q_0, but that of Q. If a nucleus is axial-symmetrically deformed, the component of the angular momentum projected on the symmetry axis, K, becomes a good quantum number in addition to the magnitude of the angular momentum I and its component along the space-fixed z-axis M. In this case, for even–even nuclei or for odd nuclei, where all nucleons rotate simultaneously[5] and $I = K = 1/2$ or $K \geq 3/2$, one can prove that Q and Q_0 are related as

$$Q = Q_0 \frac{3K^2 - I(I+1)}{(2I+3)(I+1)} . \tag{4.47}$$

Especially, if $I = K$ as in the case of the ground state,

$$Q = \frac{I}{I+1}\frac{2I-1}{2I+3}Q_0 . \tag{4.48}$$

Equation (4.48) shows that $Q = 0$ irrespective of Q_0 if $I = 0$ or $1/2$. This result can be expected from the coupling rule of angular momenta or of tensors and the orthogonality of states with different values of angular momentum. The derivation of Eq. (4.47) is given in Appendix A.7.

References

1. W.N. Cottingham, D.A. Greenwood, *An Introduction to Nuclear Physics*, 2nd edn. (Cambridge University Press, Cambridge, 2001)
2. I.J. Thompson, F.M. Nunes, *Nuclear Reactions for Astrophysics: Principles, Calculation and Applications of Low-Energy Reactions* (Cambridge University Press, Cambridge, 2009)
3. J.D. Bjorken, S.D. Drell, *Relativistic Quantum Mechanics* (McGraw-Hill, New York, 1964)
4. L.D. Landau, E.M. Lifshitz, *Quantum Mechanics–Non-Relativistic Theory*, 2nd edn. (Pergamon Press, Oxford, 1965), p. 465

[5]There exist two possibilities for odd nuclei. In one case, the core part made from even–even nucleus makes a collective rotation, while the last odd nucleon moves almost independently from this motion. In the other case, both the core and the last nucleon rotate together as a whole. The former and the latter are called *weak coupling* and *strong coupling* states, respectively.

Chapter 5
Shell Structure

Abstract Although the bulk properties of nuclei such as the mass and size, and also fission and the compound nucleus reactions suggest that the nucleus behaves like a liquid, there exist characteristic properties which cannot be understood from such a point of view. The existence of magic numbers in various phenomena is the most evident example, suggesting the shell structure. The magic numbers appear in various systems in nature. The existence of the noble gases in the periodic table of the elements is the most popular example. Contrary to those magic numbers for atoms, which are associated with the long range Coulomb interaction, the magic numbers for nuclei which originate from the short range force differ in numbers. Also, the spin–orbit interaction plays a crucial role in the magic numbers for nuclei. In this chapter we discuss how the magic numbers arise, and discuss the spin and parity properties of the nuclei in the vicinity of magic numbers. Although the shell model which is based on the mean-field theory succeeds, it is suggested that it is important to take into account the effects of the residual interaction, i.e., the pairing correlation in order to explain the details of the nuclear phenomena. As an example of the current topics, we also discuss the present status of the research of superheavy elements which are stabilized by shell effects.

5.1 Magic Numbers

The saturation properties of density and binding energy per nucleon suggest the validity of the liquid-drop model for nuclei. The mass formula is also based on the picture of a nucleus as a liquid drop, and various basic properties of fission can be explained based on the liquid-drop model. Also, the *compound nucleus reaction*, one of the representative nuclear reactions, strongly supports the validity of the liquid-drop model for nuclei.

The liquid-drop model is the picture that physical quantities vary continuously or monotonically as functions of the mass and atomic numbers and that there exist

N. Takigawa and K. Washiyama, *Fundamentals of Nuclear Physics*,
DOI 10.1007/978-4-431-55378-6_5

neither special mass number nor special proton and neutron numbers. However, there exist many experimental data which contradict this picture. We mention here some of the examples.

Figure 1.7 shows that the nuclei whose neutron number N is 50, 82, or 126 exist significantly more than the surrounding nuclei. Figure 2.6 and more detailed analysis of the binding energy show that the nuclei whose proton number and/or neutron number are one of 20, 28, 50, 82, 126 have larger binding energy per nucleon than the other nuclei whose binding energy can be well represented by the mass formula and are more stable than them. Also, as stated concerning Fig. 2.17, most of the spontaneous fission and the induced fission at low energies are asymmetric fission, where the mass number and the atomic number of the fission products which have large yield are related to the special numbers mentioned above. As we learn in Chap. 7, the shape of nucleus is also intimately related to the mass and atomic numbers, and the shape of the nucleus whose proton number and neutron number are one of the special numbers mentioned above is spherical, i.e., the absolute value of their intrinsic quadrupole moment is either 0 or small. On the other hand, ${}^{4}_{2}$He and ${}^{16}_{8}$O, whose proton and neutron numbers are both 2 or 8, respectively, are much more stable than the surrounding nuclei.

In this way, *2, 8, 20, 28, 50, 82, 126* are special numbers for nuclei and are called *magic numbers*.[1]

Sidebar: The Connection Between the Magic Numbers and Life The atomic numbers 2, 10, 18, 36, 54, 86 of the noble gases in the periodic table of the elements can also be thought to be the magic numbers of atoms. The respiration of organism is guaranteed by the rich abundance of oxygen thanks to the fact that the magic number for nuclei is not 10, but 8. As we learn in the next section, the change of the magic number from 10 to 8 is caused because of the short range character of the nuclear force. If the nuclear force were not of short range, organism could not sustain respiration, i.e., could not obtain energy by reacting oxygen with glucose, and should have got into serious trouble or had to rely upon some different processes to obtain energy.

[1] The magic numbers can be thought to be the proton and/or neutron numbers for which the nucleus becomes spherical, or those at which the binding energy per nucleon, i.e., the separation energy of a nucleon, becomes particularly large compared with that in the surrounding nuclei, or the proton and neutron numbers at which the nuclear radius suddenly changes as a function of the proton and neutron numbers. If we consider in this way, the magic number is determined by the combination of the proton and neutron numbers, and the magic numbers for nuclei which are unstable with respect to β-decay such as neutron-rich unstable nuclei or neutron-deficient nuclei, so-called proton-rich nuclei, can be different from the known magic numbers for stable nuclei. The change of magic numbers or the appearance of new magic numbers in the region of unstable nuclei is one of the central research subjects in current nuclear physics.

5.2 Explanation of the Magic Numbers by Mean-Field Theory

5.2.1 The Mean Field

The magic numbers for atoms can be explained based on the picture that electrons are moving independently in the Coulomb field V_C made by the nucleus. In this picture, the central roles are played by the facts that the Coulomb field is of long ranged and the functional form is given by the inverse of the distance from the center r and the Pauli exclusion principle.

Let us derive the magic numbers for nuclei by similarly assuming that nucleons are moving independently from each other inside the nucleus, but obeying the Pauli exclusion principle. However, there exists no central particle which governs the force like the nucleus for electrons inside an atom. Instead, we assume that each nucleon is moving in the mean field made by the nucleons themselves. In this case, the field of force at the position $\mathbf{r}$ will be given as

$$V(\mathbf{r}) = \sum_{i=1}^{A} v_{NN}(\mathbf{r} - \mathbf{r}_i) = \int v_{NN}(\mathbf{r} - \mathbf{r}')\rho(\mathbf{r}')\mathrm{d}\mathbf{r}' , \tag{5.1}$$

by adding the forces exerted on the nucleon at $\mathbf{r}$ from the other nucleons. $v_{NN}(\mathbf{r} - \mathbf{r}')$ is the nuclear potential working between two nucleons at $\mathbf{r}$ and $\mathbf{r}'$, and $\rho(\mathbf{r})$ is the number density of nucleons at the position $\mathbf{r}$. We used that $\rho(\mathbf{r})$ is given by

$$\rho(\mathbf{r}) = \sum_{i=1}^{A} \delta(\mathbf{r} - \mathbf{r}_i) , \tag{5.2}$$

in transforming to the last term of Eq. (5.1).

If we assume

$$v_{NN}(\mathbf{r}) = -V_0\delta(\mathbf{r}) \quad (V_0 > 0), \tag{5.3}$$

by emphasizing the short-range character of the nuclear force, we obtain

$$V_N(\mathbf{r}) = -V_0\rho(\mathbf{r}) \tag{5.4}$$

for the nuclear potential. Equation (5.4) suggests that the nuclear part of the mean potential which nucleons feel inside the nucleus has the radial dependence similar to that of the number density of nucleons inside the nucleus.

For this reason, the nuclear potential for nucleons inside a nucleus is often assumed to be

$$V_N(r) = -\frac{V_0}{1 + \exp[(r - R)/a]} , \tag{5.5}$$

after Eq. (2.22), and is called the Woods–Saxon potential. Here, we assumed a spherical nucleus.

The value of V_0 can be estimated to be 40–50 MeV from the facts that the Fermi energy is 33–40 MeV and the binding energy per nucleon is about 8 MeV.

5.2.2 Energy Levels for the Infinite Square-Well Potential

In the following, let us argue by ignoring the Coulomb force, although one has to add the contribution of the Coulomb force for protons. One needs to perform numerical calculations in order to determine the energy levels in the Woods–Saxon potential. Let us therefore replace the potential by a square-well potential with an infinitely high wall at $r = R$ in order to gain a qualitative understanding,

$$V(r) = \begin{cases} -V_0 & \text{for } 0 \le r < R \\ +\infty & \text{for } r \ge R \,. \end{cases} \tag{5.6}$$

The wave function in the region of the potential well is then given by

$$\psi(\mathbf{r}) = A_\ell j_\ell(Kr) Y_{\ell m}(\theta, \varphi) \,, \tag{5.7}$$

$$K = \sqrt{\frac{2M_N}{\hbar^2}(E + V_0)} \,, \tag{5.8}$$

$$j_\ell(KR) = 0 \,. \tag{5.9}$$

If we denote the nth zero of the spherical Bessel function $j_\ell(x)$ by $x_{n\ell}$, then we obtain the following formula to give the energy levels

$$E_{n\ell} = -V_0 + \frac{\hbar^2}{2M_N R^2} x_{n\ell}^2 \,. \tag{5.10}$$

Table 5.1 shows the values of some of the zeros $x_{n\ell}$, and numbers them in order of magnitude.

Table 5.1 The zeros $x_{n\ell}$ of the spherical Bessel function $j_\ell(x)$

n	$j_0(x)$	$j_1(x)$	$j_2(x)$	$j_3(x)$	$j_4(x)$	$j_5(x)$	$j_6(x)$
1	① 3.14	② 4.49	③ 5.76	⑤ 6.99	⑦ 8.18	⑨ 9.36	⑫ 10.51
2	④ 6.28	⑥ 7.73	⑧ 9.10	⑪ 10.42	11.70	12.97	14.21
3	⑩ 9.42	⑬ 10.90	12.32	13.70	15.04	16.35	17.65
4	12.57	14.07	15.51	16.92	18.30	19.65	20.98

5.2.3 The Harmonic Oscillator Model

The harmonic oscillator model, where the mean field is assumed to be given by the following formula, is often used as a more convenient method,

$$V(r) = \frac{1}{2} M_N \omega^2 r^2 \,. \tag{5.11}$$

In this case, the energy eigenvalues and the quantum numbers are given by

$$E_{n_x,n_y,n_z} = \hbar\omega \left(N + \frac{3}{2} \right) \,, \tag{5.12}$$

$$N = n_x + n_y + n_z \,, \tag{5.13}$$

$$n_x, n_y, n_z = 0, 1, 2, \ldots \,, \tag{5.14}$$

in the Cartesian coordinates, i.e., in the Descartes coordinates, and by

$$E_{n\ell m} = \hbar\omega \left(N + \frac{3}{2} \right) \,, \tag{5.15}$$

$$N = 2(n-1) + \ell \,, \quad \text{principal quantum number,} \tag{5.16}$$

$$n = 1, 2, 3, \ldots \,, \quad \text{radial quantum number,} \tag{5.17}$$

$$\ell = 0, 1, \ldots, \quad \text{azimuthal quantum number (orbital angular momentum),} \tag{5.18}$$

$$m = -\ell, -\ell+1, \ldots, \ell-1, \ell \,, \quad \text{magnetic quantum number,} \tag{5.19}$$

in the spherical polar coordinates. n is the quantum number or index which shows the numerical order of the states with the same angular momentum when they are counted from the lowest energy state. $n_r \equiv n - 1 = 0, 1, 2, \ldots$ is also often used for n. n_r is the number of nodes of the radial wave function at positions other than the origin of the coordinate system, i.e., within the potential well.

The energy levels and the wave functions in the case of Cartesian coordinates can be easily obtained by the series expansion method or by the method using the creation and annihilation operators as is explained in the standard textbook of quantum mechanics, and the wave function is given by

$$\psi_{n_x}(x) = N_{n_x} H_{n_x}(\alpha x) \exp\left(-\frac{1}{2}\alpha^2 x^2 \right) \,, \tag{5.20}$$

$$\alpha = \sqrt{M_N \omega / \hbar} \,, \tag{5.21}$$

by using the Hermite polynomials H_n.[2] $N_{n_x} = (\frac{\alpha}{\sqrt{\pi}2^{n_x}(n_x)!})^{1/2}$ is the normalization constant.

One can easily guess which kinds of angular momentum states are included in each energy level when the polar coordinates are used by considering the parity and the degeneracy of each energy level, which can be easily determined in the Cartesian coordinates, and the fact that the representations in terms of the Cartesian coordinates and the polar coordinates can be transformed to each other using a unitary transformation.

Exercise 5.1 Let us write the wave function in the polar coordinates as

$$\psi_{n_r\ell m}(\mathbf{r}) \equiv R_{n_r\ell}(r)Y_{lm}(\hat{\mathbf{r}}) = \frac{u_{n_r\ell}(r)}{r}Y_{lm}(\hat{\mathbf{r}}) \,. \tag{5.22}$$

Show that the normalized radial wave function $u_{n_r\ell}$ is then given by

$$u_{n_r\ell}(r) = \frac{\sqrt{2\alpha}}{\Gamma(\ell+\frac{3}{2})}\left[\frac{\Gamma(n_r+\ell+\frac{3}{2})}{n_r!}\right]^{1/2} x^{\ell+1}\mathrm{e}^{-\frac{1}{2}x^2}F\left(-n_r;\ell+\frac{3}{2};x^2\right) \tag{5.23}$$

$$= \sqrt{2\alpha}\left[\frac{n_r!}{\Gamma(n_r+\ell+\frac{3}{2})}\right]^{1/2} x^{\ell+1}\mathrm{e}^{-\frac{1}{2}x^2}L_{n_r}^{(\ell+\frac{1}{2})}(x^2) \,, \tag{5.24}$$

where $x = \alpha r$. F is the confluent hypergeometric function,

$$F\left(-n_r;\ell+\frac{3}{2};z\right) = \sum_{k=0}^{n_r}(-1)^k 2^k \binom{n_r}{k}\frac{(2\ell+1)!!}{(2\ell+2k+1)!!}z^k \,, \tag{5.25}$$

and $L_{n_r}^{(\ell+\frac{1}{2})}$ is the generalized Laguerre polynomial.

5.2.4 *The Magic Numbers in the Static Potential Due to Short Range Force*

The first and second columns from the left in Fig. 5.1 give the single-particle energy levels together with the quantum numbers based on Eqs. (5.15) and (5.10), respectively. The magic numbers appear when one energy level is fully occupied and the energy gap to the next energy level is large. If we denote the number of states in the

[2]The parameter b which is defined by $b \equiv 1/\alpha = \sqrt{\hbar/M_N\omega}$ is also often used and is called the oscillator parameter. It gives a measure of the strength of the potential which confines nucleons and has the dimension of length. It gives the extension of the $0s$-state and the measure of the size of clusters such as the α-cluster.

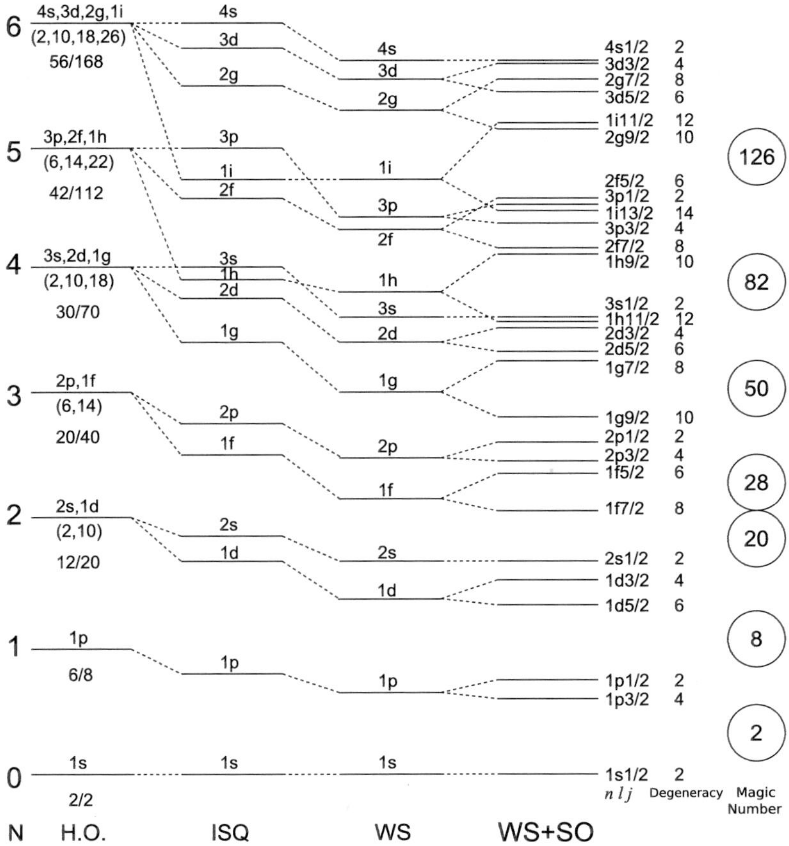

Fig. 5.1 Nucleon energy levels for several spherical mean fields. H.O. stands for the harmonic oscillator model, ISQ stands for the infinite square-well potential, WS stands for the Woods–Saxon potential. The last column represents the energy levels when the spin–orbit interaction is taken into account

energy level with the principal quantum number N by $\mathscr{N}(N)$ in order to predict the magic numbers based on the harmonic oscillator model, then

$$\mathscr{N}(N) = (N+1)(N+2) \tag{5.26}$$

regardless of the even-odd property of N. We show in Table 5.2 the $\mathscr{N}(N)$ and the total number of states whose principal quantum number is smaller than or equal to N, $\mathscr{N}_{\text{tot}}(N) \equiv \sum_{N'=0}^{N} \mathscr{N}(N')$.

Table 5.2 The degeneracy of energy levels and the magic numbers for the harmonic oscillator model

N	0	1	2	3	4	5	6	7
$\mathscr{N}(N)$	2	6	12	20	30	42	56	72
$\mathscr{N}_{\rm tot}(N)$	2	8	20	40	70	112	168	240

Table 5.2 shows that the magic numbers 2, 8 and 20 can be explained by the harmonic oscillator model. Figure 5.1 shows that the infinite square-well potential also explains the small magic numbers 2, 8 and 20. In addition, in the square-well potential, the degeneracy concerning the angular momentum is resolved and the state with larger angular momentum becomes energetically lower. This is because the state with larger orbital angular momentum among the states with the same principal quantum number has larger probability in the surface region, and hence more strongly feels the difference, i.e., a negative perturbation, between the square-well potential and the harmonic oscillator potential.

The third column in Fig. 5.1 shows the energy levels for the Woods–Saxon potential. The structure of the energy levels in this case resembles that for the square-well potential.

Exercise 5.2 Prove Eq. (5.26) by using that $\mathscr{N}(N)$ is given by $\sum_{\ell=0,2,\ldots}^{N}(2\ell+1)\times 2$ when N is an even number and by $\sum_{\ell=1,3,\ldots}^{N}(2\ell+1)\times 2$ when N is an odd number because of the degeneracies concerning the magnetic quantum number and the spin.

5.2.5 Spin–Orbit Interaction

Suppose that the spin–orbit interaction is acting on each nucleon in addition to the static central force in order to explain the magic numbers larger than 20. The Schrödinger equation to determine the energy levels of nucleons is then given by

$$\left[-\frac{\hbar^2}{2M_N}\Delta + V(r) + v_{\rm LS}(\hat{\boldsymbol{\ell}}\cdot\hat{\mathbf{s}})\right]\psi(\mathbf{r}) = \mathscr{E}\psi(\mathbf{r})\,. \tag{5.27}$$

We assume here that $v_{\rm LS} = \xi_{\rm LS}$ (constant) for simplicity.

The modified energy levels for each N can be obtained by diagonalizing the Hamiltonian with the degenerate states according to the perturbation theory for degenerate case. Here, however, we use a more convenient method. Since the operator $\hat{\boldsymbol{\ell}}\cdot\hat{\mathbf{s}}$ does not commute with any components of the orbital angular momentum and spin operator, their magnetic quantum numbers cannot be good quantum numbers separately. On the other hand, since

$$\left[\hat{\boldsymbol{\ell}}\cdot\hat{\mathbf{s}}, \hat{\boldsymbol{\ell}}^2\right] = \left[\hat{\boldsymbol{\ell}}\cdot\hat{\mathbf{s}}, \hat{\mathbf{s}}^2\right] = \left[\hat{\boldsymbol{\ell}}\cdot\hat{\mathbf{s}}, \hat{\mathbf{j}}^2\right] = \left[\hat{\boldsymbol{\ell}}\cdot\hat{\mathbf{s}}, \hat{j}_\alpha\right] = 0\,, \tag{5.28}$$

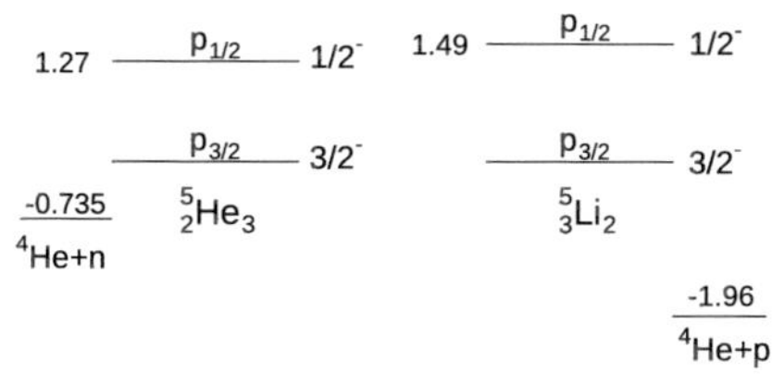

Fig. 5.2 The ground and first excited states of ^{5}He and ^{5}Li [1]. The numbers on the *left* of each first excited state are the excitation energies in MeV. Also shown are the threshold energies for the neutron and proton emissions. As seen in the figure, all the levels are above the threshold energy for nucleon emission and are resonance states with large widths. The threshold energies for the neutron and proton decays are after [2]

one can choose the magnitude of the orbital angular momentum, the magnitude of spin, the magnitude of the total angular momentum $\hat{\mathbf{j}} \equiv \hat{\boldsymbol{\ell}} + \hat{\mathbf{s}}$ and its z-component, $\ell, s = \frac{1}{2}, j, m$, as good quantum numbers to specify states. Consequently, the energy level splits into two as

$$\mathscr{E}_{n\ell j} = \begin{cases} E_{n\ell} - \xi_{LS}\frac{1}{2}(\ell+1) & (\text{for } j = \ell - \frac{1}{2}) \\ E_{n\ell} + \xi_{LS}\frac{1}{2}\ell & (\text{for } j = \ell + \frac{1}{2}) \,, \end{cases} \tag{5.29}$$

according to the magnitude of the total angular momentum. We used

$$\begin{aligned} \hat{\boldsymbol{\ell}} \cdot \hat{\mathbf{s}}|\ell s j m\rangle &= \frac{1}{2}(\hat{\mathbf{j}}^2 - \hat{\boldsymbol{\ell}}^2 - \hat{\mathbf{s}}^2)|\ell s j m\rangle \\ &= \frac{1}{2}\left[j(j+1) - \ell(\ell+1) - \frac{1}{2}\left(\frac{1}{2}+1\right)\right]|\ell s j m\rangle \end{aligned} \tag{5.30}$$

to derive Eq. (5.29).

We show in Fig. 5.2 the ground and first excited states of ^{5}He and ^{5}Li in order to learn the strength and the sign of the spin–orbit interaction. The figure suggests that the sign of ξ_{LS} is negative,[3] hence the energy of the $j = \ell + 1/2$ state shifts downward, while that of $j = \ell - 1/2$ state upward.[4]

[3]For electrons in atoms, the spin–orbit interaction is repulsive. Hence the energy of the $j = \ell - 1/2$ state gets lowered as shown in Fig. 4.1.

[4]It is difficult to determine the resonance parameters such as the level position for a resonance state with a large width. In fact, [3] gives values which are very different from those in Fig. 5.2 for the energy splitting between the $p_{1/2}$ and $p_{3/2}$ states. For example, the excitation energy of the $p_{1/2}$ state of ^{5}He is 4.6MeV. Reference [1] also gives two different parameter sets depending on the method of analysis. The resonance position and the width given in Fig. 5.2 and mentioned also in the following exercise have been obtained from the pole position of the S-matrix on the complex energy plane (extended R-matrix method; after [4]). On the other hand, the conventional method, i.e., standard R-matrix method, determines the parameters from the poles of the cross section by restricting the energy to be real as experiments are performed. The former has the advantage that the results do not depend on the choices of the boundary condition and the channel radius which appear in the R-matrix theory. See [5, 6] for details of the R-matrix theory.

Exercise 5.3 Estimate the magnitude of ξ_{LS} for protons and neutrons from Fig. 5.2.[5] The width of the level is $\Gamma = 648$ keV for the $p_{3/2}$ state and $\Gamma = 5.57$ MeV for the $p_{1/2}$ state for ^{5}He, while $\Gamma = 1.23$ MeV for $p_{3/2}$ state and $\Gamma = 6.60$ MeV for $p_{1/2}$ state for ^{5}Li [1]. Estimate the half-life and the mean life of the ground and excited states of ^{5}He and ^{5}Li from these values.

The last column of Fig. 5.1 gives the nucleon energy levels when the effect of the spin–orbit interaction is added based on these considerations. Each energy level is designated by the set of the radial quantum number n, the orbital angular momentum ℓ and the total angular momentum j. The figure shows that the magic numbers 28, 50, 82, 126 arise due to the spin–orbit interaction.[6] Note that there are $2j+1$ degenerate states with different magnetic quantum numbers in a single energy level with the angular momentum j.

Note that the magic numbers 50, 82 and 126 are caused by the penetration of the $j = \ell + 1/2$ levels with a large orbital angular momentum into the energy region of the energy levels with the principal quantum number N smaller by one due to the spin–orbit interaction and the surface effect, i.e., the effect that the realistic potential is deeper in the surface region than that in the harmonic oscillator model. These states are called *intruder states*. The $1g_{9/2}$, $1h_{11/2}$ and $1i_{13/2}$ levels play the principal role as the intruder states for the magic numbers 50, 82 and 126, respectively.

Exercise 5.4 Answer the following questions concerning the transformation from the $|\ell\, s = \frac{1}{2}\; m_\ell m_s\rangle$ basis to the $|\ell s j m_j\rangle$ basis.

1. Show that the total number of states is $(2\ell + 1) \times 2$ in both bases.
2. The states are given by $|Y_{\ell m}\alpha\rangle$ and $|Y_{\ell m}\beta\rangle$ in the former basis, where α and β represent the spin-up and spin-down states, respectively. Write the general expression of the Unitary transformation which represents the $|\ell s j = \ell + 1/2\; m_j\rangle$ and $|\ell s j = \ell - 1/2\; m_j\rangle$ states in the $|\ell s m_\ell m_s\rangle$ basis. Also, write explicitly the $|f_{7/2}\; m = 7/2\rangle$ and $|f_{5/2}\; m = 5/2\rangle$ in terms of Y_{3m}, α, β.

Sidebar: Japanese Contribution to the Development of Shell model—Pioneers of History: Tadayoshi Hikosaka and Takahiko Yamanouchi The independent particle model, i.e., the shell model, thus succeeds in explaining the existence and the actual numbers of magic numbers through the shell layer structure of the energy levels of nucleons, and the Pauli exclusion principle. Historically, pioneering studies of the independent particle model (shell model) had been developed in Japan too by Tadayoshi Hikosaka, Tohoku University, and Takahiko Yamanouchi, University of Tokyo. However, their pioneering works had been forgotten before their theory is accepted as the correct effective model for many nuclear properties.

[5] It is necessary to handle both the ground and excited states as resonance states in order to achieve a quantitatively accurate estimate. Here, however, use the harmonic oscillator model and ignore the radial dependence of the spin–orbit force in order to obtain a rough estimate of ξ_{LS}.

[6] The tensor force also strongly affects the magic numbers and the shape of nucleus, and is one of the active research subjects in recent years in connection with, e.g., the structure of unstable nuclei such as the change of the magic number. For example, ^{40}Ca is spherical, but ^{32}Mg is deformed among the $N = 20$ isotones (see [7]).

There remain historical episodes that both Hikosaka and Yamanouchi received strong objections when they discussed with Niels Bohr based on the success of the liquid-drop model to explain fission and the compound nucleus reactions. On the contrary, the existence of magic numbers was a strong experimental evidence in the cases of Mayer and Jensen who established the shell model later. There remains also an episode that Jensen got the inspiration of the spin–orbit interaction while he was attending a dance party. Incidentally, Hikosaka is the same person as who has proposed the heterogeneous reactor.

Exercise 5.5 The puzzle of the magic numbers: Comparison with the cases for atoms and microclusters. Discuss the magic numbers for many-body systems of fermions when the interactions among the constituent particles can be approximated by the mean fields listed below in the zeroth order approximation.

1. An attractive potential proportional to $1/r$ when the distance from the center is denoted by r (periodic table of the elements).
2. Spherical Woods–Saxon potential (magic numbers for Alkali metals [8]).
3. Three-dimensional harmonic oscillator plus attractive spin–orbit interaction (magic numbers for nuclei).
4. Three-dimensional harmonic oscillator plus repulsive spin–orbit interaction.
5. Two-dimensional harmonic oscillator.

5.3 The Spin and Parity of the Ground and Low-Lying States of Doubly-Magic ±1 Nuclei

The nucleus in which both proton and neutron numbers are magic numbers such as ${}^{16}_{8}\mathrm{O}_8$ and ${}^{208}_{82}\mathrm{Pb}_{126}$ is called *doubly-magic nucleus* or *doubly-closed shell nucleus*. One can predict the spin and parity of the ground state of the nuclei whose proton or neutron number is smaller or larger than those of a doubly-magic nucleus by one by using the shell model. The spin and parity of low-lying states of those nuclei can also be predicted to some extent.

Consider ${}^{209}_{82}\mathrm{Pb}$ for example. For the ground state, the single-particle energy levels are successively filled with nucleons from the lowest energy level according to the Pauli exclusion principle. Figure 5.1 suggests that all the 82 protons are just filled in the states from the $n\ell_j = 1s_{1/2}$ level to the $3s_{1/2}$ level. The 126 neutrons occupy all the states from the $1s_{1/2}$ to the $2f_{5/2}$ levels. Let us call this part, which corresponds to doubly-magic numbers, the *core* part. One then fills the last 127th neutron, called the valence neutron, in the next energy level $2g_{9/2}$. Since there is only one way to fill all the $2j+1$ degenerate states for a given $n\ell_j$ with the same number of particles, $J_{\mathrm{tot}} = 0$ because of $2J_{\mathrm{tot}}+1 = 1$ if we denote the total spin of these $2j+1$ nucleons by J_{tot}. Also, the total parity of these nucleons is plus irrespective of the even–odd property of the orbital angular momentum ℓ, since $2j+1$ is always an even number. Consequently, the total spin and parity of the nucleons forming the core are 0^+. As

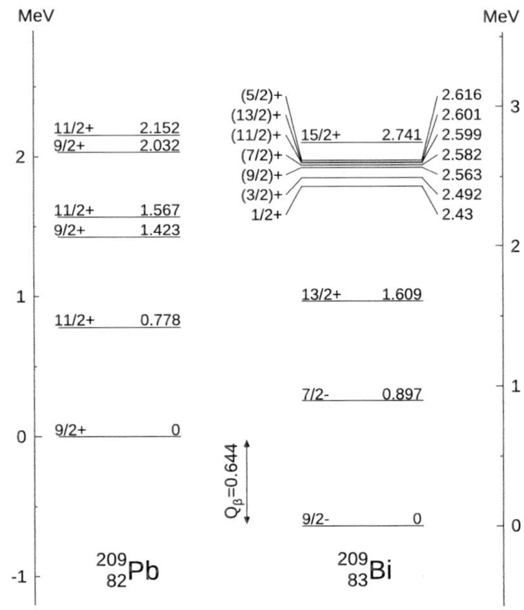

Fig. 5.3 The ground and low-lying excited states of ^{209}Pb and ^{209}Bi (the energies are in MeV) [3]

the result, the spin and parity of the whole nucleus is determined by those of the valence neutron. Hence the spin and parity of the ground state of $^{209}_{82}$Pb are expected to be $9/2^+$.

The left side of Fig. 5.3 shows the experimental energy-level diagram for $^{209}_{82}$Pb (Data from [3]). The spin and parity of the ground state agree with the prediction of the shell model described above. One can keep the core part as it is and change the level to fill the valence neutron from the $2g_{9/2}$ state to another state successively from the low energy level in order to predict the spin and parity of the other states. As an alternative, one can consider the state, where the core is polarized and the nucleon in the energy level $n_h\ell_{hj_h}$ in the core, where the index h stands for hole, is promoted to the $n_p\ell_{pj_p}$ level outside the core, where the index p stands for particle. If the original valence nucleon is in the $n\ell_j$ state, the corresponding wave function can be written, for example, as $\left|\left[(n_h\ell_{hj_h})^{-1} \otimes [(n_p\ell_{pj_p}) \otimes (n\ell_j)]^{[J']}\right]^{[j']}; j'm'\right\rangle$.[7] The state of this kind of configuration is called two-particle one-hole (2p1h) state. Although there are many possibilities depending on the way to make a hole, the choice of the particle state to promote the original core nucleon, and the way to couple the angular momenta, the state, where the core neutron is promoted to the same level as that of the valence neutron and where the total angular momentum of two valence neutrons

[7] We used the convention to omit the core part in writing the configuration, and to make the core part as the reference configuration.

J' couples to zero, appears at low energy because of the pairing interaction between the valence nucleons. In this case, the spin and parity of the whole nucleus become identical with those of the hole state, i.e., j_h and $(-1)^{\ell_h}$.

An interesting another possibility is that a series of characteristic states appear in low energy region by the process that the core part makes a low energy collective excitation and weakly couples with the valence nucleon. An example can be seen in ^{209}Bi shown on the right side of Fig. 5.3 [3]. The ground state and the first through third excited states can be understood as the single-particle states or 2p1h state in the shell model. The septet states above them, $3/2^+, 5/2^+, 7/2^+, 9/2^+, 11/2^+, 13/2^+$, and $15/2^+$, are the states, where the core of ^{209}Bi makes a vibrational excitation of the octupole type (see Chap. 7) and weakly couples with the valence proton in the $h_{9/2}$ orbit, and can be expressed as $|[3^- \otimes h_{9/2}]^J; J^+\rangle$ [9] (see also [10]).

Exercise 5.6 1. Predict the configurations for the low-lying excited states of $^{209}_{82}$Pb.
2. Predict the configurations as well as the spin and parity of the ground state and low-lying excited states of ^{17}O, ^{15}O and $^{209}_{83}$Bi, and compare the predicted spin and parity with those in the experimental data in [3].

Incidentally, in the case of odd–odd nuclei in which both proton and neutron numbers are odd numbers, the parity and the spin of the ground state and low-lying excited states are expected to be given by $(-1)^{\ell_p+\ell_n}$ and one of the values between $|j_p - j_n|$ and $j_p + j_n$, where ℓ_p, j_p, ℓ_n.j_n are the orbital and total angular momenta of the orbits of the last proton and neutron. For example, in the case of $^{14}_{7}$N, $\ell_p = \ell_n = 1$, $j_p = j_n = 1/2$ and the experimental data for the spin and parity of the ground state are 1^+.

5.4 The Magnetic Dipole Moment in the Ground State of Odd Nuclei: Single Particle Model

The idea which attributes the properties of not only the doubly-magic ± 1 nuclei, but all nuclei with an odd mass number, i.e., the even–odd or odd–even nuclei, or simply the odd nuclei, in general entirely to the last proton or neutron whose total number is an odd number is called the single particle model. The single particle model often well predicts the spin and parity of the ground state and some states in the vicinity of the ground state of odd nuclei. This happens because nucleons move around pairwise by forming a pair of total angular momentum and parity 0^+ due to the pairing interaction (see Sect. 5.8), and consequently the total spin and parity of the nucleus is determined by the last unpaired nucleon or hole regardless of the actual number of valence nucleons outside the core or the number of holes inside the core.

Exercise 5.7 Predict the configurations and the spin and parity of the ground and some low-lying excited states of $^{45}_{21}$Sc and compare the predicted spin and parity with the experimental data [3].

Here, let us learn that the single particle model provides a rough estimate of the magnetic dipole moment of odd nuclei, and further study how the discrepancy between the experimental data and the prediction of the single particle model is explained by the concept of *core polarization* or the *configuration mixing*.

5.4.1 The Schmidt Lines

The magnetic moment is defined by $\mu \equiv \langle jj|\hat{\mu}_z|jj\rangle$ in the single particle model following Eq. (4.39), where the operator of the magnetic moment $\hat{\boldsymbol{\mu}}$ is given by

$$\hat{\boldsymbol{\mu}} = \mu_N(g_s\hat{\mathbf{s}} + g_\ell\hat{\boldsymbol{\ell}}) \,, \tag{5.31}$$

according to Eqs. (4.15)–(4.17).

Here let us derive the explicit expression of μ by using the projection theorem for the expectation value of vector operator (see Appendix A.6.4). From Eq. (A.106),

$$\mu = \frac{1}{j(j+1)}\langle jj|\hat{j}_Z|jj\rangle\langle jj|(\hat{\boldsymbol{\mu}}\cdot\hat{\mathbf{j}})|jj\rangle \tag{5.32}$$

$$= \frac{1}{j+1}\langle jj|(\hat{\boldsymbol{\mu}}\cdot\hat{\mathbf{j}})|jj\rangle \,. \tag{5.33}$$

The expectation values of the operators $\hat{\mathbf{s}}\cdot\hat{\mathbf{j}}$ and $\hat{\boldsymbol{\ell}}\cdot\hat{\mathbf{j}}$, which are needed to evaluate $\langle jj|(\hat{\boldsymbol{\mu}}\cdot\hat{\mathbf{j}})|jj\rangle$, can be easily obtained in the same way as that was used to obtain the expectation value of the operator $\hat{\boldsymbol{\ell}}\cdot\hat{\mathbf{s}}$ in Sect. 5.2.5. Finally, μ is given by

$$\mu = \begin{cases} \mu_N\left[g_\ell\left(j-\frac{1}{2}\right)+\frac{1}{2}g_s\right] & \text{for } j=\ell+1/2 \\ \mu_N\left[g_\ell\left(j+\frac{3}{2}\right)-\frac{1}{2}g_s\right]\frac{j}{j+1} & \text{for } j=\ell-1/2\,. \end{cases} \tag{5.34}$$

Thus one can expect that the magnetic moments are given by the solid lines in Fig. 5.4 depending on whether $j=\ell-1/2$ or $j=\ell+1/2$ by remarking that $g_\ell = 1.0$, $g_s = g_p \approx 5.58$ in the case of odd-Z nuclei and $g_\ell = 0.0$, $g_s = g_n \approx -3.82$ in the case of odd-N nuclei. These lines are referred to as the *Schmidt lines*. The experimental values lie between the two Schmidt lines except for a small number of nuclei such as ^{3}H, ^{3}He, ^{13}C and $^{189}_{76}$Os .

Exercise 5.8 Derive Eq. (5.34) for the Schmidt lines by using that the wave function of the $|jj\rangle$ state is given by

$$|jj\rangle = \begin{cases} Y_{\ell\ell}|\alpha\rangle & \text{for } j=\ell+1/2 \\ \frac{1}{\sqrt{2(\ell+1/2)}}\left(-Y_{\ell\ell-1}|\alpha\rangle+\sqrt{2\ell}\,Y_{\ell\ell}|\beta\rangle\right) & \text{for } j=\ell-1/2\,. \end{cases} \tag{5.35}$$

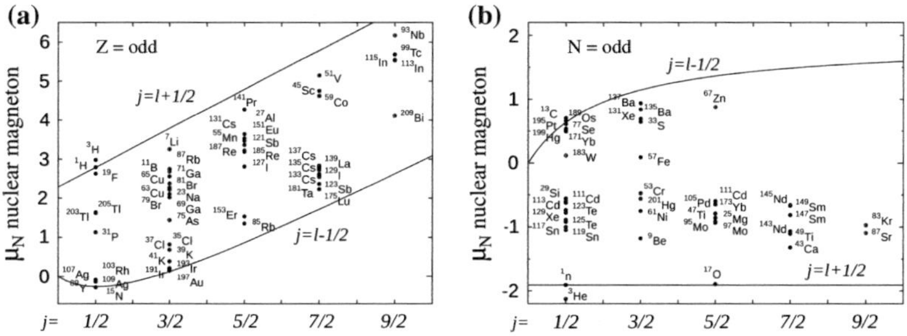

Fig. 5.4 The magnetic moment for odd nuclei. **a** Odd-Z nuclei. **b** Odd-N nuclei. After [11]

5.4.2 Configuration Mixing and Core Polarization

Figure 5.4 shows the success of the shell model in the sense that it well explains the gross behaviour of the magnetic moment, while it also shows the limitation of the single particle model in reproducing the magnitude. A large part of the deviation of the experimental values from the corresponding Schmidt values can be explained by considering the *configuration mixing* due to the residual interaction which is not incorporated in the mean field [12]. Let us denote the residual interaction by V, and write the wave function which takes into account the effect of the residual interaction as

$$|jj\rangle\rangle = \alpha_0|jj\rangle + \sum_{ph}\alpha_{ph}\left|\left[j'\otimes\left[j_h^{-1}\otimes j_p\right]^{[J]}\right]^{[j]};jj\right\rangle + \cdots \tag{5.36}$$

$$= \alpha_0|jj\rangle + \sum_{i=1}^{n}\alpha_i|C_i\rangle\ . \tag{5.37}$$

by assuming that V is a two-body interaction. $|jj\rangle = |jm_j = j\rangle$ is the zeroth order wave function in the shell model (single particle model). Figure 5.5 illustrates the

Fig. 5.5 Configuration mixing by the core polarization

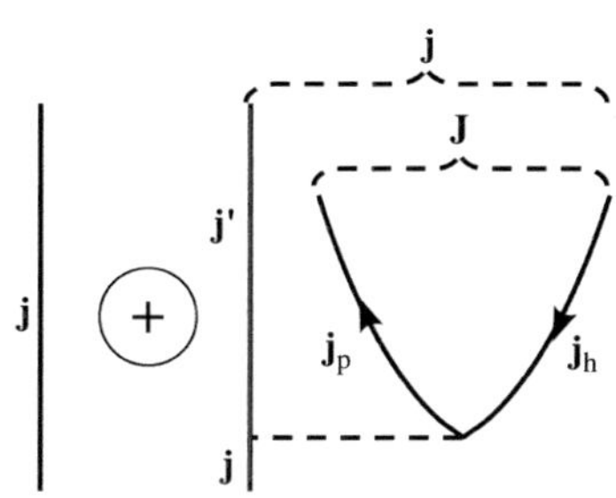

state given by Eq. (5.36). In Eq. (5.37), we used an abbreviated notation $|C_i\rangle$ for the additional terms due to the configuration mixing.

The effect of configuration mixing is often called the effect of *core polarization*. In fact, for the nuclei in the vicinity of doubly-magic nuclei, the correction terms mixed by the configuration mixing or a part of them are caused by a process where the core part is excited, i.e., polarized. However, it will be more appropriate to understand that the effect of configuration mixing is a wider concept than that of simple-minded core polarization. In Eq. (5.36), we used the indices p and h suggesting the particle or hole states as indices of quantum numbers in order to facilitate imagining the physical process.

The magnetic moment is given by

$$\mu = |\alpha_0|^2 \langle jj|\hat{\mu}_Z|jj\rangle + 2\mathrm{Re}\sum_{i=1}^{n} \alpha_0^* \alpha_i \langle jj|\hat{\mu}_Z|C_i\rangle + \mathcal{O}(\alpha_i \alpha_{i'}) , \tag{5.38}$$

when there exists configuration mixing, or core polarization. If we assume, $n = 20, \alpha_i = 0.1 (i = 1 - 20)$ and $\alpha_0^2 = 0.80$, i.e., $\alpha_0 = 0.89$, then $\alpha_0 \sum_{i=1}^{20} \alpha_i = 1.78$. Hence a large correction can arise if there are many different ways to polarize the core and if they contribute cooperatively even if the mixing amplitude of each core polarization is small. In fact, Arima and Horie have succeeded in reproducing the experimental values of the magnetic moment by taking into account the effect of core polarization [12].

Various types of configuration mixing can be considered depending on the residual interaction. For example, if $V = (\boldsymbol{\sigma}_1 \cdot \boldsymbol{\sigma}_2) V_\sigma(r)$ or $V = (\boldsymbol{\sigma}_1 \cdot \boldsymbol{\sigma}_2)(\boldsymbol{\tau}_1 \cdot \boldsymbol{\tau}_2) V_{\sigma\tau}(r)$, then J in Eq. (5.36) becomes 1, and one can expect that the effect of core polarization becomes large for the nuclei in which the $j = \ell + 1/2$ level among the group of levels split by the spin–orbit interaction from a degenerate energy level in the static mean field is occupied and the counter $j = \ell - 1/2$ level is empty.

So far, we used the values of free nucleons for the g-factor of nucleons. As mentioned in Sect. 4.1.2, those values can change inside nuclei e.g., by a quenching effect. More generally, the effects of meson and the excited state of nucleon Δ_{33} also contribute to the deviation from the Schmidt values. Especially, the magnetic moments of ^{3}H and ^{3}He, which lie outside the Schmidt lines, are explained by the meson effect. It is suggested from the detailed studies on the interaction between pions and nucleons and from the electromagnetic form factors of these particles that the anomalous magnetic moment of nucleons is weakened inside the nucleus by about 10 %. In this connection, we wish to mention that the study of the $I^\pi = 11^-$ isomer state of ^{210}Po shows that the g_ℓ-factor of proton is larger by about 10 % as an example of the meson effect [13].

5.4.3 [Addendum] The Anomalous Magnetic Moments of Nucleons in the Quark Model

In Chap. 1, we attributed the anomalous magnetic moments of nucleons to the effects of meson cloud. Alternatively, here we show that the anomalous magnetic moments of nucleons can be naturally derived from the point of view of the quark model.

In the quark model, proton is made up of two u quarks and one d quark. It is considered that all the baryons existing in nature are in the color-singlet state, which is antisymmetric concerning the color degree of freedom, and are spatially symmetric in low-energy states. Hence the wave function for proton in the combined spaces of the spin and flavor degrees of freedom must be symmetric in total, and is given by [14]

$$|\mathrm{p}\uparrow\rangle = \sqrt{\frac{1}{18}}\Big[\mathrm{uud}(\uparrow\downarrow\uparrow + \downarrow\uparrow\uparrow - 2\uparrow\uparrow\downarrow) + \mathrm{udu}(\uparrow\uparrow\downarrow + \downarrow\uparrow\uparrow - 2\uparrow\downarrow\uparrow) \\ + \mathrm{duu}(\uparrow\downarrow\uparrow + \uparrow\uparrow\downarrow - 2\downarrow\uparrow\uparrow)\Big] . \tag{5.39}$$

Similarly, the wave function for $|\mathrm{n}\uparrow\rangle$ state is given by interchanging u and d quarks.

On the other hand, the operator of the magnetic moment of the ith quark is given by

$$\hat{\boldsymbol{\mu}}_i^{(\mathrm{q})} = Q_i^{(\mathrm{q})} \frac{e\hbar}{2m_i^{(\mathrm{q})}c} \hat{\boldsymbol{\sigma}}_i^{(\mathrm{q})} \equiv \mu_i \hat{\boldsymbol{\sigma}}_i^{(\mathrm{q})} \tag{5.40}$$

if we assume that quarks are structureless Dirac particles, where $Q_i^{(\mathrm{q})}e$ and $m_i^{(\mathrm{q})}$ are the charge and the mass of the ith quark. The magnetic dipole moment of proton is given by

$$\mu_\mathrm{p} = \sum_{i=1}^{3} \langle \mathrm{p}\uparrow | \hat{\mu}_{iz}^{(\mathrm{q})} | \mathrm{p}\uparrow\rangle , \tag{5.41}$$

following the definition (4.39). Using Eq. (5.39) in Eq. (5.41), we obtain

$$\mu_\mathrm{p} = \frac{1}{3}(4\mu_\mathrm{u} - \mu_\mathrm{d}) . \tag{5.42}$$

Similarly, we obtain

$$\mu_\mathrm{n} = \frac{1}{3}(4\mu_\mathrm{d} - \mu_\mathrm{u}) \tag{5.43}$$

for the magnetic dipole moment of neutron. If we identify the quarks with constituent quarks and assume that $m_\mathrm{u} = m_\mathrm{d}$, then it follows

$$\frac{\mu_\mathrm{n}}{\mu_\mathrm{p}} = -\frac{2}{3} . \tag{5.44}$$

This value well agrees with the experimental value $\mu_\mathrm{n}/\mu_\mathrm{p} \approx -0.685$.

Exercise 5.9 Estimate the value of μ_p by assuming $m_u c^2 = m_d c^2 \approx 300\,\text{MeV}$, and compare the estimated value with the experimental data.

5.5 Mass Number Dependence of the Level Spacing $\hbar\omega$

The spacing between two successive energy levels, i.e., $\hbar\omega$ in the harmonic oscillator model, plays the central role in the shell model. Here, we estimate the value of $\hbar\omega$.

We learnt in Sect. 2.1 that the mean-square radius for stable nuclei is given by

$$\langle r^2 \rangle = \frac{3}{5} R_{\text{eq}}^2 = \frac{3}{5} r_0^2 A^{2/3} \tag{5.45}$$

by using a radius parameter $r_0 \sim 1.2\,\text{fm}$ and the mass number A.

We have

$$\langle r^2 \rangle \equiv \frac{1}{A} \sum_{k=1}^{A} \langle r_k^2 \rangle \tag{5.46}$$

$$= \frac{1}{A} \sum_{N=0}^{N_{\text{max}}} \frac{\hbar}{M_N \omega} \left(N + \frac{3}{2} \right) (N+1)(N+2) \times 2 \tag{5.47}$$

$$\approx \frac{1}{A} \frac{\hbar}{M_N \omega} \frac{1}{2} (N_{\text{max}} + 2)^4 \tag{5.48}$$

from the definition of the mean-square radius. We used the following virial theorem,

$$\frac{1}{2} M_N \omega^2 \langle r^2 \rangle_N = \langle V \rangle_N = \frac{1}{2} E_N = \frac{1}{2} \hbar\omega \left(N + \frac{3}{2} \right) , \tag{5.49}$$

to transform from Eq. (5.46) to (5.47).

On the other hand, we have

$$A = \sum_{N=0}^{N_{\text{max}}} (N+1)(N+2) \times 2 \approx \frac{2}{3} (N_{\text{max}} + 2)^3 . \tag{5.50}$$

We assumed $N_{\text{max}} + 2 \gg 1$ in deriving Eqs. (5.48) and (5.50). Equations (5.48) and (5.50) lead to

$$\langle r^2 \rangle = \frac{1}{2} \left(\frac{3}{2} \right)^{4/3} \frac{\hbar}{M_N \omega} A^{1/3} . \tag{5.51}$$

Thus we obtain

$$\hbar\omega = \frac{1}{2}\left(\frac{3}{2}\right)^{4/3} \frac{(\hbar c)^2}{M_N c^2} \frac{5}{3} r_0^{-2} A^{-1/3} \sim 41.1 A^{-1/3}\,\text{MeV} \tag{5.52}$$

from Eq. (5.45).

5.6 The Magnitude and Origin of Spin–Orbit Interaction

It will be worth commenting on the magnitude and origin of the spin–orbit interaction [15]. For nuclei with large mass numbers, the spin–orbit interaction is suggested to be

$$V_{\text{LS}} \sim -20\hat{\boldsymbol{\ell}} \cdot \hat{\mathbf{s}} A^{-2/3}\,\text{MeV} \tag{5.53}$$

from the experimental data for the energy difference between the $j = \ell + 1/2$ and $j = \ell - 1/2$ states split by the spin–orbit interaction. Since $\ell \sim k_F R = (9\pi/8)^{1/3} A^{1/3}$ for nucleons in the vicinity of the Fermi surface, V_{LS} becomes non-negligible compared with $\hbar\omega$ given by Eq. (5.52), and plays an important role in determining magic numbers.

The spin–orbit interaction used in Eq. (5.27) is the one-body spin–orbit interaction. It originates partly from the two-body spin–orbit interaction, which we studied in Sect. 3.2.2, by averaging over the coordinates of one of the nucleons. The tensor force is another source of the one-body spin–orbit interaction. In the relativistic mean-field theory which we learn in Sect. 6.3, the spin–orbit interaction originates from the coupling of nucleons to the σ-, ρ-, ω-mesons, and depends on the isospin of nucleons due to the contribution of the ρ-meson.

5.7 Difference Between the Potentials for Protons and for Neutrons: Lane Potential

The potential given by Eq. (5.1) corresponds to the Hartree potential which we learn in Sect. 6.1. More precisely, the strength of the potential V_0 depends on the isospin of nucleons reflecting, e.g., the exchange effects, and is replaced by

$$V = V_0 - \frac{1}{2}\hat{t}_z \frac{N-Z}{A} V_1 \,, \tag{5.54}$$

if we take the definition that $\hat{t}_z|\text{n}\rangle = \frac{1}{2}|\text{n}\rangle$, and $\hat{t}_z|\text{p}\rangle = -\frac{1}{2}|\text{p}\rangle$. Hence the potential for nucleons depends on the isospin of nucleons. The isospin dependent term is called the *Lane potential*. One can estimate the strength of V_1 as

$$V_1 \sim 4b_{\mathrm{sym}}^{(\mathrm{pot})} = 4\left(b_{\mathrm{sym}} - b_{\mathrm{sym}}^{(\mathrm{kin})}\right) \approx 100\,\mathrm{MeV}$$

from the magnitude of the contribution of the potential energy to the symmetry energy term in the mass formula.

Exercise 5.10 Show that

$$E_{\mathrm{sym}}^{(\mathrm{pot})} \approx \frac{1}{8}\frac{(N-Z)^2}{A}V_1 \tag{5.55}$$

if we express the magnitude of the potential energy given by the Lane potential as $E_{\mathrm{sym}}^{(\mathrm{pot})}$, hence that $V_1 \sim 4b_{\mathrm{sym}}^{(\mathrm{pot})}$.

The Lane potential can be derived in the following way. Suppose that the interaction between two nucleons is given by

$$V(x_1, x_2) = -V_0(w + mP_{\mathrm{M}})v(\mathbf{r}_1 - \mathbf{r}_2) = -V_0(w + mP_{\mathrm{M}})v(\mathbf{r}_{12}) \tag{5.56}$$

$$= -V_0\left[w - m\frac{1}{4}(1 + \boldsymbol{\tau}_1 \cdot \boldsymbol{\tau}_2)(1 + \boldsymbol{\sigma}_1 \cdot \boldsymbol{\sigma}_2)\right]v(\mathbf{r}_{12}) , \tag{5.57}$$

where P_{M} is the Majorana operator, and w and m are the strength parameters of the Wigner and Majorana forces, respectively. By ignoring the correlation between the nucleons, and adding on average with respect to the second nucleon, we obtain

$$V(\mathbf{r}_1) = -V_0\left(w - \frac{m}{4}\right)\left[1 - \frac{m}{w - \frac{m}{4}}\frac{1}{A}(\hat{\mathbf{t}} \cdot \hat{\mathbf{T}})\right]\mathscr{V}(\mathbf{r}_1) , \tag{5.58}$$

$$\mathscr{V}(\mathbf{r}_1) \equiv \int \rho(\mathbf{r}_2)v(\mathbf{r}_{12})d\mathbf{r}_2 , \tag{5.59}$$

for the nucleon potential. $\hat{\mathbf{T}}$ represents the total isospin operator of nucleons excluding the nucleon under consideration. The isospin dependent term is nothing but the Lane potential. Let us consider two cases in order to estimate the magnitude. One of the cases is the Serber exchange potential, where $m = w = 1/2$. In this case, $w - m/4 = 3/8$, $m/(w - m/4) = 4/3$. The other case is the Volkov force, which is often used for the study of cluster structure of light nuclei. In this case, $m = 0.6$, $w = 1 - m = 0.4$. Hence $w - m/4 = 0.25$, $m/(w - m/4) = 2.4$. Since V_0 and V_1 in Eq. (5.54) are about 40 and 100 MeV, respectively, the Volkov force well agrees with the phenomenological strength of the Lane potential.

Physically, the Lane potential can be considered to reflect the fact that the force between proton and neutron is stronger than the forces between protons and between neutrons. Since the difference is intimately related to the effect of the tensor force, one can suppose that the Volkov interaction is a force which effectively incorporates the effect of the tensor force.

5.8 The Spin and Parity of Low-Lying States of Doubly-Magic ±2 Nuclei and the Pairing Correlation

5.8.1 *The Spin and Parity of the Ground and Low Excited States of* $^{210}_{82}Pb$

The spin and parity of the ground and low excited states of magic number ±2 nuclei provide a good example of showing the important role of the *pairing correlation* in nuclei.

Consider $^{210}_{82}$Pb as an example. Figure 5.1 suggests that the configuration for low-lying states including the ground state is $(2g_{9/2})^2_{\mathrm{n}}$. The power 2 means to fill two nucleons.[8] Hence, from the law of addition of angular momentum, we expect that there appear $0^+, 1^+, \ldots, 7^+$, and 8^+ states in low-energy region.

Figure 5.6 shows the energy-level diagrams for $^{210}_{82}\mathrm{Pb}_{128}$ and $^{210}_{84}\mathrm{Po}_{126}$ in low-energy region [3]. There appear indeed the $0^+, 2^+, 4^+, 6^+, 8^+$ states in low-energy region in the figure for $^{210}_{82}$Pb as expected.

One can prove that the odd-spin states do not appear because of the Pauli exclusion principle as follows. Let us use the total spin j to represent the quantum number of the single-particle level. The state in which two neutrons or two protons fill the j_α and j_β states is then given by

$$|(j_\alpha j_\beta)JM\rangle_a = N\left[|(j_\alpha j_\beta)JM\rangle - P_{12}|(j_\alpha j_\beta)JM\rangle\right] \tag{5.60}$$

by making antisymmetrization in order to satisfy the Pauli exclusion principle. N is the normalization constant. The lower index a denotes antisymmetrization and P_{12} is the operator to interchange the particles 1 and 2. We used that the wave function in the isospin space is symmetric concerning the exchange of two nucleons, since we are considering the case where the two nucleons are the same kind of nucleons.

One can rewrite the second term of Eq. (5.60) as

$$\begin{aligned}
P_{12}|(j_\alpha j_\beta)JM\rangle &= P_{12}\sum_{m_\alpha,m_\beta}|j_\alpha j_\beta m_\alpha m_\beta\rangle\langle j_\alpha j_\beta m_\alpha m_\beta|JM\rangle \\
&= \sum_{m_\alpha,m_\beta}|j_\beta j_\alpha m_\beta m_\alpha\rangle\langle j_\alpha j_\beta m_\alpha m_\beta|JM\rangle \\
&= (-1)^{j_\alpha+j_\beta-J}\sum_{m_\alpha,m_\beta}|j_\beta j_\alpha m_\beta m_\alpha\rangle\langle j_\beta j_\alpha m_\beta m_\alpha|JM\rangle \\
&= (-1)^{j_\alpha+j_\beta-J}|(j_\beta j_\alpha)JM\rangle\ .
\end{aligned} \tag{5.61}$$

We used the following property of the Clebsch–Gordan coefficient,

[8] We used the lower index n in order to indicate that it is the energy level for neutrons.

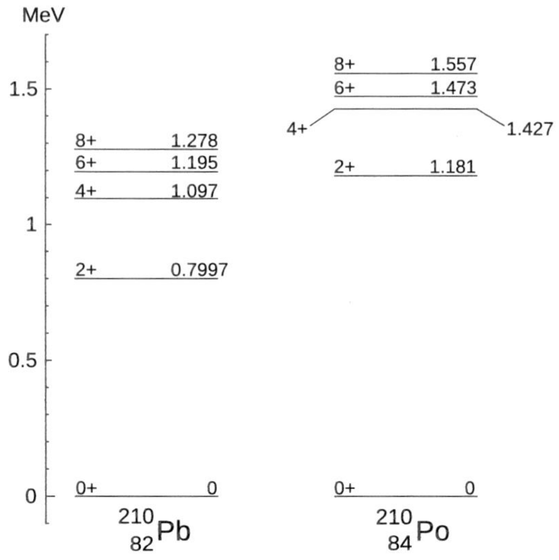

Fig. 5.6 The energy-level diagrams for ^{210}Pb and ^{210}Po (the energy is in MeV)

$$\langle j_\alpha j_\beta m_\alpha m_\beta | JM \rangle = (-1)^{j_\alpha + j_\beta - J} \langle j_\beta j_\alpha m_\beta m_\alpha | JM \rangle \ , \tag{5.62}$$

in transforming from the second to the third lines in Eq. (5.61). Finally, we obtain

$$|(j_\alpha j_\beta)JM\rangle_a = N\left[|(j_\alpha j_\beta)JM\rangle - (-1)^{j_\alpha + j_\beta - J}|(j_\beta j_\alpha)JM\rangle\right] \ . \tag{5.63}$$

Equation (5.63) shows that only even numbers are allowed for the total spin J when one fills two identical nucleons in the same $j_\alpha = j_\beta = j$ orbit, because $2j$ is an odd number.

5.8.2 The Effect of δ-Type Residual Interaction: Pairing Correlation

The shell model which assumes that nucleons move independently from each other in a mean field under the constraint of Pauli exclusion principle well explains the basic features of nuclei such as the magic numbers. However, it has to be refined by taking into account the effects of residual interaction in order to explain the details appearing in the experimental data.

We rewrite the Schrödinger equation

$$\left[-\sum_i \frac{\hbar^2}{2M_N} \nabla_i^2 + \sum_{i>j} v_{ij} \right] \Psi(x_1, \ldots, x_A) = E\Psi(x_1, \ldots, x_A) \tag{5.64}$$

for an A-nucleon system, where the nucleons are interacting with each other through the two-body interaction v, into

$$\left\{\sum_i \left[-\frac{\hbar^2}{2M_N}\nabla_i^2 + U(x_i)\right] + \sum_{i>j} V_{ij}\right\} \Psi(x_1,\ldots,x_A) = E\Psi(x_1,\ldots,x_A) \tag{5.65}$$

by introducing a mean field $U(x)$. V is the residual interaction if we consider the $\hat{h}^{(0)} \equiv -\hbar^2\nabla^2/2M_N + U$ as the unperturbed single-particle Hamiltonian of the shell model which uses, e.g., the harmonic oscillator potential for U.

We learnt in Sect. 5.4.2 that the residual interaction plays an important role in the magnetic moment. The degenerate $0^+, 2^+, \ldots, 8^+$ levels in $^{210}_{82}$Pb are also resolved by the residual interaction.

Let us assume that the residual interaction between two nucleons is given by

$$V(\mathbf{r}_1 - \mathbf{r}_2) = -V_0\delta(\mathbf{r}_1 - \mathbf{r}_2)\ . \tag{5.66}$$

Using the perturbation theory, one can then show (see Appendix A.11) that the energy of two-nucleon system, where each of the single-particle levels $(n_\alpha \ell_\alpha j_\alpha)$ and $(n_\beta \ell_\beta j_\beta)$ is occupied by one nucleon and their angular momenta couple to the total angular momentum J, changes by the amount of

$$\Delta E_J = -\frac{V_0}{2}\left[1 + (-1)^{\ell_\alpha+\ell_\beta-J}\right] F(n_\alpha\ell_\alpha, n_\beta\ell_\beta) A(j_\alpha j_\beta; J)\ , \tag{5.67}$$

where

$$F(n_\alpha\ell_\alpha, n_\beta\ell_\beta) = \int_0^\infty \frac{1}{r^2} u^2_{n_\alpha\ell_\alpha}(r) u^2_{n_\beta\ell_\beta}(r) \mathrm{d}r\ , \tag{5.68}$$

$$A(j_\alpha j_\beta; J) = \frac{1}{1+\delta_{n_\alpha n_\beta}\delta_{\ell_\alpha\ell_\beta}\delta_{j_\alpha j_\beta}} \frac{1}{4\pi} \frac{(2j_\alpha+1)(2j_\beta+1)}{2J+1} \left\langle j_\alpha j_\beta \frac{1}{2} - \frac{1}{2} \middle| J0 \right\rangle^2 , \tag{5.69}$$

where $u_{n\ell}(r)$ is the radial wave function defined according to Eq. (5.22). Equations (5.67)–(5.69) show that the energy shift by the residual interaction depends on the total spin J only through the factor A via the Clebsch–Gordan coefficient and $2J + 1$. Table 5.3 shows the factor A as a function of J for the case when we fill two identical nucleons in the same $j_\alpha = j_\beta = 9/2$ orbit. The table suggests that the energy split among the $0^+, 2^+, \ldots, 8^+$ states shown in Fig. 5.6 has been induced by the short-range residual interaction of δ-function type. The $J = 0$ state appears as the lowest energy level, because in that state two nucleons move in the same orbit with opposite directions, so that the spatial overlap is large. Consequently, the correlation energy of attractive δ-function type is maximized. The spatial overlap of two nucleons gets smaller with increasing J. Accordingly the energy shift from the energy in the independent particle model becomes smaller. Note that the energy shift

Table 5.3 The angular momentum dependence of the energy shift factor A for the case of $j_\alpha = j_\beta = 9/2$

J	0	2	4	6	8
$8\pi A$	10	80/33 (2.42)	180/143 (1.26)	320/429 (0.75)	980/2439 (0.40)

for the $J = 0$ state is especially large, and that the interval between the successive energy levels gets narrower with increasing J. The J dependence shown in Table 5.3 can be qualitatively understood from the following asymptotic formula which holds when both j and J are large [16],

$$\left\langle jj\frac{1}{2} - \frac{1}{2}\right| J0\left.\right\rangle^2 \sim \frac{2}{\pi}\frac{1}{j+\frac{1}{2}}\sqrt{\frac{4j(j+1)}{(2j+1)^2}\frac{(J+\frac{1}{2})^2}{J(J+1)} - \frac{(J+\frac{1}{2})^2}{(2j+1)^2}}$$
$$\sim \frac{2}{\pi j}\sqrt{1-\left(\frac{J}{2j}\right)^2}\,. \tag{5.70}$$

In this way, in the ground state of nuclei, two identical nucleons occupy two orbits, which are time-reversed orbits to each other, as a pair of the total angular momentum $J = 0$. This is called the *pairing correlation* in nuclei and corresponds to the *Cooper pair* and *superfluidity* in condensed matter physics.[9]

5.9 Column: Superheavy Elements (SHE)

As stated in Sect. 2.3.4 concerning the stability with respect to fission, the liquid-drop model suggests that the fission barrier disappears when the atomic number exceeds about 100, because the Coulomb force which prefers to fission then overcomes the surface tension which resists fission. Hence it predicts that there exist no heavy nuclei beyond $Z \sim 100$. However, theoretical calculations predict that there appear fission barriers due to shell effect and that the atomic number $Z = 114$ and the neutron number $N = 184$[10] are the magic numbers next to the known magic numbers $Z = 82$ and $N = 126$ and that there exists a region of relatively stable nuclei around the set of these numbers in the nuclear chart. They are called superheavy nuclei and the corresponding elements are called superheavy elements (SHE). Extensive activities to examine such theoretical prediction and to explore the limit of existence of nuclei in the region of large mass number are continued.

Figures 5.7 and 5.8 show reports of the Joint Institute for Nuclear Research (JINR) in Dubna, Russia and the RIKEN at Wako, Japan, which are currently the world centers of experimental activities in this subject. In Fig. 5.7, the superheavy nuclei synthesized by the so-called *hot fusion* which uses actinides such as Cf and Cm as the target and ^{48}Ca as the projectile are shown by circles and the superheavy nuclei synthesized by the *cold fusion* which

[9]One can rigorously discuss the level structure in the presence of the pairing correlation in an algebraic way by using the quasi-spin formalism if one simplifies that only the pair whose total angular momentum J is zero is affected by the attractive residual interaction [17].

[10]These numbers vary somewhat depending on the theoretical models.

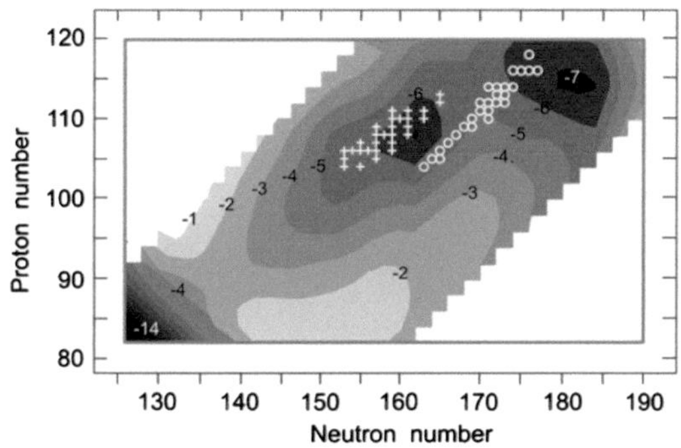

Fig. 5.7 Superheavy elements and the shell correction energy (in MeV), taken from [18]

bombards Pb or Bi target with, e.g., Ni by crosses.[11] The numbers attached to the contour lines are the magnitudes of the shell correction energy.

As the names cold fusion and hot fusion suggest, superheavy elements are synthesized by making two heavy nuclei with atomic numbers Z_1 and Z_2 collide to fuse to form a nucleus with a larger atomic number $Z_1 + Z_2$. A feature of heavy-ion fusion reactions used to synthesize superheavy elements with large atomic number is that the fission barrier locates inside the fusion barrier for the entrance channel. Therefore, it is not guaranteed that a compound nucleus with a certain lifetime is formed even if the colliding nuclei succeed to approach inside the fusion barrier for the entrance channel. Thus the formation cross section of superheavy elements gets very small.

Together with the experimental studies, many theoretical studies have been undertaken to elucidate the reaction mechanism and to suggest suitable combinations of the projectile and target nuclei as well as experimental conditions for the synthesis of superheavy nuclei using transport theories such as the stochastic differential equations and various other methods [19].

Superheavy elements have short lifetimes, and the cross section to synthesize them is so small as a few events in the experiments of a few months. Hence it is a hard task to experimentally confirm that the superheavy nuclei have been synthesized. The elements whose atomic numbers are 107–112 have been synthesized at the GSI Helmholtz Centre for Heavy Ion Research in Darmstadt, Germany, by the cold fusion reactions using Bi and Pb as the target nuclei. The electromagnetic separator called SHIP (Separator for Heavy Ion reaction Products) was used in a series of experiments, and the identification of the nuclear species of the synthesized superheavy nuclei has been done by confirming that the synthesized superheavy nuclei reach to a known nucleus after repeating α decays. As Fig. 5.8 shows, the experiments at RIKEN in 2004 and 2005 also used the fact that the α-decay chains from the superheavy nucleus have been observed and that the lifetime of the daughter nucleus $^{262}_{105}$Db which caused fission at the last stage does not contradict the known lifetime of the spontaneous fission as the evidence of the synthesis of element 113. The synthesis of the isotope $^{278}113$ of the element 113 was confirmed by the experiment in August, 2012, where six successive α-chains

[11]The reactions which have a large reaction Q-value and hence synthesize a superheavy nucleus as the evaporation residue after emitting about three neutrons by the evaporation process are called hot fusion, while the reactions which have a small Q-value and emit one neutron in synthesizing a superheavy nucleus is called cold fusion. Since the average binding energy of nucleon is about 8 MeV, one neutron is emitted for each 8 MeV of the excitation energy.

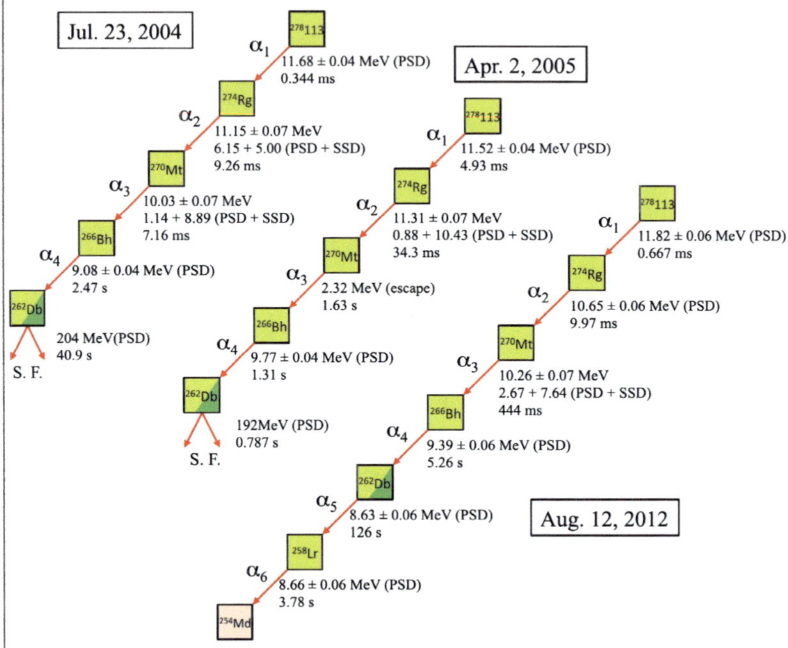

Fig. 5.8 The observed decay chains which show the synthesis of the element $^{278}113$ by the $^{209}\text{Bi} + ^{70}\text{Zn} \rightarrow {}^{278}113 + \text{n}$ reactions [20]

leading to $^{254}_{101}\text{Md}$ were observed [20]. Of the observed six α-decays the last two, i.e., the α-decays of Db and Lr, agreed with the known α-decays. Thus, the synthesis of the element 113 became unquestionable. The right to name the element $Z = 113$ was then given to Morita et al. at RIKEN by IUPAP/IUPAC in December, 2015. The name was announced as Nh (Nihonium) on June 8, 2016. By comparison, for the experiments at JINR using the hot fusion it has been an obstacle to confirmation of the synthesis of superheavy nuclei that the decay chains from the synthesized nuclei do not connect to the known region. However, those elements have been finally recognized by IUPAP/IUPAC, since the synthesis of the 112th element (^{283}Cn) by hot fusion has been reproduced at GSI, and in addition the number of events of the synthesis of 114th and 116th elements has been increased by the double-check experiments by the collaboration between JINR and Lawrence Livermore National Laboratory, USA and it was accepted that their decay can be reasonably interpreted as the decay through ^{283}Cn. The $Z = 114$ and $Z = 116$ elements have been named Fl (Flerovium) and Lv (Livermorium), respectively, in May, 2012. Furthermore, the $Z = 115$, 117 and 118 elements, which were synthesized by hot fusion, were announced as Mc (Moscovium), Ts (Tennessine), Og (Oganesson), respectively, on June 8, 2016.

References

1. D.R. Tilley et al., Nucl. Phys. A **708**, 3 (2002)
2. National Nuclear Data Center. http://www.nndc.bnl.gov/nudat2/
3. C.M. Lederer, V. Shirley, *Table of Isotopes*, 7th edn. (Wiley, New York, 1978)
4. A. Csótó, G.M. Hale, Phys. Rev. C **55**, 536 (1997)
5. M.A. Preston, *Physics of the Nucleus* (Addison-Wesley, London, 1962)
6. A.M. Lane, R.G. Thomas, Rev. Mod. Phys. **30**, 257 (1958)
7. T. Otsuka, T. Suzuki, R. Fujimoto, H. Grawe, Y. Akaishi, Phys. Rev. Lett. **95**, 232502 (2005)
8. W.D. Knight et al., Phys. Rev. Lett. **52**, 2141 (1984)
9. I. Hamamoto, Nucl. Phys. A **126**, 545 (1969); **141**, 1 (1970); **155**, 362 (1970)
10. A. Bohr, B.R. Mottelson, *Nuclear Structure*, vol. II (Benjamin, New York, 1975)
11. N.J. Stone, At. Data Nucl. Data Tables **90**, 75 (2005)
12. A. Arima, H. Horie, Prog. Theor. Phys. **12**, 623 (1954)
13. T. Yamazaki, T. Nomura, S. Nagamiya, T. Katou, Phys. Rev. Lett. **25**, 547 (1970)
14. F. Halzen, A.D. Martin, *QUARKS and LEPTONS* (Wiley, New York, 1984)
15. A. Bohr, B.R. Mottelson, *Nuclear Structure*, vol. I (Benjamin, New York, 1969)
16. A. Molinari, M.B. Johnson, H.A. Bethe, W.M. Alberico, Nucl. Phys. A **239**, 45 (1975)
17. G.E. Brown, *Unified Theory of Nuclear Models and Forces* (North-Holland, Amsterdam, 1967)
18. Yu. Oganessian, J. Phys. G **34**, R165 (2007)
19. Here, we quote some of the related theoretical papers. (a) Papers concerning stochastic differential equations: Y. Aritomo, T. Wada, M. Ohta, Y. Abe, Phys. Rev. C **59**, 796 (1999); C. Shen, G. Kosenko, Y. Abe, Phys. Rev. C **66**, 061602 (2002); Y. Abe, D. Boilley, G. Kosenko, C. Shen, Acta Phys. Pol. B **34**, 2091 (2003); Y. Aritomo, Phys. Rev. C **80**, 064604 (2009); V. Zagrebaev, W. Greiner, Phys. Rev. C **78**, 034610 (2008); (b) Paper on the quantum diffusion theory: N. Takigawa, S. Ayik, K. Washiyama, S. Kimura, Phys. Rev. C **69**, 054605 (2004); (c) Paper using coupled-channels calculations: N. Rowley, N. Grar, K. Hagino, Phys. Lett. B **632**, 243 (2006)
20. K. Morita et al., J. Phys. Soc. Jpn. **81**, 103201 (2012)

Chapter 6
Microscopic Mean-Field Theory (Hartree–Fock Theory)

Abstract In the last chapter we discussed the shell structure of nuclei by assuming a simplified mean field such as the Wood–Saxon potential or the harmonic oscillator potential, which appears realistic. However, properly speaking, it is necessary to self-consistently determine the mean field and the whole nucleus as well as the individual states of nucleons inside the nucleus, because the mean field itself is determined based on the interaction between constituent nucleons. In this chapter we describe the basic ingredients of the Hartree–Fock theory, which is a representative of such self-consistent approaches, and of some related microscopic theories. It will not be an overstatement to say that the self-consistent calculations from microscopic points of view are at the center of current theoretical studies of nuclear structure. An important aspect in performing microscopic calculations such as the Hartree–Fock calculations is that the force between the constituent particles, i.e., the nuclear force, is not necessarily well established or involves some features which make the theoretical treatment difficult in contract to the systems which are objects of, e.g., condensed matter physics, where the interaction between the constituent particles is the well-known Coulomb interaction. The studies based on microscopic theories utilizes interdependence of theory and experiments. One analyzes experimental data with theories using the new nuclear force which looks more realistic. One then improves the nuclear force so as to incorporate the results of the analysis, then reconfirms thus obtained new theoretical predictions with experimental data.

6.1 Hartree–Fock Equation

In the Hartree–Fock theory, we approximate the solution of Eq. (5.64) by a single Slater determinant such that

$$\Psi(x_1, \ldots, x_A) = \frac{1}{\sqrt{A!}} \begin{vmatrix} \psi_1(x_1) & \psi_1(x_2) & \cdots & \psi_1(x_A) \\ \psi_2(x_1) & \psi_2(x_2) & \cdots & \psi_2(x_A) \\ \vdots & \vdots & \ddots & \vdots \\ \psi_A(x_1) & \psi_A(x_2) & \cdots & \psi_A(x_A) \end{vmatrix}, \tag{6.1}$$

N. Takigawa and K. Washiyama, *Fundamentals of Nuclear Physics*,
DOI 10.1007/978-4-431-55378-6_6

and determine the single-particle wave function $\{\psi\}$ by the variational principle. From the condition of the variational principle

$$\frac{\delta}{\delta\psi_i^*(x_j)}(\langle\Psi|H|\Psi\rangle - E\langle\Psi|\Psi\rangle) = 0 , \tag{6.2}$$

we obtain

$$\left[-\frac{\hbar^2}{2M_N}\Delta + \Gamma_{\mathrm{H}}(\mathbf{r})\right]\psi_k(\mathbf{r}) + \int \Gamma_{\mathrm{EX}}(\mathbf{r},\mathbf{r}')\psi_k(\mathbf{r}')d\mathbf{r}' = \varepsilon_k\psi_k(\mathbf{r}) , \tag{6.3}$$

$$\Gamma_{\mathrm{H}}(\mathbf{r}) = \int v(\mathbf{r}-\mathbf{r}')\sum_{j=1}^{A}|\psi_j(\mathbf{r}')|^2 d\mathbf{r}' = \int v(\mathbf{r}-\mathbf{r}')\rho(\mathbf{r}')d\mathbf{r}' , \tag{6.4}$$

$$\Gamma_{\mathrm{EX}}(\mathbf{r},\mathbf{r}') = -v(\mathbf{r}-\mathbf{r}')\sum_{j=1}^{A}\psi_j^*(\mathbf{r}')\psi_j(\mathbf{r}) = -v(\mathbf{r}-\mathbf{r}')\rho(\mathbf{r},\mathbf{r}') , \tag{6.5}$$

$$\hat{\rho} \equiv \sum_{j=1}^{A}|j\rangle\langle j| . \tag{6.6}$$

We assumed a central potential of the Wigner-type for the nucleon–nucleon interaction v. Equation (6.3) is called the Hartree–Fock equation, and Γ_{H} and Γ_{EX} are the Hartree and Fock or exchange potentials, respectively. Note that the Hartree potential matches the potential assumed in Eq. (5.1) and that the Fock potential is non-local if the range of the nuclear force is finite.

Equation (6.3) which determines single-particle wave functions and the corresponding single-particle energies has to be solved self-consistently, since the Hartree and Fock potentials depend on the density of the nucleus at each position. The historically standard method for that aim is to expand the single-particle wave functions $\{\psi\}$ in terms of the wave functions of the harmonic oscillator model which we learnt in the previous section and determine the expansion coefficients by the iteration method. Recently, however, it is getting popular also to use an alternative method which directly solves the Hartree–Fock equation by numerical integration without introducing the expansion basis by dividing the whole space into small meshes. Since the accuracy of the expansion method depends on the size of the expansion basis, the direct integration method is useful, especially for the currently active area of study of unstable nuclei where the binding energy of valence nucleons is small.

6.1.1 Equivalent Local Potential, Effective Mass

As we learnt in the previous chapter, the shell model usually assumes a local mean field. One can convert the integro-differential equation due to non-local potential into an equivalent local equation if certain conditions are fulfilled. One can then understand the characteristic effects caused by the non-local potential in different physical terms.

We show in Appendix A.5 that the effect of non-locality can be represented by the *energy or momentum dependence* of the effective local potential if the change of the potential within a wave length is small and hence if the wave function can be well approximated in the WKB approximation, i.e., by an extension of the plane wave using the local wave number. Furthermore, if the non-locality is moderate, the non-local effect can be represented in terms of a position-dependent effective mass.

6.1.2 Nuclear Matter and Local Density Approximation

6.1.2.1 Local Density Approximation

As Eq. (6.5) shows, the Fock potential depends on the non-diagonal matrix elements of the density operator. Its primary effect can be understood by considering the nucleus locally as a nuclear matter[1] and by approximating the wave functions of nucleons with the plane waves distributed up to the Fermi surface corresponding to the density $\rho((\mathbf{r}+\mathbf{r}')/2)$ at each position. This is called the *local density approximation.* Following this idea, $\rho(\mathbf{r},\mathbf{r}')$ is given by

$$\rho_{\mathrm{LDA}}(\mathbf{r},\mathbf{r}') \equiv \left(\frac{1}{\sqrt{V}}\right)^2 \int_0^{k_{\mathrm{F}}(R)} \mathrm{e}^{-\mathrm{i}\mathbf{k}\cdot\mathbf{r}'}\mathrm{e}^{\mathrm{i}\mathbf{k}\cdot\mathbf{r}}\mathrm{d}\mathbf{k} \times \frac{V}{(2\pi)^3} \times 2 \times 2 \tag{6.7}$$

$$= 3\rho(R)\frac{1}{k_{\mathrm{F}}(R)s}\, j_1(k_{\mathrm{F}}(R)s)\;, \tag{6.8}$$

where $\mathbf{s} \equiv \mathbf{r}-\mathbf{r}'$, $\mathbf{R} = (\mathbf{r}+\mathbf{r}')/2$.

Exercise 6.1 Using Eqs. (6.3) and (6.8), show that the contribution of the Fock term to the Coulomb energy is given by

$$\mathscr{E}^{\mathrm{ex}}_{\mathrm{Coul}} = -\frac{1}{2}e^2 \iint \frac{\left[\langle \mathbf{r}_1|\hat{\rho}^{(\mathrm{p})}|\mathbf{r}_2\rangle\right]^2}{|\mathbf{r}_1-\mathbf{r}_2|}\mathrm{d}\mathbf{r}_1\mathrm{d}\mathbf{r}_2 \times \frac{1}{2} = -\frac{27}{16}\frac{Z^2e^2}{R^3}\frac{1}{(k_{\mathrm{F}}^{(\mathrm{p})})^2}\;, \tag{6.9}$$

when the density is uniform. The last factor 1/2 in the second term is the factor which takes into account the orthogonality in the spin space when two protons are exchanged.

[1] Such a medium made of nucleons which has uniform density and an infinite spatial extension is called nuclear matter.

The Coulomb energy is thus given by

$$\mathscr{E}_{\text{Coul}} \approx \frac{3}{5}\frac{e^2 Z^2}{R}\left[1 - 5\left(\frac{3}{16\pi Z}\right)^{2/3}\right] \tag{6.10}$$

including the Fock term for nuclei with a large mass number in the local density approximation.

6.1.2.2 Effective Mass in the Local Density Approximation

Let us study how the effective mass depends on the parameters of the nuclear force and the nuclear density by using the local density approximation.[2]

We begin by noting that the potential energy U in the Hartree–Fock approximation is given in general by

$$U = \frac{1}{2}\sum_{\alpha,\beta=\text{occupied}} \langle\alpha\beta|v|\alpha\beta\rangle_a \,, \tag{6.11}$$

where α and β denote the total quantum numbers in the usual, the spin and the isospin spaces, and the ket state $|\alpha\beta\rangle_a$ represents the antisymmetrized state $(|\alpha\beta\rangle - |\beta\alpha\rangle)$. The sum is carried over all states below the Fermi surface. We assume that the nuclear force is the central force of the Yukawa type given by Eq. (3.62),

$$v(r) = V_0\frac{\mathrm{e}^{-\mu r}}{\mu r}(w + mP_M) = v_0(r)(1 - m + mP_M) \,. \tag{6.12}$$

If we express v as Eq. (3.29) by introducing the projection operators, then the potential in each projected state is given by

$$V_{tt}(r) = V_{ss}(r) = V_0(1-2m)\frac{\mathrm{e}^{-\mu r}}{\mu r} \,, \quad V_{st}(r) = V_{ts}(r) = V_0\frac{\mathrm{e}^{-\mu r}}{\mu r} \,. \tag{6.13}$$

Let us consider especially the infinite nuclear matter consisting of equal numbers of protons and neutrons, where both spin-up and down states are occupied. In that case, the total potential energy can be expressed as

$$\begin{aligned} U = \frac{1}{2}\left[\frac{\Omega}{(2\pi)^3}\right]^2 \int_{k<k_{\mathrm{F}}} \mathrm{d}^3k \int_{k'<k_{\mathrm{F}}} \mathrm{d}^3k' \Big[&(9V_{tt}^{(0)}+3V_{ts}^{(0)}+3V_{st}^{(0)}+V_{ss}^{(0)})\langle\mathbf{kk'}|\hat{v}_0|\mathbf{kk'}\rangle \\ &- (9V_{tt}^{(0)} - 3V_{ts}^{(0)} - 3V_{st}^{(0)} + V_{ss}^{(0)})\langle\mathbf{kk'}|\hat{v}_0|\mathbf{k'k}\rangle\Big] \end{aligned} \tag{6.14}$$

by taking into account the statistical weight due to the degree of degeneracy and the symmetry property of the spatial wave function for each state of two-nucleon system.

---

[2]Similar consideration can be found in [1, 2].

We expressed the interaction in each projected state as the product of a common factor $v_0(r) = V_0 e^{-\mu r}/\mu r$ and a state dependent coefficient $V_{tt}^{(0)}$, $V_{ts}^{(0)}$, $V_{st}^{(0)}$ and $V_{ss}^{(0)}$, which are $1 - 2m$, 1, 1 and $1 - 2m$, respectively. Equation (6.14) shows that the weights of the direct and exchange terms are given by

$$\begin{aligned} 9V_{tt}^{(0)} + 3V_{ts}^{(0)} + 3V_{st}^{(0)} + V_{ss}^{(0)} &= 16 - 20m \ , \\ 9V_{tt}^{(0)} - 3V_{ts}^{(0)} - 3V_{st}^{(0)} + V_{ss}^{(0)} &= 4 - 20m \ . \end{aligned} \tag{6.15}$$

The following expressions can be obtained for the local and non-local terms in Eq. (6.3) by performing the functional differentiation of U with respect to $\psi_k^*(\mathbf{r})$.

$$\Gamma_{\mathrm{H}}(\mathbf{r}) = (4 - 5m) \int v_0(|\mathbf{r} - \mathbf{r}'|)\rho'(\mathbf{r}')\mathrm{d}\mathbf{r}' \ , \tag{6.16}$$

$$\Gamma_{\mathrm{EX}}(\mathbf{r}, \mathbf{r}') = -(1 - 5m)v_0(|\mathbf{r} - \mathbf{r}'|)\rho'(\mathbf{r}, \mathbf{r}') \ , \tag{6.17}$$

$$\hat{\rho}' \equiv \sum_{\mathbf{k}'=0}^{k_{\mathrm{F}}} |\mathbf{k}'\rangle\langle\mathbf{k}'| = \sum_{\mathbf{k}'=0}^{k_{\mathrm{F}}} |\psi_{\mathbf{k}'}\rangle\langle\psi_{\mathbf{k}'}| \ . \tag{6.18}$$

We divided each term by 4 by specifying spin and isospin in taking the functional differentiation.

Let us now calculate the Fock term in the Hartree–Fock equation by using the explicit form of the nuclear force $v_0(\mathbf{r} - \mathbf{r}') = V_0 e^{-\mu|\mathbf{r}-\mathbf{r}'|}/\mu|\mathbf{r} - \mathbf{r}'|$ and the result of Eq. (6.8). In that procedure, we use the plane wave for the wave function according to the nuclear matter approximation. The Fock term can then be expressed as $\Gamma_{\mathrm{EX}}^{\mathrm{ELP}}(\mathbf{r})\psi_k(\mathbf{r})$, where the equivalent local potential (ELP) for the Fock term $\Gamma_{\mathrm{EX}}^{\mathrm{ELP}}(\mathbf{r})$ is given by

$$\Gamma_{\mathrm{EX}}^{\mathrm{ELP}}(\mathbf{r}) = -\frac{1}{4}(1 - 5m)V_0 3\rho(\mathbf{r}) \int \frac{e^{-\mu s}}{\mu s} \frac{j_1(k_{\mathrm{F}}(\mathbf{r})s)}{k_{\mathrm{F}}(\mathbf{r})s} e^{i\mathbf{k}\cdot\mathbf{s}}\mathrm{d}\mathbf{s} \tag{6.19}$$

$$\begin{aligned} &= -(1 - 5m)V_0 \left[3\pi\rho(\mathbf{r}) \frac{1}{\mu k_{\mathrm{F}}(\mathbf{r})} \int_0^\infty e^{-\mu s} j_1(k_{\mathrm{F}}(\mathbf{r})s)\mathrm{d}s \right. \\ &\qquad \left. -\pi\rho(\mathbf{r}) \frac{1}{\mu(\mu^2 + k_{\mathrm{F}}^2)^2} k^2 + \cdots \right] \end{aligned} \tag{6.20}$$

by noting that $\rho'(\mathbf{r}, \mathbf{r}') = \rho(\mathbf{r}, \mathbf{r}')/4$. This leads to the following expression for the effective mass,

$$\frac{1}{M^*} = \frac{1}{M_N} \left[1 + 2\pi(1 - 5m)V_0 \frac{M_N c^2}{(\hbar c)^2} \rho(\mathbf{r}) \frac{1}{\mu(\mu^2 + k_{\mathrm{F}}^2)^2} \right] . \tag{6.21}$$

Equation (6.21) shows that the contribution of the Fock term to the effective mass is zero, i.e., $M^* = M_N$, in the limit when the range of the nuclear force is zero, i.e., when $\mu \to \infty$, and gets larger with increasing range. The same equation as Eq. (6.21) can be obtained also from the evaluation of the energy of nucleons with momentum $\hbar k$. If we take $k_{\rm F} = 1.27\,{\rm fm}^{-1}$, $\rho = 0.138\,{\rm fm}^{-3}$ (Table 2.2), $m = 0.5$ (Serber force), $V_0 = -48.1\,{\rm MeV}$, $a = 1/\mu = 1.17\,{\rm fm}$ (see Eq. (3.64)), then $M_N^* = 0.76 M_N$. This well agrees with the results of the standard Hartree–Fock calculations such as the Skyrme Hartree–Fock calculations, which we learn in the next section.

Incidentally, the Hartree potential given by Eq. (6.16) becomes

$$\Gamma_{\rm H}(\mathbf{r}) = V_0(4 - 5m)\frac{2}{3\pi}\left(\frac{k_{\rm F}}{\mu}\right)^3 \tag{6.22}$$

in the nuclear matter approximation.

6.1.3 Saturation Property in the Well-Behaved Potential, Constraint to the Exchange Property

A characteristic of the potential given by Eq. (6.12) in comparison with potentials with repulsive core which have been introduced in Chap. 3 is that one can perform Hartree–Fock calculations as it is, since the matrix elements of the nuclear force do not diverge.[3] We comment here on how the saturation property, i.e., the fact that nuclei do not collapse or that nuclei exist with a finite density, is guaranteed within the framework of the Hartree–Fock theory for such a well-behaved potential. Similarly to Sect. 6.1.2.2 we consider the infinite nuclear matter consisting of the same numbers of neutrons and protons, where both spin-up and down states are occupied, and assume that the radial dependence of the nuclear force is given by $v_0(r) = V_0 {\rm e}^{-\mu r}/\mu r$ similarly to Eq. (6.12). Then, the condition

$$16 - 20m < 0, \quad \text{i.e.,} \quad m > \frac{4}{5}\,, \tag{6.23}$$

has to be satisfied in order for nuclei to exist with a finite density as we can see from Eq. (6.22).

The condition given by Eq. (6.23) contradicts the requirement of two-nucleon scattering, which suggests $m \sim 1/2$ (for example, Serber force). In this way, it is difficult to satisfy the requirement of the saturation property with well-behaved potentials such as that given by Eq. (6.12) solely by the exchange property of the nuclear force. Hence it is necessary to attribute the saturation property to some effective

[3] Such potentials are often called well-behaved potentials.

many-body interaction, which takes into account the medium effects. The Skyrme interaction, which we describe in the following, is one example of such interactions.[4]

6.2 Skyrme Hartree–Fock Calculations for Finite Nuclei

6.2.1 Skyrme Force

It is not so easy to systematically study nuclear structure from the point of view of a quantum many-body problem based on the Brueckner theory (see Sect. 3.8). On the other hand, the Hartree–Fock calculations using a phenomenological nuclear force began in early 1970 s and have been very popular up to now. Especially, the Hartree–Fock calculations using the simple force invented by Skyrme are known under the name of the Skyrme Hartree–Fock calculations [3] and have become a standard calculation for nuclear structure.

Skyrme invented the extreme nuclear force of δ-function type to emphasize the short-range property of the nuclear force. In addition, he introduced a three-body force to guarantee the saturation property,

$$V = \sum_{i<j} V(i,j) + \sum_{i<j<k} V(i,j,k) \,. \tag{6.24}$$

(A) Two-Body Force The force currently used as the Skyrme force, which is an extension of the force introduced originally by Skyrme himself, reads

$$\begin{aligned} V(1,2) &= t_0(1+x_0P_\sigma)\delta(\mathbf{r}_1-\mathbf{r}_2) \\ &+ \frac{1}{2}t_1(1+x_1P_\sigma)\left[\delta(\mathbf{r}_1-\mathbf{r}_2)\mathbf{k}_R^2 + \mathbf{k}_L^2\delta(\mathbf{r}_1-\mathbf{r}_2)\right] \\ &+ t_2(1+x_2P_\sigma)\mathbf{k}_L\cdot\delta(\mathbf{r}_1-\mathbf{r}_2)\mathbf{k}_R \\ &+ \mathrm{i}W_0(\boldsymbol{\sigma}_1+\boldsymbol{\sigma}_2)\cdot\mathbf{k}_L\times\delta(\mathbf{r}_1-\mathbf{r}_2)\mathbf{k}_R \,, \end{aligned} \tag{6.25}$$

where P_σ is the spin exchange operator defined by Eq. (3.18), and $\mathbf{k}_R$ and $\mathbf{k}_L$ are defined by

$$\mathbf{k}_R \equiv \frac{1}{2\mathrm{i}}(\nabla_1-\nabla_2) \,, \tag{6.26}$$

$$\mathbf{k}_L \equiv -\frac{1}{2\mathrm{i}}(\nabla_1-\nabla_2) \,, \tag{6.27}$$

and operate on the right and the left, respectively.

[4]The study of equation of state using the simple perfect rigid sphere model shows that the saturation property is guaranteed by the repulsive core in the high density region higher than about six times the standard density $\rho_0 \sim 0.14\,\mathrm{fm}^{-3}$ if one assumes the radius of the repulsive core to be 0.4 fm.

(B) Three-body force The three-body force[5] which was originally introduced by Skyrme reads

$$V(1,2,3) = t_3\delta(\mathbf{r}_1 - \mathbf{r}_2)\delta(\mathbf{r}_2 - \mathbf{r}_3) \,. \tag{6.28}$$

However, currently, it is more popular to introduce an additional two-body term $\sum_{i<j} V_{\mathrm{DD}}(i,j)$ using the density-dependent two-body force

$$V_{\mathrm{DD}}(1,2) = \frac{1}{6}t_3(1 + x_3P_\sigma)\rho^\alpha\left(\frac{\mathbf{r}_1 + \mathbf{r}_2}{2}\right)\delta(\mathbf{r}_1 - \mathbf{r}_2) \,, \tag{6.29}$$

instead of the term of the three-body force $\sum_{i<j<k} V(i,j,k)$. This model is designed to guarantee the saturation and control the nuclear incompressibility through the value of α, which is often assumed to be either $1/3$ or $1/6$. The latter gives, of course, a smaller incompressibility.

The W_0 term in Eq. (6.25) represents the two-body spin–orbit interaction. The t_0 term corresponds to the limit of zero-range nuclear force, and is the attractive term which is indispensable for the nucleus to exist as a self-bound system. The t_1 and t_2 terms have been introduced to take into account the effects of the finite range property of the nuclear force. In order to understand this, let us consider a two-body central force with a finite range,[6,7]

$$v(r) = V_0\mathrm{e}^{-(r/a)^2} \,. \tag{6.30}$$

In this case, the matrix elements using the eigenstates of the momentum operator are given by

$$\langle\mathbf{k}|\hat{v}|\mathbf{k}'\rangle = \frac{1}{\Omega}(a\sqrt{\pi})^3V_0\mathrm{e}^{-a^2(\mathbf{k}-\mathbf{k}')^2/4} \tag{6.31}$$

$$= \frac{1}{\Omega}(a\sqrt{\pi})^3V_0\left[1 - \frac{1}{4}a^2(\mathbf{k}-\mathbf{k}')^2 + \cdots\right] \,. \tag{6.32}$$

Equation (6.31) can be derived by inserting the completeness relation $\int \mathrm{d}\mathbf{r}|\mathbf{r}\rangle\langle\mathbf{r}| = 1$ on both sides of the operator $\hat{v}$ and using $\langle\mathbf{r}|\mathbf{k}\rangle = \frac{1}{\sqrt{\Omega}}\mathrm{e}^{\mathrm{i}\mathbf{k}\cdot\mathbf{r}}$ and also that the matrix element of a local potential $\hat{v}$ in the Dirac notation is given by

[5]This term is an effective interaction, which represents the many-body effects and the effect of repulsive core appearing in the G-matrix theory and has been introduced to avoid the collapse of nuclei, and should be distinguished from the genuine three-body force appearing in the discussion of nuclear force based on the meson theory as the force involving three nucleons [1]. The magnitude of the latter has been discussed through detailed calculations of the binding energy and the form factor of light nuclei such as the triton and ^{3}He using, e.g., the Faddeev equation.

[6]Instead of the Yukawa type, we assumed the Gauss type as a two-body central force to facilitate calculations.

[7]Here, we eliminate the center-of-mass motion and formulate only for the relative motion, and treat it as a one-body problem.

$$\langle \mathbf{r}|\hat{v}|\mathbf{r}'\rangle = \delta(\mathbf{r}-\mathbf{r}')v(\mathbf{r}) \,. \tag{6.33}$$

Since Eq. (6.32) is an expansion with respect to the range of the nuclear force a, let us keep up to the first term representing the finite range effect, i.e., up to the second order with respect to a in Eq. (6.32), and transform back to the coordinate representation. We then have

$$\begin{aligned}\langle \mathbf{r}|\hat{v}|\mathbf{r}'\rangle &= (a\sqrt{\pi})^3 V_0 \Big\{ \delta(\mathbf{r})\delta(\mathbf{r}') \\ &\quad + \frac{1}{4}a^2 \left[\nabla_r^2\delta(\mathbf{r})\delta(\mathbf{r}') + \delta(\mathbf{r})\nabla_{r'}^2\delta(\mathbf{r}') + 2\nabla_r\delta(\mathbf{r})\cdot\nabla_{r'}\delta(\mathbf{r}') \right] \Big\} \,,\end{aligned} \tag{6.34}$$

which can be obtained by inserting the completeness relation $\frac{\Omega}{(2\pi)^3}\int d\mathbf{k}|\mathbf{k}\rangle\langle\mathbf{k}| = 1$ on both sides of $\hat{v}$. Correspondingly, one obtains

$$\begin{aligned}&\iint \mathrm{d}\mathbf{r}\mathrm{d}\mathbf{r}' F(\mathbf{r})\langle \mathbf{r}|\hat{v}|\mathbf{r}'\rangle G(\mathbf{r}') \\ &= (a\sqrt{\pi})^3 V_0 \Big\{ \iint \mathrm{d}\mathbf{r}\mathrm{d}\mathbf{r}'\delta(\mathbf{r})F(\mathbf{r})\delta(\mathbf{r}')G(\mathbf{r}') \\ &+ \frac{1}{4}a^2 \iint \mathrm{d}\mathbf{r}\mathrm{d}\mathbf{r}'\delta(\mathbf{r}) \left[(\nabla_r^2 F(\mathbf{r}))\delta(\mathbf{r}')G(\mathbf{r}') + F(\mathbf{r})\delta(\mathbf{r}')(\nabla_{r'}^2 G(\mathbf{r}')) \right] \\ &+ \frac{1}{2}a^2 \iint \mathrm{d}\mathbf{r}\mathrm{d}\mathbf{r}'\delta(\mathbf{r})\nabla_r F(\mathbf{r})\delta(\mathbf{r}')\cdot\nabla_{r'}G(\mathbf{r}') \Big\}\end{aligned} \tag{6.35}$$

for functions $F(\mathbf{r})$ and $G(\mathbf{r}')$. This means that the $v(\mathbf{r})$ approximated up to the second order of the range of the force is given by

$$\begin{aligned}v(\mathbf{r}) \approx (a\sqrt{\pi})^3 V_0 \Bigg\{ \delta(\mathbf{r}) &- \frac{1}{4}a^2 \left[\left(-\frac{1}{\mathrm{i}}\nabla\right)_L^2 \delta(\mathbf{r}) + \delta(\mathbf{r}) \left(\frac{1}{\mathrm{i}}\nabla\right)_R^2 \right] \\ &+ \frac{1}{2}a^2 \left(-\frac{1}{\mathrm{i}}\nabla\right)_L \delta(\mathbf{r}) \left(\frac{1}{\mathrm{i}}\nabla\right)_R \Bigg\} \,.\end{aligned} \tag{6.36}$$

By comparing Eq. (6.25) with Eq. (6.36), one can expect that $t_0 < 0$, $t_1 > 0$, and $t_2 < 0$. Also, the coefficient $(a\sqrt{\pi})^3$ in Eq. (6.36) can be understood by considering one of the representations of the δ-function, $\delta(x) = \lim_{a\to 0} \mathrm{e}^{-(x/a)^2}/\sqrt{\pi}a$.

6.2.2 Skyrme Hartree–Fock Equation

In the Skyrme Hartree–Fock calculations, thanks to the feature of the interaction that it is of the δ-function type one can explicitly write down the Hartree–Fock equation in terms of the parameters of the nuclear force and the nucleon density ρ [3]. We give here the Hartree–Fock equation for the $N = Z$ system in the case when the Coulomb force is ignored and when we assume the original interaction by Skyrme, i.e., when we set $x_1 = x_2 = 0$ for the two-body force, and assume the effective three-body force given by Eq. (6.28),

$$\left[-\nabla \cdot \frac{\hbar^2}{2M_N^*(\mathbf{r})}\nabla + U(\mathbf{r}) - \mathrm{i}\mathbf{W}(\mathbf{r}) \cdot (\nabla \times \boldsymbol{\sigma})\right]\psi_i = \varepsilon_i \psi_i \ , \tag{6.37}$$

where the effective mass $M_N^*(\mathbf{r})$, scalar $U(\mathbf{r})$ and spin–orbit $\mathbf{W}(\mathbf{r})$ potentials are given as

$$\frac{\hbar^2}{2M_N^*(\mathbf{r})} = \frac{\hbar^2}{2M_N} + \frac{1}{16}(3t_1 + 5t_2)\rho(\mathbf{r}) \ , \tag{6.38}$$

$$U(\mathbf{r}) = \frac{3}{4}t_0\rho + \frac{3}{16}t_3\rho^2 + \frac{1}{16}(3t_1 + 5t_2)\tau + \frac{1}{32}(5t_2 - 9t_1)\nabla^2\rho - \frac{3}{4}W_0\nabla \cdot \mathbf{J} \ , \tag{6.39}$$

$$\mathbf{W}(\mathbf{r}) = \frac{3}{4}W_0\nabla\rho + \frac{1}{16}(t_1 - t_2)\mathbf{J} \ , \tag{6.40}$$

with

$$\tau(\mathbf{r}) = \sum_{q=\mathrm{p,n}}\sum_{i,s}|\nabla\psi_i(\mathbf{r}, s, q)|^2 \sim \frac{3}{5}k_F^2\rho \ , \tag{6.41}$$

$$\mathbf{J}(\mathbf{r}) = -\mathrm{i}\sum_{q=\mathrm{p,n}}\sum_{i,s,s'}\psi_i^*(\mathbf{r}, s, q)\left[\nabla\psi_i(\mathbf{r}, s', q) \times \langle s|\boldsymbol{\sigma}|s'\rangle\right] \ . \tag{6.42}$$

The τ and $\mathbf{J}$ are the kinetic energy and the spin–orbit densities, respectively. There appears the effective mass $M_N^*(r)$ in the Hartree–Fock equation reflecting the momentum-dependent terms in the Skyrme force (see Eq. (6.25)). In accord with the result of Eq. (6.21), the effective mass stays the same as the bare nucleon mass in the limit of zero-range force, i.e., in the limit of $t_1 = 0$, $t_2 = 0$, and gets smaller with increasing range of the nuclear force.

6.2.3 Energy Density and Determination of Parameters

In the case of the Skyrme force we can express the interaction energy as a functional of the nucleon density. On the other hand, in the Thomas–Fermi approximation, the kinetic energy, e.g., for protons, is given by

$$E_{\mathrm{kin}}^{(\mathrm{p})} = \frac{1}{(2\pi\hbar)^3}\int^{p_{\mathrm{F}}^{(\mathrm{p})}(r)}\frac{p^2}{2M_N}\mathrm{d}\mathbf{p}\mathrm{d}\mathbf{r} \times 2 = \int\frac{3}{5}\frac{(\hbar k_{\mathrm{F}}^{(\mathrm{p})}(r))^2}{2M_N}\rho_{\mathrm{p}}(\mathbf{r})\mathrm{d}\mathbf{r} \ . \tag{6.43}$$

Consequently, if we represent the total energy as

$$E = E_{\text{Skyrme}} = \int \mathscr{H}(\mathbf{r})d\mathbf{r} \tag{6.44}$$

using the energy density $\mathscr{H}$, then $\mathscr{H}$ can be expressed as a functional of the proton and neutron densities. Especially, if we consider the case of $N = Z$ and ignore the Coulomb force, then $\mathscr{H}$ is given by

$$\begin{aligned}\mathscr{H} &= \frac{\hbar^2}{2M_N}\tau + \frac{3}{8}t_0\rho^2 + \frac{1}{16}t_3\rho^3 + \frac{1}{16}(3t_1 + 5t_2)\rho\tau \\ &+ \frac{1}{64}(9t_1 - 5t_2)(\nabla\rho)^2 - \frac{3}{4}W_0\rho\nabla\cdot\mathbf{J} + \frac{1}{32}(t_1 - t_2)\mathbf{J}^2 ,\end{aligned} \tag{6.45}$$

corresponding to Eqs. (6.37)–(6.42). Let us next move to the problem of determining the values of parameters such as t_0 and t_1. To that end, let us consider the case of $N = Z$ and the nuclear matter, where the surface effect can be ignored. In this case, the derivative terms in Eq. (6.45) can be ignored, and the total energy per nucleon is given by

$$\frac{E}{A} = \frac{\mathscr{H}V}{\rho V} = \frac{3}{5}T_F + \frac{3}{8}t_0\rho + \frac{1}{16}t_3\rho^2 + \frac{3}{80}(3t_1 + 5t_2)\rho k_F^2 , \tag{6.46}$$

where $T_F = \hbar^2 k_F^2/2M_N$ is the Fermi energy. Equation (6.46) leads to

$$K \equiv k_F^2 \frac{\partial^2}{\partial k_F^2}\left(\frac{E}{A}\right) = \frac{6}{5}T_F + \frac{9}{4}t_0\rho + \frac{15}{8}t_3\rho^2 + \frac{3}{4}(3t_1 + 5t_2)\rho k_F^2 \tag{6.47}$$

for the incompressibility of the nuclear matter.[8] On the other hand, the pressure is given by

$$P = -\frac{\partial}{\partial V}E = \rho^2 \frac{\partial}{\partial \rho}\left(\frac{E}{A}\right) \tag{6.48}$$

by definition when the temperature is 0.

Exercise 6.2 Express the pressure as a function of the nucleon density ρ and the parameters of the nuclear force.

It is known from the mass formula that the binding energy per nucleon is about 15 MeV for the nuclear matter. On the other hand, the incompressibility K is suggested to be about 200 MeV[9] from the excitation energy [4] of the *isoscalar*

[8] Alternatively, the nuclear incompressibility is defined also by $k \equiv -V\frac{\partial}{\partial V}P = \rho\frac{\partial P}{\partial \rho}$. The two nuclear incompressibilities are related as $K = \frac{9k(\rho_0)}{\rho_0}$ if we denote the equilibrium density, which gives $P(\rho_0) = 0$, by ρ_0.

[9] It is, however, not so easy to determine the incompressibility of nuclear matter to a good accuracy from the data of actual nuclei with finite sizes.

monopole vibration (*breathing mode of vibration*; the excitation motion of the nucleus, where the protons and neutrons make in phase collective motion, and the nucleus repeats expansion and contraction of the volume without changing the shape) and the study of astrophysical phenomena such as neutron stars and supernova explosions. Also, the pressure has to be 0 at the standard density, i.e., at the saturation density $\rho = \rho_0 \sim 0.17\,\text{fm}^{-3}$, since nuclei are isolated systems. The parameters of the Skyrme force are determined so as to satisfy these conditions and to reproduce the radii and the binding energies of some of the representative nuclei such as ^{16}O and ^{208}Pb, and several different parameter sets have been proposed depending on the experimental data, which have been taken into account. Table 6.1 lists some examples.

SIII [5] is one of the frequently used parameter sets, where the values of parameters have been determined so as to reproduce the experimental data of the binding energy and the radius of spherical nuclei ^{16}O, ^{40}Ca, ^{48}Ca, ^{56}Ni, ^{90}Zr, ^{140}Ce, and ^{208}Pb. SkM* [6] determines the parameters to reproduce the fission barrier of ^{240}Pu as well. Alternatively, SLy4 [7] determines the parameters so as to be suitable for the study of neutron-rich nuclei and the neutron matter by paying attention to the isospin degree of freedom. Table 6.1 shows also the properties of the nuclear matter calculated with each parameter set. Since the nuclear incompressibility K is suggested to be about 200 MeV from the excitation energy of the compressional vibration, i.e., the monopole vibration, the size of the neutron star and also supernova explosion as stated

Table 6.1 The values of parameters of the Skyrme force and the properties of the nuclear matter obtained by using them. ρ_0 is the baryon number density, E/A is the binding energy per nucleon, K is nuclear incompressibility and m_N^* is the effective mass of nucleon in the central region

	SIII	SkM*	SLy4
t_0 (MeV fm^3)	−1128.75	−2645.00	−2488.91
t_1 (MeV fm^5)	395.0	410.00	486.82
t_2 (MeV fm^5)	−95.0	−135.00	−546.39
t_3 (MeV fm$^{3+\alpha}$)	14000.0	15595.00	13777.0
x_0	0.45	0.09	0.834
x_1	0.00	0.00	−0.344
x_2	0.00	0.00	−1.000
x_3	1.00	0.00	1.354
α	1	1/6	1/6
W_0 (MeV fm^5)	120.0	130.0	123.0
Properties of the nuclear matter			
ρ_0 (fm^{-3})	0.145	0.1603	0.160
E/A (MeV)	−15.87	−15.78	−15.969
m_N^*/M_N	0.76	0.79	0.70
K (MeV)	356	216.7	229.9

before, SkM* and SLy4 which give soft equation of state (EOS) look reasonable, while SIII gives too hard EOS.[10]

6.2.4 Comparison with the Experimental Data

The Hartree–Fock calculations assuming the Skyrme force well reproduce the experimental data of fundamental observables such as the nuclear radii and the charge distributions. As an example of the comparison with the experimental data, Fig. 6.1 compares the energy levels for nucleons obtained by the Hartree–Fock calculations using SkM* and SLy4 forces with the experimental data. The figure shows that the Hartree–Fock calculations with the Skyrme force well reproduce the features of the experimental data for the single-particle levels.

However, if one examines closely, it is noticed that the Hartree–Fock calculations give too large level spacing, i.e., smaller level density, compared with the experimental data. That trend is more significant for the SLy4 force than for the SkM* force, especially, for neutrons. The two forces give similar results for protons. In order to see the origin of this problem and its feature, we compare in Fig. 6.2 the effective mass and potentials of nucleons obtained for each calculation. The figure shows that

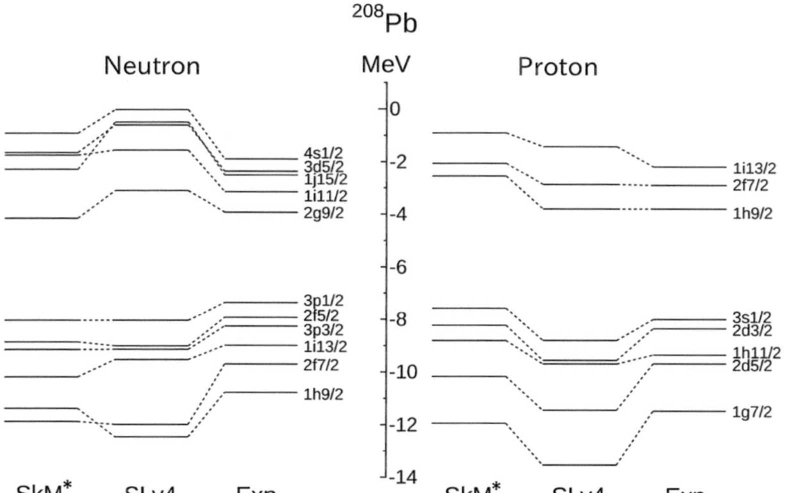

Fig. 6.1 Nucleon energy levels near the Fermi surface. Comparison of the experimental data with the Hartree–Fock calculations using the SkM* and SLy4 forces

---

[10]It is known that the fragmentation reactions observed in medium energy nucleus–nucleus collisions also reflect the hard or soft property of the EOS [8].

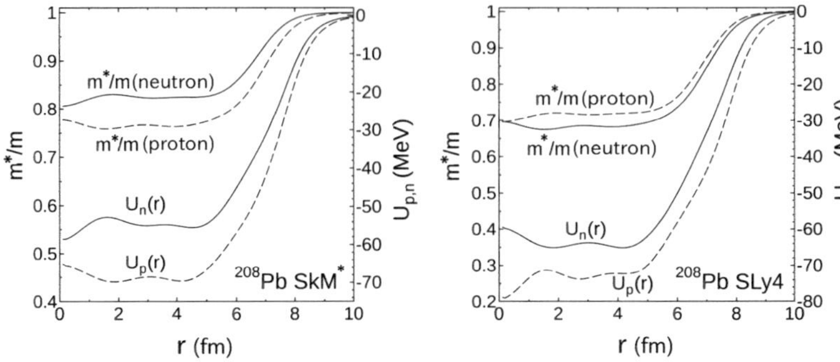

Fig. 6.2 The effective mass and the potential for nucleons in ^{208}Pb for SkM* (*left*) and for SLy4 (*right*)

the effective mass in the Hartree–Fock calculations is significantly smaller than the bare mass. Moreover, it shows that the effective mass for neutrons is considerably smaller for SLy4 force than that for SkM* force. On the other hand, the effective mass for protons is nearly the same for two forces. Since the level spacing in the square well potential increases with decreasing mass (see Eq. (5.10)), the above mentioned problem of the Hartree–Fock calculations concerning the nucleon level density could be sought one of its origins in the incorrect treatment of the effective mass, i.e., in that they ignore the *e*-mass, which we discuss later.

6.2.5 *The Equation of State, Saturation, Spinodal Line, Surface Thickness*

6.2.5.1 The Equation of State and Saturation

The equation of state is one of the fundamental properties of a physical system and is often expressed by representing the pressure as a function of density. One can analytically write down the equation of state if one assumes the Skyrme force. The equation of state plays an important role in various problems such as the nuclear saturation, compressional vibrational excitation, nuclear fragmentation reactions[11] and astrophysical phenomena such as neutron stars and supernova explosions. Here, we discuss the equation of state for the nuclear matter with $N = Z$ based on the Skyrme force. We ignore the Coulomb force.

[11]Nuclear reactions observed in the nucleus-nucleus collisions at medium energies of several tens MeV per nucleon, where the colliding nuclei dissociate into nucleons and fragments with various small mass numbers.

The pressure at finite temperature is given by the derivative of the *Helmholtz* free energy $F = E - TS$:

$$P = -\left(\frac{\partial}{\partial V}F\right)_T . \tag{6.49}$$

Since the internal energy E is given by the sum of the interaction energy E_{int} and the kinetic energy E_{kin}, we divide the pressure into the part originating from the interaction energy and that from the kinetic energy

$$P = P_V + P_T . \tag{6.50}$$

(1) Pressure due to the Interaction Energy: P_V We estimate the pressure originating from the interaction by replacing the total energy E with the interaction energy E_{int} in the formula (6.48) at zero temperature. Corresponding to Eq. (6.46), let us assume that the interaction energy per nucleon is given by

$$\frac{E_{\text{int}}}{A} = -C_1\rho + C_2\rho^{1+\alpha} . \tag{6.51}$$

We have changed the power in the second term on the right-hand side of Eq. (6.51) by considering the general case when the nuclear force is given by Eq. (6.29) instead of Eq. (6.28), although Eq. (6.46) corresponds to the case when $\alpha = 1$.[12] The pressure due to the interaction is then given by

$$P_V = -C_1\rho^2 + (1+\alpha)C_2\rho^{2+\alpha} . \tag{6.52}$$

(2) Pressure due to the Kinetic Energy: P_T In order to obtain the pressure due to the kinetic energy, we first calculate the kinetic energy per unit volume based on

$$\frac{E_{\text{kin}}}{V} = \frac{4}{(2\pi)^3}\int d^3k \frac{\hbar^2k^2}{2M_N}\frac{1}{\exp[(\frac{\hbar^2k^2}{2M_N} - \lambda)/T] + 1} , \tag{6.53}$$

then obtain P_T by inserting E_{kin} in the place of E in Bernoulli's formula [9, 10],

$$PV = \frac{2}{3}E , \tag{6.54}$$

which holds for both classical and quantum mechanical ideal gases irrespective of the statistical property. λ in Eq. (6.53) is the Fermi level or Fermi potential.
(2a) The Degenerate Limit: the Case of $T = 0$ When $T = 0$, one can easily show that

---

[12]The prescription which takes into account the finite range property of the nuclear force through the t_1 and t_2 terms following Skyrme would need to be reconsidered when we use α smaller than 2/3 in discussing the saturation, since the last term on the right-hand side of Eq. (6.46) is proportional to $\rho^{5/3}$.

$$P_T = \frac{2}{5}\rho\varepsilon_{\mathrm{F}} = \frac{2}{5}\rho\frac{\hbar^2 k_{\mathrm{F}}^2}{2M_N} \propto \rho^{5/3} , \tag{6.55}$$

where ε_{F} is the Fermi level or Fermi energy at $T = 0$.

(2b) The Case of Non-degenerate System at High Temperature: At high temperatures, one can approximate the Fermi distribution by the Boltzmann distribution. We then have

$$\begin{aligned} \frac{E_{\mathrm{kin}}}{V} &\sim \frac{4}{(2\pi)^3}4\pi\frac{\hbar^2}{2M_N}\int_0^\infty \mathrm{d}k k^4 \mathrm{e}^{-(\frac{\hbar^2k^2}{2M_N}-\lambda)/T} \\ &\sim \frac{4}{(2\pi)^3}4\pi\frac{1}{2M_N}\frac{1}{\hbar^3}\mathrm{e}^{\lambda/T}\frac{3}{8}\sqrt{\pi}(2M_N T)^{5/2} , \end{aligned} \tag{6.56}$$

$$\begin{aligned} \frac{N}{V} &\sim \frac{4}{(2\pi)^3}4\pi\int_0^\infty \mathrm{d}k k^2 \mathrm{e}^{-(\frac{\hbar^2k^2}{2M_N}-\lambda)/T} \\ &\sim \frac{4}{(2\pi)^3}4\pi\frac{1}{\hbar^3}\mathrm{e}^{\lambda/T}\frac{1}{4}\sqrt{\pi}(2M_N T)^{3/2} , \end{aligned} \tag{6.57}$$

and obtain

$$\frac{E_{\mathrm{kin}}}{V} = \frac{3}{2}\rho T . \tag{6.58}$$

Using Bernoulli's formula, the pressure due to the kinetic energy is finally given by

$$P_T = \rho T , \tag{6.59}$$

which is well known for the ideal gas.

(2c) The Case of Fermi Gas at Low Temperature: The kinetic energy at temperature T is given by

$$E_{\mathrm{kin}}(T) = E_{\mathrm{kin}}(T=0) + \frac{1}{4}\frac{\pi^2}{\varepsilon_F}AT^2 \tag{6.60}$$

when the temperature T satisfies the condition $T \ll \lambda$. The pressure due to the kinetic energy is therefore given by

$$P = \frac{2}{5}\rho\varepsilon_{\mathrm{F}}\left[1 + \frac{5}{12}\pi^2\left(\frac{T}{\varepsilon_{\mathrm{F}}}\right)^2\right] \tag{6.61}$$

by using Eq. (6.54).

Let us guess the behaviour of the equation of state based on these results. We first consider the case when the temperature is 0. Equations (6.52) and (6.55) suggest that the effect of kinetic energy, i.e., the degeneracy pressure, dominates in the region of low density and the pressure is positive and increases with density. Eventually, the pressure due to attractive two-body interaction becomes to overcome and the

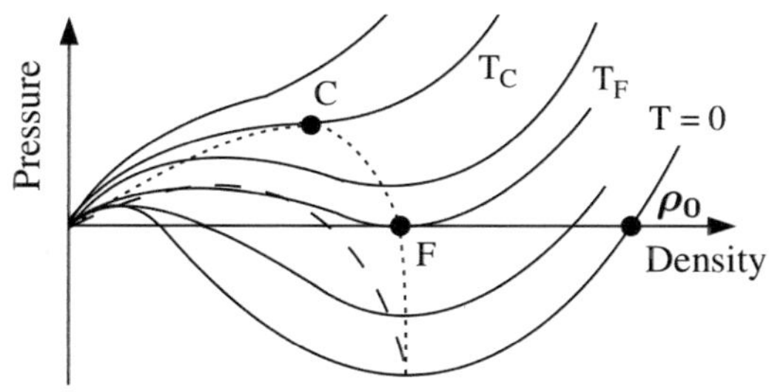

Fig. 6.3 The conceptual illustration (*solid lines*) of the equation of state of nuclear matter (*isothermal lines*). The *dotted* and the *dashed lines* are the isothermal and isentropic spinodal lines, which connect the points where $(\frac{\partial P}{\partial \rho})|_T = 0$ and $(\frac{\partial P}{\partial \rho})|_S = 0$, respectively (taken from [13])

pressure decreases with density and turns to be negative. For even higher density, the pressure turns to increase with density because of the three-body force or the density-dependent force, i.e., the second term on the right-hand side of Eq. (6.52), which have been introduced to guarantee the saturation, and becomes 0 at the normal density $\rho_0 \sim 0.17\,\text{fm}^{-3}$, for which the nucleus exists as a stable isolated system, and then continues to increase. However, it is necessary to refine the theory by taking into account the effect of repulsive core and various effects, which are ignored in Eq. (6.55), in order to discuss the region of very high density.[13] As Eq. (6.61) shows, the pressure increases with increasing temperature and eventually converges to the equation of state for the classical ideal gas given by Eq. (6.59).

Figure 6.3 shows the equation of state drawn based on these considerations. Roughly speaking, the pressure behaves like a cubic function of the density at low temperatures, and becomes a monotonically increasing function of density at temperatures higher than the critical temperature T_C.

Finally, we express the equation of state in the form of the energy density $\mathscr{H}(\rho)$. To that end, we remark that the incompressibility defined by Eq. (6.47) can be expressed also as

$$K = 9\rho^2 \frac{\partial^2}{\partial \rho^2} \left(\frac{E}{A} \right) \bigg|_{\rho=\rho_0}, \tag{6.62}$$

[13]The equation of state at high density is one of the crucial issues in the problems of the neutron stars and the supernova explosions. However, it is not precisely known yet. Concerning neutron stars, it has been reported that the pressure gets lower if one takes into account baryons other than protons and neutrons, that is, hyperons. This is natural, because for a given baryon density the degeneracy pressure must decrease if more different kinds of baryon exist. A problem is that the softening of the equation of state by the presence of hyperons contradicts the experimental observation of massive neutron stars with the mass as large as about two times the solar mass. This problem is often referred to as the hyperon puzzle or the hyperon crisis. Various possibilities such as the recovery of the stiffness of the equation of state at the high density side due to the repulsive effect of a three-body force working universally among all baryons [11] are now under investigation to resolve this puzzle and to establish a theory which is compatible with the data of both neutron stars and hypernuclei [12]. In this connection, we note that the density at deep inside of neutron stars is thought to be significantly higher than the normal nuclear density, although the actual value is not known and varies over a wide range between 4 and 8 times the normal nuclear density depending on the theoretical models.

where ρ_0 is the density of nucleus in the ground state. Equation (6.62) shows that the incompressibility corresponds to the curvature in the case when we expand the energy per nucleon as a function of the density ρ in the vicinity of ρ_0. Hence the $\mathscr{H}(\rho)$ can be expressed as

$$\mathscr{H}(\rho) = \frac{E}{A}\rho = \rho\left[\frac{K}{18}\left(1 - \frac{\rho}{\rho_0}\right)^2 + \varepsilon_0\right] + \frac{1}{2}B(\nabla\rho)^2 , \tag{6.63}$$

with

$$B = \frac{1}{32}[9t_1 - (5 + 4x_2)t_2] + \delta B_W , \tag{6.64}$$

$$\delta B_W = \frac{1}{18}\frac{\hbar^2}{2M_N}\frac{1}{\rho} , \tag{6.65}$$

where ε_0 is the binding energy per nucleon $\sim -15\,\text{MeV}$. We used the extended Thomas–Fermi approximation,

$$\tau(\rho) = \tau_{\text{TF}}(\rho) + \tau_2(\rho) = \frac{3}{5}(3\pi^2)^{2/3}\rho^{5/3} + \tau_2(\rho) , \tag{6.66}$$

$$\tau_2(\rho) = \frac{1}{36}\frac{(\nabla\rho)^2}{\rho} + \frac{1}{3}\Delta\rho , \tag{6.67}$$

for the kinetic energy, since we have taken into account the $(\nabla\rho)^2$ term, which is related to the surface energy, in the interaction energy. The first term on the right-hand side of Eq. (6.67) is called the *Weizsäcker correction*. The last term δB_W in Eq. (6.64) is the term representing the *Weizsäcker correction*.

6.2.5.2 Liquid–Gas Phase Transition, Spinodal Line

The behaviour shown in Fig. 6.3 resembles the van der Waals equation of state. Like van der Waals, the problem of liquid–gas phase transition in nuclei, which are self-bound systems with finite number of constituents interacting through strong interactions, has been discussed based on this equation of state in connection with the fragmentation reactions in medium energy nucleus–nucleus collisions [14].

The nucleus once becomes a state of high temperature and high density after the collision at high energy, then cools down nearly isentropically with expansion, and eventually is observed as fragments of various masses including nuclei with relatively large mass number. In this procedure, the region of negative nuclear incompressibility plays an important role as the region of mechanical instability. This can be understood in the following way.

Let us consider the Helmholtz free energy density $\mathscr{F} \equiv \frac{F}{A}\rho$ corresponding to the Hamiltonian density $\mathscr{H} \equiv \frac{E}{A}\rho$ introduced by Eq. (6.44). By taking the second

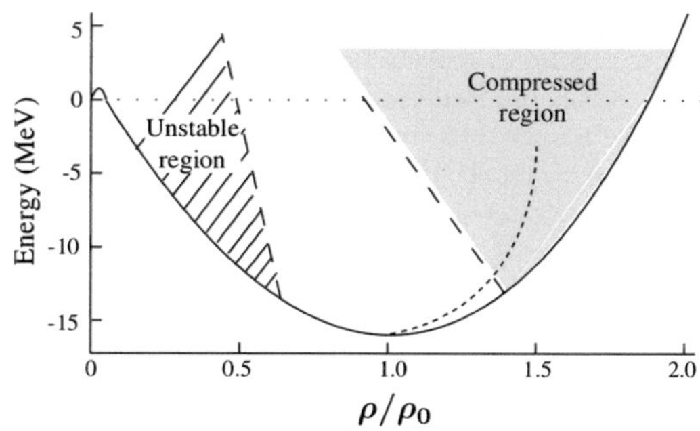

Fig. 6.4 The unstable region (the *oblique lines*: the boundary is called the *spinodal line*) and the compressed region (*shaded* region) which is expected to lead to the fragmentation reaction in heavy-ion collisions. The *dotted line* is an example of the trajectory in heavy-ion collisions. The figure is after D.K. Scott

derivative of $\mathscr{F}$ with respect to the density, we have

$$\frac{\partial^2 \mathscr{F}}{\partial \rho^2} = 2\frac{\partial}{\partial \rho}\left(\frac{F}{A}\right) + \rho\frac{\partial^2}{\partial \rho^2}\left(\frac{F}{A}\right) . \tag{6.68}$$

On the other hand, from the definition of pressure (6.49), we have

$$\frac{\partial P}{\partial \rho} = 2\rho\frac{\partial}{\partial \rho}\left(\frac{F}{A}\right) + \rho^2\frac{\partial^2}{\partial \rho^2}\left(\frac{F}{A}\right) . \tag{6.69}$$

Equations (6.68) and (6.69) lead to the relationship,

$$\frac{\partial^2 \mathscr{F}}{\partial \rho^2} = \frac{1}{\rho}\frac{\partial P}{\partial \rho} . \tag{6.70}$$

This relationship shows that the negative gradient of the pressure, i.e., negative incompressibility, is equivalent to that the second derivative of the free energy density with respect to density is negative.

$$\frac{\partial P}{\partial \rho} < 0 \quad \leftrightarrow \quad \frac{\partial^2 \mathscr{F}}{\partial \rho^2} < 0 . \tag{6.71}$$

The condition $\frac{\partial^2 \mathscr{F}}{\partial \rho^2} < 0$ means that the system is unstable with respect to the density fluctuation, i.e., the *mechanical instability*. It is because the change of the free energy density due to a local fluctuation of the density is given by

$$\begin{aligned}\Delta\mathscr{F} &= \frac{1}{2}[\mathscr{F}(\rho+\Delta\rho) + \mathscr{F}(\rho-\Delta\rho)] - \mathscr{F}(\rho)\\ &= \frac{1}{2}\frac{\partial^2 \mathscr{F}}{\partial \rho^2}(\Delta\rho)^2 ,\end{aligned} \tag{6.72}$$

and it becomes $\Delta\mathscr{F} < 0$ if $\frac{\partial^2\mathscr{F}}{\partial\rho^2} < 0$. The former, i.e., $\Delta\mathscr{F} < 0$, means that the free energy gets lower in the state with density fluctuation.

Figure 6.4 shows the mechanically unstable region, whose boundary is called the *spinodal line*, and the region which is compressed immediately after the collision and is expected to be required in order to reach instability region after cooling in the nucleus–nucleus collision at medium energies.

6.2.5.3 Surface Diffuseness

The surface property of nuclei is one of the intriguing physical quantities concerning nuclear structure as well as nuclear reactions. Equation (6.63) suggests that the parameter B given by Eq. (6.64) is related to the surface thickness. In fact, using Eq. (6.63) one can show [15] that the density distribution of nuclei is given by

$$\rho = \rho_0 \tanh^2\left(\frac{r-R}{2b}\right), \tag{6.73}$$

$$b^2 = 9\frac{B}{K}\rho_0 . \tag{6.74}$$

If we approximate the functional form of Eq. (6.73) by the standard Woods–Saxon type, then the surface diffuseness parameter a is given by $a \approx 2b/3$. Equations (6.74) and (6.64) suggest that the surface thickness of nuclei is governed by the nuclear incompressibility and the Weizsäcker term in the asymptotic surface region, where the density is extremely small, or by the range of nuclear force through the parameter $9t_1 - (5+4x_2)t_2$ in more inside region.

6.2.6 *Beyond the Hartree–Fock Calculations: Nucleon–Vibration Coupling; ω-Mass*

We remarked in Sect. 6.2.4 that the single-particle level density obtained by the Hartree–Fock calculations is too small compared with the experimental data, and suggested the possibility that one of the reasons of this failure is related to the effective mass.

The effective mass hitherto discussed within the framework of the Hartree–Fock theory arises from the antisymmetrization of nucleons. In revised theories which go beyond the Hartree–Fock approximation, the single-particle potential has an energy dependent correction term in addition to the potential $U_{\rm HF}$ in the Hartree–Fock theory due to, e.g., the coupling of nucleons to the surface vibration, which is one of the nuclear collective excitations. In consequence, the energy of the single-particle state is given by

$$e = \frac{p^2}{2m} + U_{\rm HF}(r, p(r)) + U_{\rm PVC}(r, e) . \tag{6.75}$$

The lower index PVC in the third term on the right-hand side means *particle–vibration coupling*. We denoted the bare nucleon mass M_N by m. Also, we represented the Hartree–Fock potential by an equivalent momentum dependent local potential in the spirit of the WKB approximation, although it is in general a non-local potential.

By taking the derivative of both sides of Eq. (6.75) with respect to p, and using the definition of the effective mass $\frac{1}{m^*} \equiv \frac{1}{p}\frac{\mathrm{d}e}{\mathrm{d}p}$, we obtain

$$\frac{m^*}{m} = \frac{m_k}{m}\frac{m_\omega}{m}, \tag{6.76}$$

$$\frac{m_k}{m} = \left(1 + \frac{m}{p}\frac{\partial U_{\rm HF}}{\partial p}\right)^{-1}, \tag{6.77}$$

$$\frac{m_\omega}{m} = 1 - \frac{\partial U_{\rm PVC}}{\partial e}, \tag{6.78}$$

where m_k stands for *k-mass* and m_ω for *ω-mass* or *e-mass*.

The ω-mass m_ω is larger than the bare mass m, and is concentrated in the surface region reflecting the coupling between nucleons and the surface vibrational motion [16]. Thanks to the ω-mass, the final effective mass becomes larger than the k-mass introduced in the Hartree–Fock approximation. Hence the agreement between the theoretical value and the experimental data concerning the single-particle level density is improved by taking into account the effect of particle–vibration coupling.

Incidentally, this is related to the fact that the parameter a,[14] which is called the level density parameter, takes the value $A/8\,\mathrm{MeV}^{-1}$ in the region of low excitation energy in contradiction to the expectation. The ω-mass approaches 1 with increasing excitation energy. Consequently, the level density parameter a is expected to approach the expected value $A/15\,\mathrm{MeV}^{-1}$ as the excitation energy increases. It has been confirmed that it is the case indeed from the study of, e.g., the excitation energy dependence of the evaporation spectrum of α particles from excited nuclei [18].

6.3 Relativistic Mean-Field Theory ($\sigma\omega\rho$ Model)

Theoretical studies which use the relativistic theory and introduce mesons to mediate the nuclear force have also been extensively developed to determine nuclear properties such as the binding energy and the shape of nuclei within the framework of the mean-field theory in parallel with the non-relativistic Hartree–Fock theory with the Skyrme force. The relativistic approach has the advantage that the spin–orbit force is

[14] If one considers the nucleus as a Fermi gas, then the level density ρ_L exponentially increases with the excitation energy E and can be roughly given by $\rho_L(E) \sim \exp(aE)^{1/2}$ (see [17] for details).

naturally introduced.[15,16] In this section, we learn the $\sigma\omega\rho$ model, which is a widely used relativistic mean-field theory[17] [20, 21].

6.3.1 Lagrangian

In the $\sigma\omega\rho$ model, the Lagrangian density is assumed to be given by[18]

$$\begin{aligned}\mathscr{L} = \bar{\psi}\left(\mathrm{i}\gamma_\mu\partial^\mu - M_N - g_\sigma\sigma - g_\omega\gamma_\mu\omega^\mu - g_\rho\gamma_\mu\tau_a\rho^{a\mu}\right)\psi \\ + \frac{1}{2}\partial_\mu\sigma\partial^\mu\sigma - U(\sigma) - \frac{1}{4}\Omega_{\mu\nu}\Omega^{\mu\nu} + \frac{1}{2}m_\omega^2\omega_\mu\omega^\mu - \frac{1}{4}R^a_{\mu\nu}R^{a\mu\nu} + \frac{1}{2}m_\rho^2\rho^a_\mu\rho^{a\mu} \\ - \frac{1}{4}F_{\mu\nu}F^{\mu\nu} - e\bar{\psi}\gamma_\mu\frac{1-\tau_3}{2}\psi A^\mu \end{aligned} \tag{6.79}$$

in terms of the field operators of nucleons ψ and mesons σ, ω, ρ.[19] $U(\sigma)$ is the potential energy term of the σ meson. We assume that it is given by

$$U(\sigma) = \frac{1}{2}m_\sigma^2\sigma^2 + \frac{1}{3}g_2\sigma^3 + \frac{1}{4}g_3\sigma^4 \,. \tag{6.80}$$

The second and third terms on the right-hand side of Eq. (6.80), which are often called non-linear terms, are the self-interaction terms of the σ mesons and have been introduced in order to reproduce the experimental data of the binding energy per particle and Fermi momentum of the nuclear matter as well as the surface energy and surface thickness of nuclei.[20]

---

[15] Another attractive point of the relativistic treatment is that one can reproduce the saturation property of nuclei within the framework of the mean-field theory. Although it is different from the relativistic mean-field theory which we describe in this section, in the case of the Brueckner–Hartree–Fock calculations using the nuclear force obtained from the experimental data of the nucleon–nucleon scattering, the relativistic treatment can simultaneously reproduce the Fermi momentum and the binding energy per nucleon of the nuclear matter. On the other hand, if one uses non-relativistic theories, there exists a correlation between the Fermi momentum and the binding energy per nucleon, called the Coester line, and one cannot simultaneously reproduce both of them in the way to match with the mass formula irrespective of the nuclear force.

[16] The relativistic approach reproduces the experimental data of the polarization phenomena in the proton–nucleus scattering much better than the non-relativistic approach.

[17] The π mesons are not introduced, because one tries to establish the theory under the mean-field approximation as will be described later. Recently, the theory which explicitly includes the π mesons is also being developed (see, e.g., [19]).

[18] In this section, we use the natural units, where one sets $\hbar = 1, c = 1$.

[19] μ and ν denote the four-dimensional time–space components (see Appendix A.10), while a the component in the isospin space.

[20] As Eq. (6.74) shows, the surface thickness of nuclei is intimately related to the nuclear incompressibility. The incompressibility K takes a large value of about 500 MeV if the non-linear terms are not introduced, and the surface properties of nuclei are not well reproduced. The value $K \sim 500$ MeV is too large also in comparison with about 210 MeV, which is suggested from the experimental value of the excitation energy of the breathing mode.

The field tensors are given by

$$\Omega^{\mu\nu} = \partial^\mu \omega^\nu - \partial^\nu \omega^\mu \,, \tag{6.81}$$

$$R^{a\mu\nu} = \partial^\mu \rho^{a\nu} - \partial^\nu \rho^{a\mu} - g_\rho \varepsilon^{abc} \rho^{b\mu} \rho^{c\nu} \,, \tag{6.82}$$

$$F^{\mu\nu} = \partial^\mu A^\nu - \partial^\nu A^\mu \,, \tag{6.83}$$

where ε^{abc} is the Levi-Civita symbol.

6.3.2 Field Equations

Following the Euler–Lagrange equation,

$$\frac{\partial}{\partial x^\mu}\left[\frac{\partial \mathscr{L}}{\partial(\partial q_i/\partial x^\mu)}\right] - \frac{\partial \mathscr{L}}{\partial q_i} = 0 \,, \tag{6.84}$$

where q_i is a field operator, the equations for the field operators are obtained as

$$\left[\gamma^\mu \left(-\mathrm{i}\partial_\mu + g_\omega \omega_\mu + g_\rho \tau_a \rho^a_\mu + e\frac{1-\tau_3}{2} A_\mu\right) + (M_N + g_\sigma \sigma)\right] \psi = 0 \,, \tag{6.85}$$

$$\partial^\nu \partial_\nu \sigma + \frac{\mathrm{d}U(\sigma)}{\mathrm{d}\sigma} = -g_\sigma \rho_S \,, \tag{6.86}$$

$$\left(\partial^\nu \partial_\nu + m^2_\omega\right) \omega^\mu = g_\omega j^\mu \,, \tag{6.87}$$

$$\left(\partial^\nu \partial_\nu + m^2_\rho\right) \rho^{a\mu} = g_\rho j^{a\mu} \,, \tag{6.88}$$

$$\partial^\nu \partial_\nu A^\mu = e j^\mu_\mathrm{p} \,. \tag{6.89}$$

Equation (6.85) is the Dirac equation, Eqs. (6.86)–(6.88) are the Klein-Gordon equations and Eq. (6.89) is the d'Alembert equation. The source terms for each meson and electromagnetic field are given by

$$\rho_S(x) = \sum_{i=1}^{A} \bar{\psi}_i(x) \psi_i(x) \,, \tag{6.90}$$

$$j^\mu(x) = \sum_{i=1}^{A} \bar{\psi}_i(x) \gamma^\mu \psi_i(x) \,, \tag{6.91}$$

$$j^{a\mu}(x) = \sum_{i=1}^{A} \bar{\psi}_i(x) \gamma^\mu \tau_a \psi_i(x) \,, \tag{6.92}$$

$$j^\mu_\mathrm{p}(x) = \sum_{i=1}^{A} \bar{\psi}_i(x) \gamma^\mu \frac{1-\tau_3}{2} \psi_i(x) \,. \tag{6.93}$$

6.3.3 The Mean-Field Theory

Equations (6.85)–(6.89) are the equations for the field operators, so that it is not easy to solve them as they are. One therefore usually introduces the approximation that replaces the operators by their expectation values, which are c-numbers,

$$\sigma \rightarrow \langle\sigma\rangle = \sigma(\mathbf{r}) \tag{6.94}$$

$$\omega^\mu \rightarrow \langle\omega^\mu\rangle = \omega^\mu(\mathbf{r}). \tag{6.95}$$

The resulting theory is called the mean-field theory.

Consequently, we obtain

$$\left[\boldsymbol{\alpha}\cdot(-\mathrm{i}\nabla - \mathbf{V}(\mathbf{r})) + \beta M^*(\mathbf{r}) + V(\mathbf{r})\right]\psi_i(\mathbf{r}) = \varepsilon_i\psi_i(\mathbf{r}) , \tag{6.96}$$

$$M^*(\mathbf{r}) = M_N + g_\sigma\sigma(\mathbf{r}) = M_N + S(\mathbf{r}) , \tag{6.97}$$

$$V(\mathbf{r}) = g_\omega\omega^0(\mathbf{r}) + g_\rho\tau_a\rho^{a0}(\mathbf{r}) + e\frac{1-\tau_3}{2}A^0(\mathbf{r}) , \tag{6.98}$$

$$\mathbf{V}(\mathbf{r}) = g_\omega\boldsymbol{\omega}(\mathbf{r}) + g_\rho\tau_a\boldsymbol{\rho}^a(\mathbf{r}) + e\frac{1-\tau_3}{2}\mathbf{A}(\mathbf{r}) , \tag{6.99}$$

for the nucleon field, and

$$(-\Delta + m_\sigma^2)\sigma(\mathbf{r}) = -g_\sigma\rho_S(\mathbf{r}) - g_2\sigma^2(\mathbf{r}) - g_3\sigma^3(\mathbf{r}) , \tag{6.100}$$

$$(-\Delta + m_\omega^2)\omega^\mu(\mathbf{r}) = g_\omega j^\mu(\mathbf{r}) , \tag{6.101}$$

$$(-\Delta + m_\rho^2)\rho^{a\mu}(\mathbf{r}) = g_\rho j^{a\mu}(\mathbf{r}) , \tag{6.102}$$

$$-\Delta A^\mu(\mathbf{r}) = e j_p^\mu(\mathbf{r}) , \tag{6.103}$$

for the meson and electromagnetic fields, as the equations for the stationary mean fields. See Appendix A.10 for the notations and the variables that appear in the above equations.

Equations (6.96)–(6.99) show that a scalar and a vector fields act on nucleons due to the interaction with the meson and the electromagnetic fields, and that the effect of the σ meson appears as a space dependent effective mass.

The relativistic mean-field theory usually assumes the time-reversal symmetry and is applied to even–even nuclei. In this case, the total current density becomes zero, so that the spatial components of the vector potential becomes zero ($\mathbf{V}(\mathbf{r}) = 0$) as seen from Eqs. (6.101)–(6.103).

6.3.4 Prologue to How to Solve the Mean-Field Equations

Here, we briefly explain the practical method to solve the equations of the relativistic mean-field theory. For spherical nuclei [22, 23], we express the wave function of a nucleon in the ith state as

$$\psi_i(\mathbf{r}, s, t) = \begin{pmatrix} \mathrm{i}\frac{G_i^{\ell j}(r)}{r}\mathscr{Y}_{jm}^{\ell}(\theta, \phi) \\ \frac{F_i^{\ell j}(r)}{r}(\boldsymbol{\sigma}\cdot\hat{\mathbf{r}})\mathscr{Y}_{jm}^{\ell}(\theta, \phi) \end{pmatrix} \chi_i(t) , \tag{6.104}$$

by noting that the good quantum number is not $(\ell, s = \frac{1}{2}, m_\ell, m_s)$, but $(\ell, s, j, m_j = m)$ due to the spin–orbit interaction, which we will describe in Sect. 6.3.5. The $\mathscr{Y}_{jm}^{\ell}$ are the spherical spinors representing the angular momentum states, and are defined by

$$\mathscr{Y}_{jm}^{\ell} = \sum_{m_\ell m_s} \langle \ell \frac{1}{2} m_\ell m_s | jm \rangle Y_{\ell m_\ell} | \frac{1}{2} m_s \rangle . \tag{6.105}$$

The $\chi(t)$ are the wave functions in the isospin space.

By inserting Eq. (6.104) into Eq. (6.96), we obtain the following coupled differential equations for the radial wave functions $G_i^{\ell j}(r)$ and $F_i^{\ell j}(r)$,

$$\varepsilon_i G_i^{\ell j}(r) = \left(-\frac{\partial}{\partial r} + \frac{\kappa_i}{r}\right) F_i^{\ell j}(r) + [M_N + S(r) + V(r)] G_i^{\ell j}(r) , \tag{6.106}$$

$$\varepsilon_i F_i^{\ell j}(r) = \left(+\frac{\partial}{\partial r} + \frac{\kappa_i}{r}\right) G_i^{\ell j}(r) - [M_N + S(r) - V(r)] F_i^{\ell j}(r) , \tag{6.107}$$

where κ is defined by

$$\kappa = \begin{cases} -(j+1/2) & \text{for } j = \ell + \frac{1}{2} \\ +(j+1/2) & \text{for } j = \ell - \frac{1}{2} . \end{cases} \tag{6.108}$$

On the other hand, the meson and the electromagnetic fields obey the radial Laplace equations, which are given by

$$\left(-\frac{\partial^2}{\partial r^2} - \frac{2}{r} + m_\Phi^2\right) \Phi = S_\Phi(r) , \tag{6.109}$$

where

$$S_\Phi(r) = \begin{cases} -g_\sigma \rho_S(r) - g_2\sigma^2(r) - g_3\sigma^3(r) & \text{for the } \sigma \text{ field,} \\ g_\omega \rho_B(r) & \text{for the } \omega \text{ field,} \\ g_\rho \rho_3(r) & \text{for the } \rho \text{ field,} \\ e\rho_C(r) & \text{for the Coulomb field.} \end{cases} \tag{6.110}$$

$$\begin{cases} 4\pi r^2 \rho_S(r) = \sum_{i=1}^{A}(|G_i(r)|^2 - |F_i(r)|^2) \ , \\ 4\pi r^2 \rho_B(r) = \sum_{i=1}^{A}(|G_i(r)|^2 + |F_i(r)|^2) \ , \\ 4\pi r^2 \rho_3(r) = \sum_{n=1}^{N}(|G_n(r)|^2 + |F_n(r)|^2) - \sum_{p=1}^{Z}(|G_p(r)|^2 + |F_p(r)|^2) \ , \\ 4\pi r^2 \rho_C(r) = \sum_{p=1}^{Z}(|G_p(r)|^2 + |F_p(r)|^2) \ . \end{cases} \tag{6.111}$$

The m_Φ is taken to be the mass of the corresponding meson if Φ is either σ or ω or ρ mesons, and zero if Φ is photon. One then solves these coupled equations self-consistently by an iteration method with either the expansion method using the harmonic oscillator wave functions as the expansion basis or with the direct integration method by discretizing the space into small lattices.

We describe actual applications in Sect. 7.6.2 by taking the study of nuclear shape as an example after we learn pairing correlations.

6.3.5 Non-relativistic Approximation and the Spin–Orbit Coupling

We learnt in Sect. 5.2.5 that the spin–orbit coupling plays a crucial role in determining the magic numbers and the stability of nuclei. Here, we introduce the non-relativistic approximation by using the *Tani–Foldy–Wouthuysen transformation* (TFW transformation), and obtain the information on the one-body spin–orbit coupling from the relativistic mean-field theory.

The TFW transformation gradually removes the odd power terms of α which admix the small and large components from the equation by repeating the unitary transformation which makes the small components, i.e., the third and fourth components, contained in the relativistic wave function small, and thus increases the accuracy of the approximation in the power series of the inverse of the mass [22, 24].

By performing the TFW transformation on Eq. (6.96), we obtain the following Hamiltonian for the spin–orbit coupling,

$$\hat{H}_{\ell s} = \frac{1}{2}\frac{1}{M_N^2}\left(\frac{1}{r}\frac{\partial V}{\partial r} - \beta\frac{1}{r}\frac{\partial S}{\partial r}\right)\boldsymbol{\ell}\cdot\mathbf{s} \tag{6.112}$$

$$\sim \frac{1}{2}\frac{1}{M_N^2}\left[\frac{1}{r}\frac{\partial}{\partial r}(V(r) - S(r))\right]\boldsymbol{\ell}\cdot\mathbf{s} \ . \tag{6.113}$$

Equation (6.113) is the equation for the large components, i.e., the first and second components, of the Dirac spinor.[21, 22]

[21] Note that the central force is governed by $V + S$ in contrast to the $\boldsymbol{\ell}\cdot\mathbf{s}$ force (see Eq. (6.106)).

[22] The case of electrons in atoms corresponds to the case when there exists only the V term. Thanks to the S term, the sign of the spin–orbit potential for nucleons inside a nucleus becomes opposite to that for the electrons in atoms [25].

If we express the $V(r) - S(r)$ in terms of the meson fields by using Eqs. (6.97) and (6.98), and determine each meson field by approximating the left-hand side of Eq. (6.109) with the mass term, then we have

$$V(r) - S(r) \sim \frac{1}{m_\sigma^2} g_\sigma^2 \rho_S + \frac{1}{m_\omega^2} g_\omega^2 \rho_B + \frac{1}{m_\rho^2} g_\rho^2 \tau_3 \rho_3 \tag{6.114}$$

$$\approx \frac{1}{m_\sigma^2} g_\sigma^2 \rho_B + \frac{1}{m_\omega^2} g_\omega^2 \rho_B + \frac{1}{m_\rho^2} g_\rho^2 (\rho_E - \rho_{NE}) , \tag{6.115}$$

where ρ_B is the baryon density (nucleon density), ρ_E and ρ_{NE} are the densities of nucleons which have the equal or non-equal isospin to that of the nucleon under consideration, respectively. In transforming from Eqs. (6.114) to (6.115), we remarked Eq. (6.111) and assumed that the main components of the Dirac spinor are much larger than the small components ($|G| \gg |F|$).

Since the nucleon density decreases with r, the results shown in Eqs. (6.113) and (6.115) match the attractive nature of the spin–orbit coupling, i.e., that v_{LS} is negative as we learnt phenomenologically in Chap. 5. Also, they show that the spin–orbit coupling is strong in the surface region where the density variation is large. Furthermore, Eq. (6.115) suggests that the spin–orbit coupling depends on the isospin due to the spin–orbit coupling stemming from the coupling to ρ mesons, hence that it is necessary to revise the isospin-independent spin-dependent term (W_0 term) assumed in the original Skyrme force. This effect is important, e.g., in understanding the isotope shift of nuclear radius over a wide range across a magic number.[23]

The spin–orbit coupling is expected to have different effects in stable nuclei and in unstable nuclei in the vicinity of the limit of existence through the difference of the surface properties and the isospin dependence. The exploration of such effects will be an interesting subject.

6.3.6 Parameter Sets

Table 6.2 shows several parameter sets which have been used for actual calculations. The common notation NL stands for non-linear.

[23] Sharma [26] succeeded in explaining the kink phenomenon which the radii of Pb isotopes show when they cross the neutron magic number $N = 126$ by replacing the W_0 term in Eq. (6.25) with $\mathrm{i}W_0(1 + x_w P_\tau)(\boldsymbol{\sigma}_1 + \boldsymbol{\sigma}_2) \cdot \mathbf{k}_L \times \delta(\mathbf{r}_1 - \mathbf{r}_2)\mathbf{k}_R$.

Table 6.2 Input parameter sets for the relativistic mean field calculations and the nuclear matter properties obtained by them

	NL1	NL2	NL-SH	NL3
m_N (MeV)	938.000	938.0	939.0	939.0
m_σ (MeV)	492.250	504.89	526.059	508.194
m_ω (MeV)	795.360	780.0	783.00	782.501
m_ρ (MeV)	763.000	763.0	763.00	763.000
g_σ	10.138	9.111	10.444	10.217
g_ω	13.285	11.493	12.945	12.868
g_ρ	4.976	5.507	4.383	4.474
g_2 (fm^{-1})	−12.172	−2.304	−6.9099	−10.431
g_3	−36.265	13.783	−15.8377	−28.885
Nuclear matter properties				
ρ_0 (fm^{-3})	0.1542	0.146	0.146	0.148
E/A (MeV)	−16.43	−17.016	−16.328	−16.299
m_N^*/M_N	0.571	0.670	0.60	0.60
K (MeV)	212	399.2	354.95	271.76

6.4 Pairing Correlation

6.4.1 Overview

We learnt in Sect. 2.3.1 that the binding energy per nucleon for even–even nuclei is systematically larger than that for odd nuclei and that for odd–odd nuclei. Also, by using $^{210}_{82}$Pb as an example, we have shown in Sect. 5.8 that the level structure, i.e., the spin, parity and the level spacing, of the ground state and low-lying excited states of doubly-magic ±2 nuclei can be explained by taking into account the short-range residual interaction, which is ignored in the mean field, i.e., in the zeroth order shell model.

In general, detailed examination of the role of residual interactions which are not included in the mean field is necessary to understand the nuclear properties in detail. Of the residual interactions the especially important is the pairing correlation. The ground state and low-lying excited states of nuclei become superfluid due to the pairing interaction. As stated in the beginning of this section, the pairing correlation strongly affects the binding energy of nuclei as well as the low-lying level structure, i.e., the spin, parity and the distribution of the excitation energies of low-lying states. Furthermore, as we will learn later, it plays an important role in determining the shapes of nuclei. Also, the pairing correlation has a strong influence on various phenomena and quantities such as the moment of inertia of the rotational motion of deformed nuclei and the cross section of two-nucleon transfer reactions.

The pairing interaction among electrons, which makes metals superfluid, is induced through the electron–phonon coupling [27]. In contrast, the pairing correlation in nuclei originates partly from the direct attractive interaction between nucleons and partly from the mechanism that the pair of nucleons, which occupy time-reversal states to each other in the vicinity of Fermi surface, form a Cooper pair via surface vibrations (see Ref. [28] for details).

In solid state physics, the origin of superfluidity depends on the system. For example, it is known that many metals and alloys become superfluid by the ^1S pair, while heavy electron systems and liquid ^{3}He become superfluid by ^3P pair [27]. The mechanism to become superfluid depends on the system for the nucleus as well. In this section, by having the ground and low-lying excited states of stable nuclei in mind, we discuss the role and the theoretical treatment of the main pairing correlation due to the pair of nucleons in the isospin-triplet and spin-singlet state, i.e., in the ^1S state.[24]

6.4.2 *Multipole Expansion of the Pairing Correlation, Monopole Pairing Correlation Model and Quasi-Spin Formalism*

Let us first consider the behaviour of the residual interaction and its effect in the case when a single-particle level with total angular momentum j is occupied by nucleons. Here, we use the second quantization representation. The residual interaction V is in general represented by two creation and two annihilation operators, because it is a two-body operator. Since Hamiltonian is a tensor of rank-0, we can represent V as

$$V = \sum_{J=\text{even}} E_J \sum_M C_+(JM)C_-(JM) \,, \tag{6.116}$$

$$C_+(JM) \equiv \frac{1}{\sqrt{2}} \sum_{m>0} \langle jjmM-m|JM\rangle a^\dagger_{jm} a^\dagger_{jM-m} [1 + (-1)^J] \,, \tag{6.117}$$

$$C_-(JM) = C_+^\dagger(JM) \,, \tag{6.118}$$

by separately coupling two creation and two annihilation operators into a reduced tensor.

The $J = 0$ term in Eq. (6.116) is called the monopole pairing, and the other terms multipole pairing, i.e., the quadrupole pairing etc. As Table 5.3 shows, the monopole pairing is much stronger than multipole pairing when the residual interaction is given by the δ function. Hence one often approximates V by keeping only the monopole

[24] In neutron stars, it is considered that neutrons in the outer region with low density called crust are in the $^1S_0$ superfluid state, the neutrons in more inner region with high density are in the $^3P_2$ superfluid state, and protons are in the $^1S_0$ superfluid state. Also, these superfluidities are considered to play important roles in the cooling and the phenomena called glitches of pulsars. See [29].

term. This is called the monopole pairing model. In that case, the residual interaction is given by[25]

$$V = -GS_+S_- \ , \tag{6.119}$$

$$G = -2E_0\frac{1}{2j+1} \ , \tag{6.120}$$

$$S_+ = \sum_{m>0}(-1)^{j-m}a^\dagger_{jm}a^\dagger_{j-m} \ , \tag{6.121}$$

$$S_- = S_+^\dagger = \sum_{m>0}(-)^{j-m}a_{j-m}a_{jm} \ . \tag{6.122}$$

Using the so-called quasi-spin formalism [1], one can analytically show that the degeneracy of the energy levels of the N-body system $|j^N\rangle$ is then resolved and the resulting levels can be classified according to the quantum number named seniority s, which represents the number of unpaired nucleons, i.e., the number of nucleons whose pairwise angular momentum is not coupled to zero.

Especially, in the case when $N = 2$, i.e., when the number of valence nucleons is 2, the seniority s is either $s = 0$ or $s = 2$, and all the energies for the $J = 2, 4, 6, \ldots$ states corresponding to $s = 2$ are degenerate and only the energy of the $J = 0$ state for $s = 0$ gets lower. This result well represents the main feature of the level structure of $^{210}_{82}$Pb which we learnt in Sect. 5.8 (Table 5.3) as an example of the doubly-magic ± 2 nuclei by emphasizing the characteristic property of the pairing correlation with the assumption that there exists only the monopole pairing correlation.

6.4.3 BCS Theory

The quasi-spin formalism can be applied not only to the case when the pairing interaction is weak, so that the single level approximation is applicable like the case which we described in the previous section, but to other cases as well. For example, if there exist many close lying levels, then one will be able to apply the quasi-spin formalism by considering those levels as approximately degenerate levels, and by considering them as a single energy level with a large effective angular momentum. Recently, there are attempts of the quasi-spin formalism which does not introduce the degeneracy approximation. However, more general and standard treatments which handle the energy distribution of plural single-particle energy levels as it is are the Bardeen–Cooper–Schrieffer (*BCS*) or the Hartree–Fock–Bogoliubov (*HFB*) theories. In this section, we learn the BCS theory.

---

[25] Since $\langle jjm-m|00\rangle = (-)^{j-m}/\sqrt{2j+1}$, S_+ and S_- are the creation and annihilation operators, respectively, of the pair of nucleons whose total angular momentum couples to 0.

6.4.3.1 The Time-Reversed State Representation of the Monopole Pairing Correlation Model

Before we describe the BCS theory, let us reconsider the meaning of the monopole pairing correlation model given by Eqs. (6.119)–(6.122) from the point of view of the correlation between the time-reversed states.

We fix the phase such that the time-reversed state is given by

$$|\overline{\alpha JM}\rangle = \hat{\mathscr{T}}|\alpha JM\rangle = (-1)^{J-M}|\alpha J - M\rangle \ , \tag{6.123}$$

when we represent the single-particle state as $|\alpha JM\rangle$ by using the magnitude of the angular momentum J, the magnitude of its z-component M and the additional quantum number α. The $\hat{\mathscr{T}}$ is the time-reversal operator, and is given by

$$\hat{\mathscr{T}} = -\mathrm{i}\sigma_y \hat{K} \tag{6.124}$$

with the operator $\hat{K}$, which implies to take complex conjugate of all c-numbers.[26]

Exercise 6.3 Following the time-reversal operator defined by Eq. (6.124), show that the time-reversed state is given by Eq. (6.123) if we define the single-particle state as

$$|n\ell jm\rangle = R_{n\ell}(r)\Big\langle \ell \frac{1}{2} m_\ell m_s \Big| jm \Big\rangle \mathrm{i}^\ell Y_{\ell m_\ell}|m_s\rangle \ . \tag{6.125}$$

Hint: Use the following property of the Clebsch–Gordan coefficients:

$$\langle j_1 j_2 m_1 m_2 | JM\rangle = (-1)^{j_1+j_2-J}\langle j_1 j_2 - m_1 - m_2 | J - M\rangle \ .$$

Note that the phase i^ℓ in Eq. (6.125) has been introduced to guarantee the phase in Eq. (6.123).

Using Eqs. (6.123), (6.119) can be rewritten as

$$V = -G \sum_{m>0,m'>0} a^\dagger_{jm} a^\dagger_{\overline{jm}} a_{\overline{jm'}} a_{jm'} \ . \tag{6.126}$$

Equation (6.126) shows that the monopole pairing model is the model which assumes that the residual interaction works only between the states which are time reversal to each other. This corresponds to that the Cooper pair of electrons in metals is formed for the pair of electrons which are in the time-reversed states ($\mathbf{k}\uparrow, -\mathbf{k}\downarrow$) to each other. Remark that the phase which appears in the quasi-spin to define the monopole pair and the phase which appears as the result of time-reversal operation are both $(-1)^{j-m}$ and identical with each other.

[26] For Eq. (6.123) and correspondingly also for Eq. (6.124), we determined the phase in the same way as that in Ref. [30]. In Ref. [17], the phase of Eq. (6.123) is taken to be $(-1)^{J+M}$. Correspondingly, the time-reversal operator is taken to be $\hat{\mathscr{T}} = \mathrm{i}\sigma_y \hat{K}$.

6.4.3.2 HF+BCS Theory

One of the standard methods to theoretically explore the structure of nuclei is first to determine single-particle levels in the mean field by using, for example, either the non-relativistic mean-field calculations such as the Skyrme Hartree–Fock calculations or the relativistic mean-field theory, then to take into account the effects of the residual interaction by the BCS theory. This method is often called the Hartree–Fock+BCS theory.

Here, we proceed by having a non-relativistic theory in mind. We denote the single-particle state determined by the Hartree–Fock equation by $|k\rangle$. As we discuss in Chap. 7, most nuclei except for those whose proton or neutron numbers are either one of the magic numbers or in their vicinity are deformed. The $n\ell jm$ which we used to write Eq. (6.126) are therefore not in general good quantum numbers for the eigenstates of the Hartree–Fock equation. Instead, as we learn in Sect. 7.4, $|k\rangle$ is given by a superposition of the states which have $n\ell jm$ as quantum numbers,

$$|k\rangle = \sum_{\alpha=\{n\ell jm\}} D_{\alpha k}|\alpha\rangle \ . \tag{6.127}$$

However, one can define the time-reversed state by $|\bar{k}\rangle \equiv \hat{\mathscr{T}}|k\rangle$ in the deformed case as well. In many cases, the study is performed by assuming an axially-symmetric deformation. In that case, the magnitude of the projection of the total angular momentum along the symmetry axis $m = \Omega$ is a good quantum number, and the magnitude of the projection of the total angular momentum of the time-reversed state along the symmetry axis is $-m = -\Omega$.

In the BCS theory, one assumes that the total Hamiltonian is given by

$$\hat{H} = \sum_{k>0} \varepsilon_k (a_k^\dagger a_k + a_{\bar{k}}^\dagger a_{\bar{k}}) - G \sum_{k,k'>0} a_k^\dagger a_{\bar{k}}^\dagger a_{\bar{k}'} a_{k'} \ , \tag{6.128}$$

by extending Eq. (6.126). The summation is taken over the states whose component of the angular momentum along the symmetry axis is positive.

Equation (6.128) implies that nucleons pairwise occupy the time-reversed states to each other. Hence we assume that the total wave function is given by

$$|\mathrm{BCS}\rangle = \prod_{k>0} (u_k + v_k a_k^\dagger a_{\bar{k}}^\dagger)|0\rangle \ , \tag{6.129}$$

$$u_k^2 + v_k^2 = 1 \ . \tag{6.130}$$

The v_k^2 represents the probability that the pair of nucleons occupy the $k, \bar{k}$ levels, and Eq. (6.130) is the normalization condition. The $|0\rangle$ is the vacuum state. The wave function assumed by Eq. (6.129) is in general a mixture of different number of nucleons N. Therefore, introducing the Lagrange multiplier λ in order to fix the particle number on average, and by performing the variational calculation to minimize $\hat{H}' = \hat{H} - \lambda\hat{N}$, we obtain

$$\left.\begin{array}{l} u_k^2 \\ v_k^2 \end{array}\right\} = \frac{1}{2}\left[1 \pm \frac{\varepsilon_k - \lambda}{\sqrt{(\varepsilon_k - \lambda)^2 + \Delta^2}}\right] , \tag{6.131}$$

$$\Delta \equiv G \sum_{k>0} u_k v_k . \tag{6.132}$$

The λ is called either the chemical potential or the Fermi energy, and is determined by the condition,

$$\langle \mathrm{BCS}|\hat{N}|\mathrm{BCS}\rangle = 2\sum_{k>0} v_k^2 = N . \tag{6.133}$$

Since the nucleus is a many-body system with a small number of constituents, the fluctuation of the particle number is a serious problem. Various methods such as the projection operator method which projects on the state with fixed number of particles or the double constraint method which gives constraint on both the mean value and the fluctuation of the particle number have been invented[27] [31–33].

Equations (6.131) and (6.132) lead to

$$\Delta = \frac{G}{2} \sum_{k>0} \frac{\Delta}{\sqrt{(\varepsilon_k - \lambda)^2 + \Delta^2}} . \tag{6.134}$$

Equation (6.134) is called the *gap equation*, and determines the *gap parameter* Δ. If the pairing correlation is sufficiently strong, or if there are plenty of levels in the vicinity of the Fermi surface, there exists a solution of finite value of Δ besides the trivial solution $\Delta = 0$. The former is the superfluid state. Figure 6.5 shows the occupation probability of each single-particle level v_k^2 as a function of the single-particle energy ε_k for the case when Δ is finite. The figure shows that the occupation probability of each level deviates from 1 or 0, and hence the distribution near the Fermi surface becomes blurred due to the pairing correlation.[28]

---

[27] The quasi-spin method is advantageous in the sense that it has no fluctuation of the particle number, although it is difficult to be applied to general cases because of the restriction on the degeneracy of the energy levels.

[28] Equation (6.128) corresponds to the monopole pairing correlation approximation described in the previous section. In more advanced treatments the strength of the pairing correlation G depends on the states and is generalized to $G_{kk'}$. As the result, the gap parameter becomes state dependent Δ_k, and the gap equation is generalized to [34]

$$\Delta_k = \frac{1}{2} \sum_{k'>0} \frac{G_{kk'}\Delta_{k'}}{\sqrt{(\varepsilon_{k'} - \lambda)^2 + \Delta_{k'}^2}} . \tag{6.135}$$

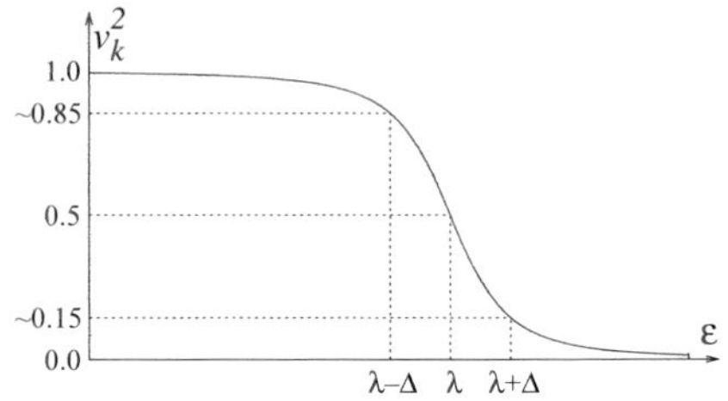

Fig. 6.5 The occupation probability of each single-particle state in the BCS theory

6.4.4 *The Magnitude of the Gap Parameter*

One of the standard ways to determine the magnitude of the gap parameter Δ is to estimate it phenomenologically from the binding energies of nuclei by remarking that the binding energies systematically differ for odd, odd–odd and even–even nuclei due to the pairing correlation (see Sect. 2.3.1). Practically, we use, for example,

$$\Delta_{\rm n} = \frac{1}{4}\left[B(A-2,Z) - 3B(A-1,Z) + 3B(A,Z) - B(A+1,Z)\right] , \quad (6.136)$$

$$\Delta_{\rm p} = \frac{1}{4}\left[B(A-2,Z-2) - 3B(A-1,Z-1) + 3B(A,Z) - B(A+1,Z+1)\right] , \quad (6.137)$$

for even–even nuclei. Figure 6.6 shows thus obtained gap parameter for neutrons $\Delta_{\rm n}$ as a function of the mass number.[29] The $\Delta_{\rm p}$ behaves in a similar way. As is written in the figure, the average behaviour of Δ can be well represented by

$$\Delta \sim \frac{12}{\sqrt{A}}\,\text{MeV} \quad (6.138)$$

for both neutrons and protons.[30]

For nuclei in the vicinity of doubly-magic ± 2 nuclei, the magnitude of the gap parameter Δ can be estimated also from the level spacing of low-lying states. For example, in the case of $^{210}_{82}$Pb, the level structure shown in Fig. 5.6 gives $2\Delta \sim 1.3$ MeV, yielding $\Delta \sim 0.7$ MeV. This value well matches the empirical formula (6.138).

One can also theoretically determine Δ so as to be consistent with Hartree–Fock calculations by solving the gap equation (6.134) once the strength of the pairing correlation G is known. According to Nilsson and Prior [38], $G_{\rm p} \approx 17/A$ MeV for protons and $G_{\rm n} \approx 25/A$ MeV for neutrons. Also, Nilsson and Ragnarsson [39]

[29] The masses of nuclei have been taken from [35].

[30] There exist also empirical formulae which take into account the isospin dependence [36].

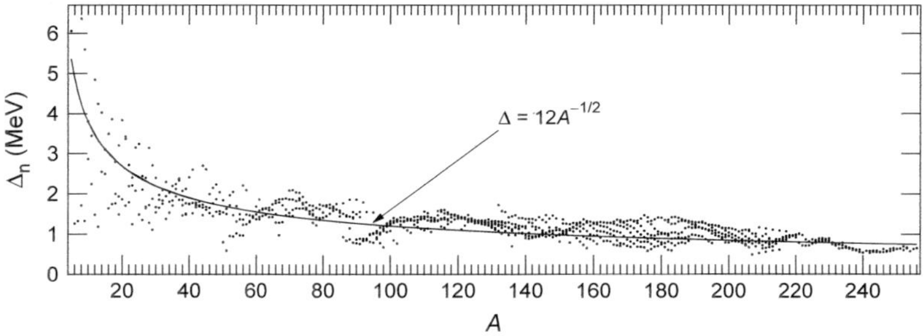

Fig. 6.6 The experimental values of Δ_n. After [37]

suggest that $G_p \approx 26/A$ MeV, $G_n \approx 20.5/A$ MeV for the nuclei in the rare-earth region. One can find some estimates also in [28].[31]

We discuss in Sect. 7.6.2 the correlation between the shape of nuclei and the nuclear superfluidity by first estimating G globally, and then by determining Δ as a function of nuclear deformation by solving the gap equation.

6.4.5 The Coherence Length

Here, we remark the spatial extension of the Cooper pair.

The *coherence length* or *the correlation length* ξ, which gives the measure of the spatial extension of the pairing correlation, is given by [27, 28]

$$\xi = \frac{\hbar v_F}{2\Delta}. \tag{6.139}$$

Exercise 6.4 Derive Eq. (6.139) based on the idea of the uncertainty relation.

Since v_F is about 30% of the speed of light, Eq. (6.138) shows that the coherence length of the pairing correlation in stable nuclei is longer than the nuclear size, i.e., the diameter $2R \sim 2 \times 1.2A^{1/3}$ fm. For example, for ^{210}Pb, $2R \sim 14.3$ fm, while $\xi \sim 35.7$ fm.

As Eq. (2.40) shows, the Fermi momentum gets smaller when the nuclear density becomes smaller. From Eq. (6.139) one can therefore expect that the coherence length becomes small in the region where the nuclear density is small. As a related topic, it is being argued recently in connection with unstable nuclei that the spatial property of the Cooper pair changes from that in the case of normal density when the nuclear density gets smaller than the nuclear saturation density ρ_0, and that there appears a spatially localized two-body correlation [41].

[31] One needs to determine the magnitude of G and the range of states k over which the sum is taken on the right-hand side of Eq. (6.134) in a consistent way [40].

References

1. G.E. Brown, *Unified Theory of Nuclear Models and Forces* (North-Holland, Amsterdam, 1967)
2. K. Sugimoto, M. Muraoka, *Nuclear Physics*, Japanese edn. (Kyouritsu, Tokyo, 1998)
3. D. Vautherin, D.M. Brink, Phys. Rev. C **5**, 626 (1972)
4. J.P. Blaizot, Phys. Rep. **64**, 171 (1980)
5. M. Beiner, H. Flocard, N. Van Giai, P. Quentin. Nucl. Phys. A **238**, 29 (1975)
6. J. Bartel, P. Quentin, M. Brack, C. Guet, H.-B. Hakansson, Nucl. Phys. A **386**, 79 (1982)
7. E. Chabanat, P. Bonche, P. Haensel, J. Meyer, R. Schaeffer, Nucl. Phys. A **627**, 710 (1997); **635**, 231 (1998)
8. A. Ono, H. Horiuchi, Prog. Part. Nucl. Phys. **53**, 501 (2004)
9. H. Ichimura, *Statistical Mechanics*, Japanese edn. (Shoukabou, 1992)
10. R. Kubo, H. Ichimura, T. Usui, N. Hashitsume, *Statistical Mechanics–An Advanced Course with Problems and Solutions* (Elsevier, Amsterdam, 1990)
11. S. Nishizaki, Y. Yamamoto, T. Takatsuka, Prog. Theor. Phys. **105**, 607 (2001); **108**, 703 (2002); Y. Yamamoto, T. Furumoto, N. Yasutake, Th. A. Rijken. Phys. Rev. C **90**, 045805 (2014)
12. T. Takatsuka, Genshikaku Kenkyu **57**, Suppl. **3**, 270 (2013); K. Masuda, T. Hatsuda, T. Takatsuka, Eur. Phys. J. A **52**, 65 (2016); I. Vidana. J. Phys. Conf. Ser. **668**, 012031 (2016)
13. P.J. Siemens, A.S. Jensen, *Elements of Nuclei: Many-Body Physics with the Strong Interaction* (Addison and Wesley, California, 1987)
14. T. Furuta, A. Ono, Phys. Rev. C **74**, 014612 (2006); **79**, 014608 (2009)
15. T.H.R. Skyrme, Philos. Mag. **1**, 1043 (1956); Proc. Phys. Soc. (Lond.) A **70**, 433 (1957); Nucl. Phys. **9**, 615 (1958)
16. J.P. Jeukenne, A. Lejeune, C. Mahaux, Phys. Rep. **25**, 83 (1976); C. Mahaux, P. F. Bortignon, R.A. Broglia, C.H. Dasso. Phys. Rep. **120**, 1 (1985)
17. Aage Bohr, Ben R. Mottelson, *Nuclear Structure*, vol. I (Benjamin, New York, 1969)
18. S. Shlomo, J.B. Natowitz, Phys. Lett. B **252**, 187 (1990)
19. H. Toki, S. Sugimoto, K. Ikeda, Prog. Theor. Phys. **108**, 903 (2002)
20. B.D. Serot, J.D. Walecka, Adv. Nucl. Phys. **16**, 1 (1986)
21. P. Ring, Y. K. Gambhir, G. A. Lalazissis, Comp. Phys. Commun. **105**, 77 (1997); Y.K. Gambhir, P. Ring, A. Thimet, Ann. Phys. (N.Y.) **198**, 132 (1990)
22. W. Greiner, *Relativistic Quantum Mechanics* (Springer, Berlin, 2000)
23. J. Meng, Nucl. Phys. A **635**, 3 (1998)
24. J.D. Bjorken, S.D. Drell, *Relativistic Quantum Mechanics* (McGraw-Hill, New York, 1964)
25. H. Kurasawa, Butsuri **49**, 628 (1994); J.V. Noble. Nucl. Phys. A **329**, 354 (1979)
26. M.M. Sharma, G. Lalazissis, J. König, P. Ring, Phys. Rev. Lett. **74**, 3744 (1995)
27. T. Tsuneto, *Superconductivity and Superfluidity* (Iwanami Modern Physics 17, Japanese edition) (Iwanami, Tokyo, 1993)
28. D.M. Brink, R.A. Broglia, *Nuclear Superfluidity-Pairing in Finite Systems* (Cambridge University Press, Cambridge, 2005)
29. R. Tamagaki, High density nuclear matter (Frontiers of Physics 15, Japanese edition) (Kyouritsu, 1986); Prog. Theor. Phys. **44**, 905 (1970)
30. W. Greiner, J.A. Maruhn, *Nuclear Models* (Springer, Berlin, 1995)
31. K. Dietrich, H.J. Mang, J.H. Pradal, Phys. Rev. **135**, B22 (1964)
32. H.J. Lipkin, Ann. Phys. (N.Y.) **9**, 272 (1960); **12**, 452 (1961)
33. Y. Nogami, Phys. Rev. **134**, B313 (1964)
34. P. Ring, P. Schuck, *The Nuclear Many-Body Problem* (Springer, Berlin, 1980)
35. G. Audi, A.H. Wapstra, C. Thibault, Nucl. Phys. A **729**, 337 (2003)
36. D.G. Madland, J.R. Nix, Nucl. Phys. A **476**, 1 (1988)
37. H. Koura, private communication
38. S.G. Nilsson, O. Prior, Mat. Fys. Medd. Dan. Vid. Selsk. **32**, 16 (1961)
39. S.G. Nilsson, I. Ragnarsson, *Shapes and Shells in Nuclear Structure* (Cambridge University Press, Cambridge, 1995)
40. A. Bulgac, Y. Yu, Phys. Rev. Lett. **88**, 042504 (2002)
41. M. Matsuo, Phys. Rev. C **73**, 044309 (2006)

Chapter 7
The Shapes of Nuclei

Abstract The nucleus behaves like a liquid drop as the mass formula and the saturation of density imply. However, a significant difference from the classical liquid drop, which is always spherical in order to make the energy by surface tension minimum, is that many nuclei are deformed except for those near the closed shells. The shape of nuclei is one of the central research subjects of nuclear structure together with the size of nuclei. The shape of nuclei is intimately related to the collective excitations of nuclei, and also strongly affects nuclear reactions including heavy-ion fusion reactions (see Balantekin and Takigawa, Rev. Mod. Phys. 70, 77 (1998); Dasgupta et al., Annu. Rev. Nucl. Part. Sci. 48, 401 (1998); Hagino et al., Comput. Phys. Commun. 123, 143 (1999); Esbensen, Nucl. Phys. A 352, 147 (1981); Hagino and Takigawa, Butsuri 57, 588 (2002); Hagino and Takigawa, Prog. Theor. Phys. 128, 1001 (2012) [1–6]). In this chapter we describe several basic concepts concerning the shape of nuclei.

7.1 The Observables Relevant to the Nuclear Shape: Multipole Moments and the Excitation Spectrum

As pointed out in Chap. 4, the magnitude of the electric quadrupole moment provides direct information on the nuclear shape, more precisely, on whether the distribution of protons inside the nucleus is spherical or deformed. We learn in Sect. 8.3 that the electromagnetic transitions give related valuable information on the shape of the nucleus. Figures 7.1a and 7.1b show the energy levels of ${}^{166}_{68}$Er and ${}^{167}_{68}$Er in the vicinity of the ground state taken from [7] (the excitation energies are in keV). The level structure of ${}^{166}_{68}$Er resembles that for ^{210}Pb and ^{210}Po shown in Fig. 5.6 in the sense that the $I^{\pi} = 0^+, 2^+, 4^+, 6^+, 8^+$ levels appear in low energy region in this order. However, it significantly differs in that the level spacing increases with increasing angular momentum I. The level spacing well agrees with the spectrum $E_I = \hbar^2 I(I+1)/2\mathscr{I}$, $\mathscr{I}$ being the moment of inertia, of the rotational motion, which is a characteristic collective motion of statically deformed nuclei. The level structure of ${}^{167}_{68}$Er is similar. Figures 7.1a and 7.1b therefore suggest that these nuclei are deformed nuclei.

N. Takigawa and K. Washiyama, *Fundamentals of Nuclear Physics*,
DOI 10.1007/978-4-431-55378-6_7

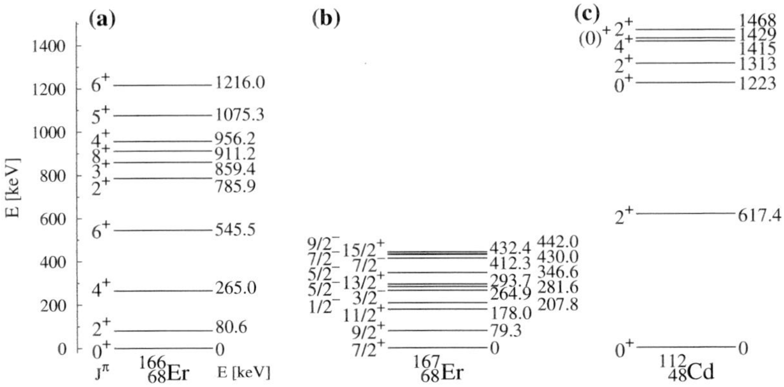

Fig. 7.1 The level structure of $^{166}_{68}$Er (**a**), $^{167}_{68}$Er (**b**) and $^{112}_{48}$Cd (**c**)

On the other hand, Fig. 7.1c shows the energy-level diagram for ^{112}Cd in the vicinity of the ground state. It is noticed that the first excited state is $I^\pi = 2^+$ state, and that there appear almost degenerate second excited group of states with $I^\pi = 0_2^+, 2_2^+$, and 4_1^+ at the excitation energy of about twice that of the first excited state. This implies that ^{112}Cd is a spherical nucleus and can be easily excited to the vibrational excitation of the quadrupole type. The 0_2^+, 2_2^+, and 4_1^+ levels can be considered to be the two phonon states.[1]

In this way, the level structure of nuclei also provides important information on the shape of nuclei.

Figure 7.2 shows the change of the energy-level diagrams for Sm isotopes in the vicinity of the ground state as the neutron number increases. The figure suggests

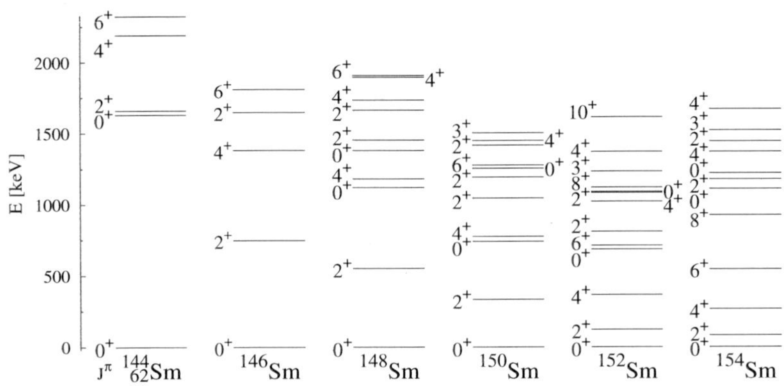

Fig. 7.2 The isotopic variation of the low-lying level structure of Sm isotopes

[1]Recently, it is argued that the 0_2^+ state has a different property from that of the two-phonon state.

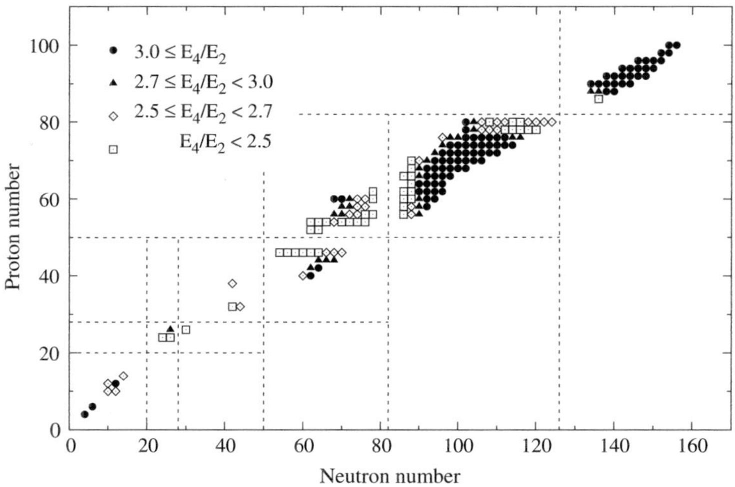

Fig. 7.3 The region of nuclei whose ground state is deformed

that the Sm isotopes change their shape from the spherical shape to the deformed shape as the neutron number increases from the magic number $N = 82$ (the phase transition of the nuclear shape, i.e., the *shape transition*).

Figure 7.3 shows the region in the nuclear chart where nuclei are deformed in their ground state. Those nuclei have been selected as the nuclei, which have either the ground-state rotational band or whose E_4/E_2 ratio is large, where E_2 is the excitation energy of the first 2^+ state and E_4 that of the first 4^+ state except for some shape coexisting nuclei.[2] They are sorted according to the value of E_4/E_2.

Typical deformed nuclei are located in the regions of rare earth such as Er isotopes and actinides such as U isotopes. In addition, ^{8}Be, ^{12}C, ^{20}Ne, and ^{24}Mg are also deformed in their ground state. Even if the ground state is spherical, some nuclei have deformed states in the region of relatively low excitation energy. Also, there exist nuclei which have states with different deformations in the same energy region. They are called the *shape coexistence*. ^{16}O and some of the Ge isotopes are representative examples of these nuclei.[3] Ge isotopes are also known to make shape transition in the ground state from the spherical shape to a deformed shape as the neutron number increases.

---

[2]The excitation energy of each level has been taken either from the Table of Isotopes [7] or from http://www.nndc.bnl.gov/chart/.

[3]Light Hg isotopes are known to be the nuclei with shape coexistence [8]. They have a rotational band of prolate shape in the region of a low excitation energy in addition to the ground-state rotational band of oblate shape. The excitation energy of the 4^+ member of the prolate band becomes lower than that of the 4^+ member of the oblate ground-state band for the isotopes whose mass number is 186 or smaller. Hence we used the excitation energy of the second 4^+ state for E_4 for the ^{182,184,186}Hg isotopes in making Fig. 7.3. We have omitted ^{180}Hg from the figure, since the level structure is presently unclear.

7.2 Deformation Parameters

Various deformation parameters are conventionally used to represent nuclear deformation. Here, we list some of them and discuss their properties and connections among them. We introduced the deformation parameters α_λ by Eq. (2.66) in order to describe fission. The parameter δ which is defined by

$$\delta \equiv \frac{3}{2}\frac{R_3^2 - R_\perp^2}{R_3^2 + 2R_\perp^2} \tag{7.1}$$

is also often used when the deformation is axially symmetric by assigning the symmetry axis to the third axis, which we sometimes call the z-axis in the following. The R_3 and $R_\perp$ represent the equivalent radii in the direction of the symmetry axis, and in the perpendicular direction, respectively.

For general deformations, one expands the radius as

$$R = R(\theta, \varphi) = R_0 \left[1 + \alpha_{00} + \sum_{\lambda=2}^{\infty} \sum_{\mu=-\lambda}^{\lambda} \alpha_{\lambda\mu}^{*} Y_{\lambda\mu}(\theta, \varphi) \right] \tag{7.2}$$

by extending Eq. (2.66). The α_{00} is not an independent parameter in the case of incompressible deformations, but is given by

$$\alpha_{00} = -\frac{1}{4\pi} \sum_{\lambda=2}^{\infty} \sum_{\mu=-\lambda}^{\lambda} |\alpha_{\lambda\mu}|^2 \tag{7.3}$$

by using the other deformation parameters due to the condition of the volume conservation.

One can derive the following properties for the deformation parameters:

$$\alpha_{\lambda-\mu}^{*} = (-1)^{\mu} \alpha_{\lambda\mu} , \tag{7.4}$$

$$\hat{P} \alpha_{\lambda\mu} \hat{P}^{-1} = (-1)^{\lambda} \alpha_{\lambda\mu} , \tag{7.5}$$

$$(\alpha_{\lambda\mu})_{ncs} = \hat{\mathscr{R}}(\omega) \alpha_{\lambda\mu} \hat{\mathscr{R}}^{-1}(\omega) = R_{\mu'\mu}^{(\lambda)}(\omega) (\alpha_{\lambda\mu'})_{ocs} , \tag{7.6}$$

by using the condition that the nuclear radius R is real and invariant under the parity and the rotational transformations and also from the properties of the spherical harmonics. $\hat{P}$ and $\hat{\mathscr{R}}$ are the parity inversion and the rotation operators, respectively. When we take the point of view to rotate the system, i.e., the nucleus, instead of the axes, $\hat{\mathscr{R}}$ is given by

$$\hat{\mathscr{R}}(\omega) = \mathrm{e}^{-\mathrm{i}\phi \hat{J}_z} \mathrm{e}^{-\mathrm{i}\theta \hat{J}_y} \mathrm{e}^{-\mathrm{i}\psi \hat{J}_z} , \tag{7.7}$$

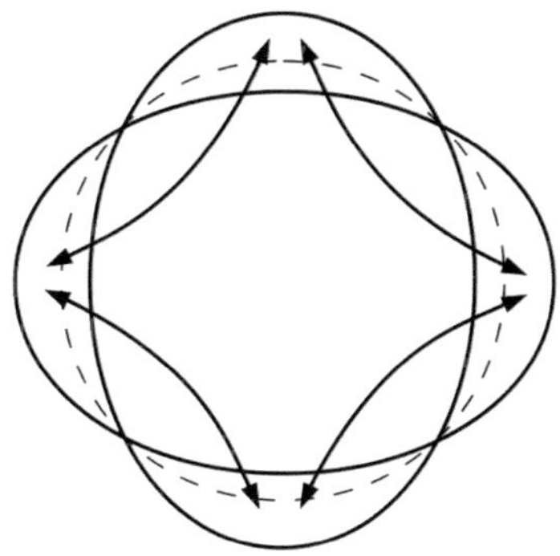

Fig. 7.4 The quadrupole vibration

in terms of the Euler angles $\omega = (\phi, \theta, \psi)$. The $R^{(\lambda)}_{\mu'\mu}(\omega)$ is the matrix which represents the rotation, and is defined by [9, 10]

$$R^{(\lambda)}_{\mu'\mu}(\omega) \equiv \langle \lambda\mu' | \hat{\mathscr{R}}(\omega) | \lambda\mu \rangle = \mathscr{D}^{\lambda\,*}_{\mu'\mu}(\omega) \ . \tag{7.8}$$

The indices ncs and ocs in Eq. (7.6) represent the components after and before the rotational transformation, respectively. Equation (7.6) shows that $\{\alpha_{\lambda\mu}\}$ forms a tensor of rank-λ and that $\alpha_{\lambda\mu}$ is its μ component (see Sect. A.6).

Exercise 7.1 Show Eqs. (7.4)–(7.6).

We have shown in Fig. 2.21 the shape of deformation for $\lambda = 2, 3$ and 4 in connection with fission.[4] The surface vibrations of spherical nuclei can be described by using $\alpha_{\lambda\mu}$ as dynamical variables (*geometrical collective model*).[5] Those vibrations correspond to the oscillations of the deviation of the shape from the spherical shape shown in Fig. 2.21 as functions of time. The vibrations for $\lambda = 2$ and 3 are called the quadrupole and octupole vibrations, respectively. Figure. 7.4 conceptually shows the quadrupole vibration as an example.

For statically deformed nuclei, it is more convenient to use the Euler angles, which show the direction of principal axes, instead of the $\alpha_{\lambda\mu}$ as a part of the variables in order to describe collective motions. As an example, let us consider the case of the quadrupole deformation. There are originally $2(2\lambda + 1)$ variables, since $\alpha_{\lambda\mu}$ are complex numbers. Among them $2\lambda + 1$ variables are independent, since there exist

[4] We assumed the prolate type, whose shape resembles that of the ball of the American football, for the quadrupole deformation $\lambda = 2$. Note that the change of the shape depends on the sign of α_{40} when the hexadecapole deformation α_{40} is added. It becomes the barrel type which has expanded peaks and belly compared to the shape in the case when there exists only the quadrupole deformation if $\alpha_{40} > 0$, while the peanuts shape when $\alpha_{40} < 0$.

[5] There exist alternative ways to describe nuclear energy level structure and excitations. One of the powerful theoretical frameworks is the *interacting boson model* (IBM) [11] which is an algebraic method using the group theory.

$2\lambda + 1$ conditions because of Eq. (7.4). Hence if $\lambda = 2$, there are two independent variables besides three Euler angles. On the other hand, since $\{\alpha_{\lambda\mu}\}$ is a tensor of rank λ, the variables $\alpha_{2\mu}$ in the space-fixed coordinate system are transformed to new variables $a_{2\nu}$ in the nucleus fixed coordinate system, i.e., the body-fixed coordinate system, where the coordinate axes are taken along the principal axes of the nucleus, by a rotation of the coordinate system:

$$\alpha_{2\mu} = \sum_{\nu} a_{2\nu} \mathscr{D}^2_{\mu\nu}(\omega) . \tag{7.9}$$

By taking the Euler angles in the direction of the principal axes, we have

$$a_{21} = a_{2-1} = 0, \quad a_{22} = a_{2-2} . \tag{7.10}$$

The Euler angles ϕ, θ, ψ and the real variables a_{22}, a_{20} become new variables. Conventionally, however, the parameters β and γ which are defined by

$$a_{20} = \beta \cos\gamma, \quad a_{22} = \frac{1}{\sqrt{2}} \beta \sin\gamma \tag{7.11}$$

are usually used for a_{22} and a_{20}. The nuclear radius in the direction of θ_b, φ_b measured in the body-fixed system is given by

$$R(\theta_b, \varphi_b) = R_0 \left\{ 1 + \beta\sqrt{\frac{5}{16\pi}} \left[\cos\gamma(3\cos^2\theta_b - 1) + \sqrt{3}\sin\gamma \sin^2\theta_b \cos 2\varphi_b \right] \right\} . \tag{7.12}$$

β is the parameter which gives the magnitude of the quadrupole deformation, and is related to $\alpha_{2\mu}$ as

$$\sum_{\mu} |\alpha_{2\mu}|^2 = a_{20}^2 + 2a_{22}^2 = \beta^2 . \tag{7.13}$$

γ is the parameter for discussing the axial symmetry. Equation (7.12) shows that

1. $\gamma = 0°, 120°, 240°$ represent the prolate type of axially symmetric deformation where the 3, 1 and 2 axes are the symmetry axes, respectively.
2. $\gamma = 180°, 300°, 60°$ represent the oblate type of axially symmetric deformation where the 3, 1 and 2 axes are the symmetry axes, respectively.
3. The γ which is not a multiple of 60° represents a triaxial deformation.
4. One can represent all the shapes by using only the region $\beta \geq 0, 0° \leq \gamma \leq 60°$ in the case of quadrupole deformation. The other regions of β, γ can be reached by suitable transformations.

Figure. 7.5 shows the connection between the deformation parameters β, γ and the nuclear shape.

The following relations hold among different deformation parameters,

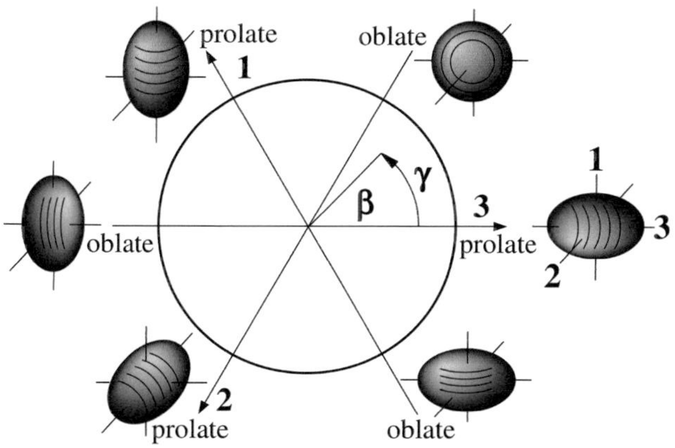

Fig. 7.5 The relation between the deformation parameters β, γ and the nuclear shape

$$\alpha_2 = \sqrt{\frac{5}{4\pi}}\alpha_{20} \sim \frac{2}{3}\delta\,, \quad \alpha_{20} = \beta \sim 1.06\delta\,, \quad \delta \sim 0.95\alpha_{20} = 0.95\beta\,, \tag{7.14}$$

when the deformation is small.

7.3 The Deformed Shell Model

Let us study the distribution of energy levels for nucleons inside a deformed nucleus. One can then understand the origin of nuclear deformation.

As Eq. (5.4) shows, in the approximation to set the range of the nuclear force to zero the single-particle potential for nucleons is proportional to the density of nucleons. This implies that the mean field is also deformed if the nucleus is deformed. This is called the *self-consistency condition*.

Suppose an axially symmetric deformation and assume that the single-particle field is given by the deformed harmonic oscillator field as

$$V = \frac{M_N}{2}\left[\omega_\perp^2(x^2+y^2) + \omega_z^2 z^2\right]\,. \tag{7.15}$$

Both oscillator parameters $\omega_\perp$ and ω_z must be functions of the magnitude of deformation δ because of the self-consistency condition. In order to determine their functional forms, we remark that the shape of the equipotential surfaces of the potential represented by Eq. (7.15) is given by the ratio of the distances in the directions of the three axes $\frac{1}{\omega_\perp} : \frac{1}{\omega_\perp} : \frac{1}{\omega_z}$. Therefore, if we denote the radius along the α ($\alpha = x, y, z$) axis by a_α, the mean radius by R_0 and the common oscillator parameter by ω_0, then it is required that

$$\omega_z(\delta) = \frac{R_0}{a_z}\omega_0(\delta) \sim \left(1 - \frac{2}{3}\delta\right)\omega_0(\delta)\ , \tag{7.16}$$

$$\omega_\perp(\delta) = \frac{R_0}{a_\perp}\omega_0(\delta) \sim \left(1 + \frac{1}{3}\delta\right)\omega_0(\delta)\ , \tag{7.17}$$

in order for the mean field to be consistent with the density distribution[6] [12]. If we denote the mean oscillator parameter by $\bar{\omega}$,[7] then we obtain

$$\omega_0(\delta) \sim \bar{\omega}\left(1 + \frac{1}{9}\delta^2\right)\ , \tag{7.18}$$

from the incompressibility condition $\omega_\perp^2\omega_z = \bar{\omega}^3$.

The potential given by Eq. (7.15) can be expressed as

$$V = \frac{1}{2}M_N\omega_0^2(\delta)r^2 - M_N\omega_0^2(\delta)r^2\beta Y_{20}(\theta,\phi) \tag{7.19}$$

by using the deformation parameter.

Since it is the problem of the three-dimensional harmonic oscillator, the energy levels can be easily obtained by using either the Cartesian or cylindrical coordinates, and are given by

$$\varepsilon(n_x, n_y, n_z) = \varepsilon(n_z, n_\rho, k_\ell) \tag{7.20}$$

$$= \hbar\omega_\perp(n_x + n_y + 1) + \hbar\omega_z\left(n_z + \frac{1}{2}\right) \tag{7.21}$$

$$= \hbar\omega_\perp(2n_\rho + |k_\ell| + 1) + \hbar\omega_z\left(n_z + \frac{1}{2}\right) \tag{7.22}$$

$$\sim \hbar\omega_0(\delta)\left[\left(N + \frac{3}{2}\right) + \frac{1}{3}\delta(N - 3n_z)\right]\ , \tag{7.23}$$

where $N = n_x + n_y + n_z = n_z + 2n_\rho + |k_\ell|$ and k_ℓ is the quantum number which represents the magnitude of the projection component of the orbital angular momentum along the symmetry axis.

As remarked concerning Eqs. (7.16) and (7.17), strictly speaking, the δ in Eq. (7.23) differs from the δ in Eq. (7.1), and while the latter is defined by

$$\delta \equiv \frac{3}{2}\frac{R_z^2 - R_\perp^2}{R_z^2 + 2R_\perp^2} = \frac{3}{2}\frac{\omega_\perp^2(\delta) - \omega_z^2(\delta)}{\omega_\perp^2(\delta) + 2\omega_z^2(\delta)}\ , \tag{7.24}$$

[6]The transformation to the last term in Eqs. (7.16) and (7.17) is correct only to the first order of the deformation δ. Strictly speaking, it is therefore different from the δ defined by Eq. (7.1).

[7]The rough estimate of $\hbar\bar{\omega}$ is given by Eq. (5.52).

the former is defined by

$$\delta \equiv 3\frac{\omega_\perp(\delta) - \omega_z(\delta)}{\omega_z(\delta) + 2\omega_\perp(\delta)} . \tag{7.25}$$

Hence some literatures, e.g., [13], distinguish it by denoting with a different notation such as $\delta_{\rm osc}$.

Figure 7.6 shows how the three levels (six levels if we include the spin), which are degenerate in the spherical shape, split with the deformation by taking the case of $N = 1$ levels as the example. Among them the $n_z = 0$ level is two-fold degenerate (four-fold degenerate if one includes the spin). For positive δ, i.e., for prolate deformation which is stretched in the direction of the symmetry axis, the state which has the finite quantum number in that direction, i.e., the $N = 1, n_z = 1$ state, in other words, the state with a smaller $|k_\ell|$ (the $k_\ell = 0$ state in the present case), has lower energy, while for negative δ, i.e., for oblate deformation which is shrunk in the direction of the symmetry axis, the state which has the finite quantum number in the direction perpendicular to the symmetry axis, i.e., the $N = 1, n_z = 0$ state, in other words, the state with a large $|k_\ell|$ value (the $k_\ell = \pm 1$ state in the present case) has lower energy. This can be easily expected from the uncertainty principle, which suggests that the kinetic energy can be smaller if the movable space gets larger.

Note also that the change of the energy due to deformation is larger on the prolate side if we compare the lower level on the prolate and oblate sides. This can be understood from Eq. (7.23) which shows that the variation of the energy is given by $2\delta/3$ for the lower $n_z = 1$ level on the prolate side, while it is given by $\delta/3$ for the $n_\perp = 1 (n_z = 0)$ level which is the lower level on the oblate side.

Figure 7.6 suggests that the deformation of nucleus occurs in close connection with the shell structure and the Pauli exclusion principle. As an example, let us consider $^8_4$Be from the point of view of the shell model. The first two protons (and neutrons) occupy the lowest 1s state ($N = 0$ state). Figure 7.6 shows that it is energetically more favorable to deform the nucleus to the prolate shape and to fill the remaining two protons (and neutrons) in the $n_z = 1$ level. For this reason, ^{8}Be becomes a deformed nucleus of prolate type.[8] As the number of nucleons increases, it becomes necessary to fill higher levels with nucleons, and the sign of deformation changes. Roughly

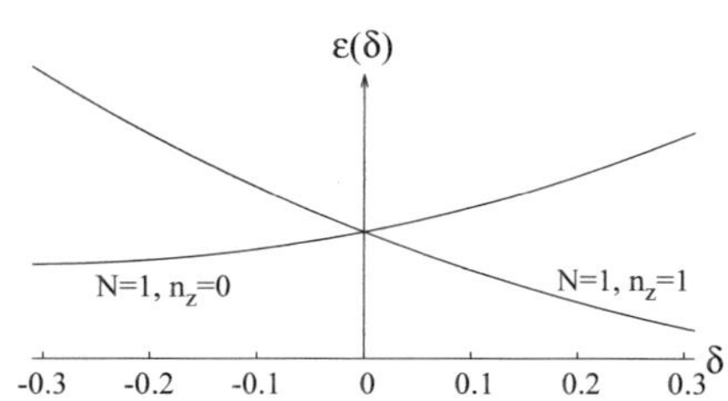

Fig. 7.6 The deformation dependence of the single-particle energy levels

[8] This is a kind of Jahn–Teller effect. There exists an alternative view for ^{8}Be in terms of the cluster model or the α-particle model, where ^{8}Be is thought to be a dumb-bell consisting of two alpha

speaking, it can be expected that the nuclei which belong to the beginning of a major shell have prolate deformation, while those towards the end of the shell oblate. For example, in the case of $N = 1$ shell, i.e., the $1p$ shell, ^{8}Be, which belongs to the first half, has a prolate deformation, while ^{12}C, which belongs to the last half, has an oblate shape. However, in reality most of the deformed nuclei are prolate, and those which are oblate in their ground state are limited to ^{12}C and small number of nuclei such as some of the Hg isotopes, which we mentioned in connection with Fig. 7.3 and will discuss again in Sect. 7.6.2.

7.4 Nucleon Energy Levels in a Deformed One-Body Field: Nilsson Levels

The spin–orbit coupling is indispensable in order to discuss real nuclei. Also, the radial dependence of the harmonic oscillator type is unrealistic. Hence we assume that the Hamiltonian for nucleons is given by

$$\hat{h} = -\frac{\hbar^2}{2M_N}\Delta + \frac{M_N}{2}\omega_\perp^2(x^2+y^2) + \frac{M_N}{2}\omega_z^2 z^2 + C\boldsymbol{\ell}\cdot\mathbf{s} + D\boldsymbol{\ell}^2 \tag{7.26}$$

by extending the deformed harmonic oscillator model introduced in the previous section. The $\boldsymbol{\ell}^2$ term has been introduced in order to effectively improve the harmonic oscillator model so as to yield the level structure for the realistic radial dependence of the Woods–Saxon type, and is designed in such a way that the states with larger orbital angular momentum feel stronger attraction (see Fig. 5.1). The single-particle levels obtained by Eq. (7.26) are called Nilsson levels or Nilsson diagram or Nilsson scheme, since this model was proposed by Nilsson [14] for the first time.

The Nilsson levels and the associated wave functions can be obtained by diagonalizing the Hamiltonian given by Eq. (7.26) by using a suitable complete set as the expansion basis. For example, using the wave functions of the deformed shell model considered in the previous section $|\alpha\rangle = |Nn_zk_\ell\Omega\rangle$ as the basis, we expand the wave function as

$$|\phi_{i=Nn_zk_\ell\Omega}\rangle = \sum_\alpha C_{i\alpha}|\alpha\rangle\ , \tag{7.27}$$

and determine the expansion coefficients $C_{i\alpha}$ and the corresponding single-particle energy ε_i by diagonalizing $\hat{h}$. $\Omega = k_\ell + m_s$ is the magnitude of the projection component of the total angular momentum along the symmetry axis, where m_s is the magnitude of the projection component of the spin along the symmetry axis.

Figure 7.7 shows an example of the Nilsson levels. The followings are some remarks concerning the Nilsson levels.

---

particles. However, the overlap between the wave functions of the shell model and of the cluster model is fairly large.

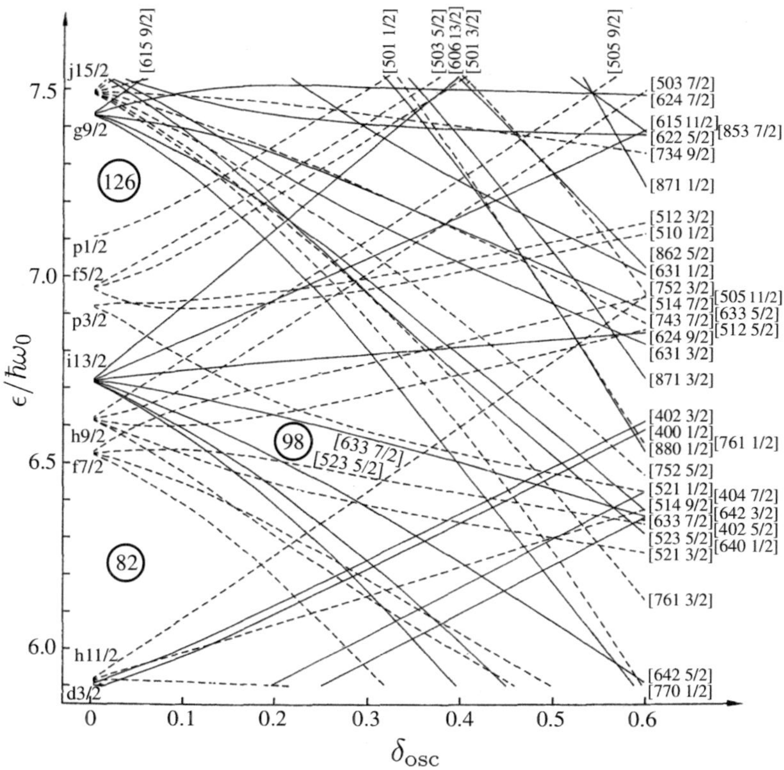

Fig. 7.7 Neutron energy levels for prolate deformation for $82 < N < 126$, taken from [15]

1. The $2j+1$ levels, which are degenerate for the spherical shape, split into $(2j+1)/2$ levels by deformation.
2. Each split energy level is twofold degenerate together with the time reversed state.
3. As the right-hand side of Eq. (7.27) shows, different N, n_z, k_ℓ values are mixed in a single energy level. Nevertheless, the set of quantum numbers $\{N, n_z, k_\ell, \Omega\}$, which is suitable in the region of large deformation and hence is sometimes called the asymptotic quantum numbers, is conventionally used as the label for each Nilsson level. However, the parity, which is determined by N is a good quantum number, and Ω is also a good quantum number as long as the deformation is axially symmetric. Note that Ω is not the component of the angular momentum along a space-fixed axis, but the projection of the angular momentum along an intrinsic, i.e., body-fixed, axis.

4. Roughly speaking, for prolate deformation ($\delta > 0$), the states with small Ω gain negative energy by deformation and the amount of the gain is larger for smaller Ω. The shift of energy by deformation gradually changes with Ω, and turns to be positive for states with large values of Ω. The trend of the effect of deformation is opposite for oblate deformation. However, the effect of deformation is more complex for oblate deformation.

7.5 The Spin and Parity of the Ground State of Deformed Odd Nuclei

One can speculate the spin and parity of the ground state of deformed odd nuclei by referring to the Nilsson diagram.

Consider $^{167}_{68}$Er as an example. From the analysis of the inelastic scattering of α particles, i.e., (α, α') scattering, it is known that the quadrupole deformation of $^{166}_{68}$Er is $\beta_2 = 0.276$ [16]. If the quadrupole moment of $^{167}_{68}$Er is assumed to be nearly equal, then the spin of the ground state of $^{167}_{68}$Er is predicted to be $7/2$ from the Ω value of the 99th energy level for neutrons in Fig. 7.7. Also, the parity is expected to be plus, since this level results from the split of the $i_{13/2}$ level. These predictions agree with the experimental results shown in Fig. 7.1b.

7.6 Theoretical Prediction of Nuclear Shape

There are various theoretical methods to confirm nuclear shapes determined or predicted by experimental data or to predict unknown nuclear shapes. Here, we explain two standard methods which determine the nuclear shape based on the calculation of binding energies.

7.6.1 The Strutinsky Method: Macroscopic–Microscopic Method

One of the methods is the method which reconciles the liquid-drop model and the shell model. It is called either the macroscopic–microscopic method or the Strutinsky shell correction method after the name of the inventor [17]. This method gives also the binding energy of nuclei and the potential surface for fission with high accuracy [18].

Leaving the detailed explanation of the method to other books [19], here we mention the basic concept of this method.

As we see in Fig. 2.6, the binding energy consists of the part E_{LDM} which smoothly depends on the mass number and can be well explained by the liquid-drop model, and the part which deviates from it and oscillates, i.e., detailed structure;

$$E_{\text{tot}} = E_{\text{LDM}} + E_{\text{osc}} . \tag{7.28}$$

The oscillating part arises because the distribution of single-particle energy levels deviates from the average level density and is non-uniform reflecting the shell structure.

In the Strutinsky method, the part which depends smoothly on the mass number is estimated macroscopically based on the liquid-drop model, while the oscillating part E_{osc} as the difference between the energy microscopically calculated by the shell model and the mean energy $\tilde{E}_{\text{sh}}$ calculated by introducing the average level density

$$E_{\text{osc}} = E_{\text{sh}} - \tilde{E}_{\text{sh}} . \tag{7.29}$$

If we denote the accurate or shell model level density by $g(\varepsilon)$, the average level density by $\tilde{g}(\varepsilon)$, the Fermi energy by λ, and the Fermi energy when the average level density is used by $\tilde{\lambda}$, then

$$g(\varepsilon) = \sum_{i=1}^{A} \delta(\varepsilon - \varepsilon_i) , \tag{7.30}$$

$$E_{\text{sh}} = \sum_{i=1}^{A} \varepsilon_i = \int_{-\infty}^{\lambda} \varepsilon g(\varepsilon) \mathrm{d}\varepsilon , \tag{7.31}$$

$$\tilde{E}_{\text{sh}} = \int_{-\infty}^{\tilde{\lambda}} \varepsilon \tilde{g}(\varepsilon) \mathrm{d}\varepsilon , \tag{7.32}$$

$$A = \int_{-\infty}^{\lambda} g(\varepsilon) \mathrm{d}\varepsilon = \int_{-\infty}^{\tilde{\lambda}} \tilde{g}(\varepsilon) \mathrm{d}\varepsilon , \tag{7.33}$$

where we denote the mass number by A.

The folding method,

$$\tilde{g}(\varepsilon) = \frac{1}{\gamma} \int_{-\infty}^{\infty} g(\varepsilon') f\left(\frac{\varepsilon' - \varepsilon}{\gamma}\right) \mathrm{d}\varepsilon' , \tag{7.34}$$

where f is a Gaussian, whose width γ is of the order of $\hbar\omega_0$, is a candidate of the method to introduce the average level density. However, one must be sure that the result does not change when the averaging procedure is repeated [19].

Incidentally, Eq. (7.29) is revised as

$$E_{\text{osc}} = E_{\text{sh}} - \tilde{E}_{\text{sh}} + E_{\text{pair}} - \tilde{E}_{\text{pair}} \tag{7.35}$$

if the pairing energy is taken into account in the BCS method.

A feature of this method is that the parameter dependence is smaller than the microscopic method, which we will discuss in the next section, because the parameters are phenomenologically determined based on, e.g., the mass formula. Hence, it has a high predictive power. As stated at the beginning of this section, it therefore provides a powerful way to calculate the binding energy and the shape of the ground state as well as the energy surface for fission. The same idea has been applied also to the theoretical study of the structure of metal clusters.

7.6.2 Constrained Hartree–Fock Calculations

The other method is the constrained mean-field calculations with the linear or the quadratic constraints. If one is interested in the quadrupole deformation, the method of linear constraint performs variational calculations for the effective Hamiltonian given by

$$\hat{H}' = \hat{H} - \lambda_Q \hat{Q}\,, \tag{7.36}$$

where $\hat{Q}$ is the operator of the quadrupole moment $\hat{Q} = \sqrt{\frac{16\pi}{5}}\hat{Q}_{20}$. λ_Q is the Lagrange multiplier and is determined such that the expectation value of the quadrupole moment becomes equal to the given value Q:

$$\langle \hat{Q} \rangle = Q \sim AR_0^2\sqrt{\frac{9}{5\pi}}\beta\,. \tag{7.37}$$

It is therefore a function of the magnitude of the quadrupole moment Q or the deformation parameter β, that is, $\lambda_Q = \lambda_Q(\beta)$.[9] In the method of quadratic constraint, one performs variational calculations to make

$$E'' \equiv \langle \hat{H} \rangle + \frac{1}{2}C(\langle \hat{Q} \rangle - \mu)^2 \tag{7.38}$$

minimum. This method corresponds to taking $\lambda_Q = C(\mu - \langle \hat{Q} \rangle)$ in the linear constraint method. If one takes a large enough value for C, E'' gives the minimum energy for the state with the deformation μ when $\mu = \langle \hat{Q} \rangle$.

The single-particle states and the corresponding energies $\varepsilon_i(\beta)$ are determined by performing mean-field calculations under these constraints. In the Hartree–Fock+BCS theory, one needs to further determine the u and v factors. There are practically two ways to that end. The one is a simplified method, which treats Δ as a deformation independent parameter. The other is to determine Δ as a function of

[9] In recent years, the study of unstable nuclei which are far from the β-stability line such as those in the vicinity of drip lines is extensively going on. One of the main concerns there is to elucidate the difference of the deformations of the proton and neutron distributions inside a nucleus. In such studies, one performs calculations by imposing separate constraints on protons and neutrons.

β by solving the gap equation (6.134) for each value of the deformation parameter β, and further to determine the corresponding u and v factors. As the Nilsson levels show, the distribution of single-particle energy levels strongly depends on β, so that the latter is, of course, much superior. As we discuss later, it is sometimes crucially important to follow the latter procedure in order to determine the shape of nuclei. In that case, one determines the strength of the pairing correlation G in advance by some way, and assumes that it does not depend on the nuclear deformation. This method is often called the *constant G method*. By comparison, the simplified method which treats Δ as a deformation independent constant is called the *constant Δ method.*

Figure 7.8 shows the energy surface for ^{184}Hg obtained by the relativistic mean-field (RMF) calculations using the method of quadratic constraint as a function of the deformation parameter β [20]. The figure shows that the ground state of ^{184}Hg is oblate and suggests that there appears prolate states at low excitation energies. Actually, these results well agree with the experimental data, i.e., that for light Hg isotopes there appears a rotational band with prolate shape at low excitation energies above the ground-state rotational band with oblate shape as we mentioned in connection with Fig. 7.3. The RMF calculations were performed by assuming the NL1 force [21], and the BCS calculations in the constant G method. In the latter, the magnitude of G has been determined in such a way that the average value of the gap parameter

$$\bar{\Delta}_{\rm n} = 4.8/N^{1/3} , \tag{7.39}$$

which has been obtained from the systematic data analyses for many nuclei [22], can be reproduced for spherical nuclei, instead of using the average value mentioned in Sect. 6.4.4 as it is for the magnitude of G. Using thus obtained $G_{\rm n}$ and the energies $\varepsilon_k(\beta)$ of the single-particle levels obtained by the constrained RMF, the gap equation

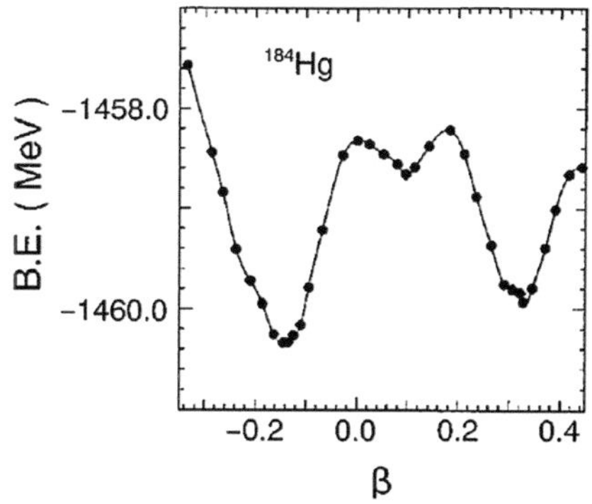

Fig. 7.8 The energy surface as a function of β. Taken from [20]

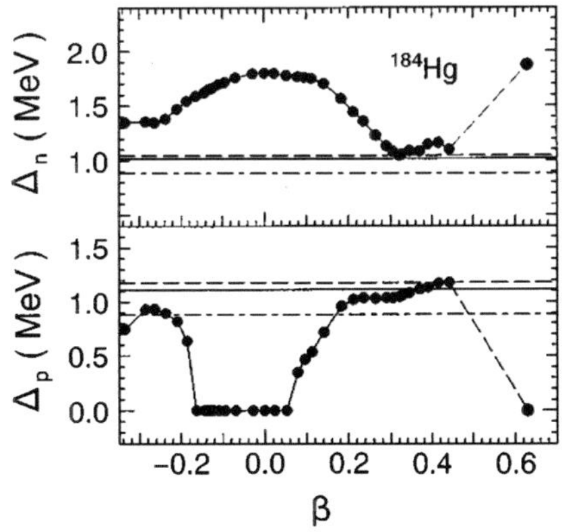

Fig. 7.9 The deformation dependence of the gap parameter. Taken from [20]

$$\frac{2}{G_n} = \sum_{k=0}^{k_{max}} \frac{1}{\sqrt{(\varepsilon_k(\beta) - \lambda)^2 + \Delta_n^2(\beta)}} \tag{7.40}$$

has been solved for each value of β and the value of the gap parameter $\Delta_n(\beta)$ has been determined. The G_p and $\Delta_p(\beta)$ for protons have also been determined by using $\bar{\Delta}_p = 4.8/Z^{1/3}$ and by the same procedure. Figure 7.9 shows thus obtained gap parameters as functions of the deformation parameter [20]. The figure clearly shows that the value of the gap parameter, hence the superfluidity of a nucleus, changes drastically with the deformation of the nucleus. This can be easily expected from the facts that the value of the gap parameter strongly depends on the level density in the vicinity of the Fermi surface as Eq. (7.40) shows, and that the distribution of the nucleon energy levels strongly reflects the degree of deformation as the Nilsson diagram studied in Sect. 7.4 shows. As mentioned in Sect. 7.3, ^{184}Hg is one of the exceptional nuclei which are oblate in the ground state. If we use the constant Δ method, ^{184}Hg is predicted to be prolate. This suggests that the pairing correlation plays a decisive role in determining the shape of nuclei, and that it is crucially important to correctly take into account the deformation dependence of the pairing correlation in order to discuss the shape of nuclei. It will be worth mentioning that this calculation well reproduces the experimental data of the isotope variation of the charge radii of Hg isotopes over a wide range of the mass number.

7.7 Column: Superdeformed States

The amount of deformation of rare-earth nuclei and actinides in their ground states is about $\delta \sim 0.3$. By comparison, the deformation of light deformed nuclei such as ^{8}Be is large.

Today, many states with large deformation have been found in the excitation energy region in the medium-mass and heavy nuclei as well. They are called superdeformed states and superdeformed rotational bands built on them, and have been extensively studied together with high spin states [23].

The fission isomers mentioned in Sect. 2.3.3 are typical examples of superdeformed states. Figure 7.10 shows the distribution of superdeformed bands and fission isomers on the nuclear chart.

We learnt in Sect. 7.3 that the deformation of nucleus is intimately related to the shell structure of single-particle levels. In this connection, Fig. 7.11 shows the distribution of single-particle energy levels in the deformed shell model. The figure shows that there appears a distinct shell structure, i.e., the degeneracy of levels, similar to that for $\delta = 0$, when the ratio between the long and short axes $R_z : R_\perp$ becomes, e.g., 2:1. This happens because in that case the ratio of the magnitudes of the energy quanta in the direction parallel to the symmetry axis and in the perpendicular direction $\hbar\omega_z : \hbar\omega_\perp$ becomes 1:2 due to the consistency condition between the potential for nucleons and the nucleon density. In this case, the deformation $\delta^{(\mathrm{SD})}$ of the superdeformed state becomes $\delta^{(\mathrm{SD})}_{\mathrm{osc}} = 0.6$ in terms of δ_{osc} defined by Eq. (7.25).

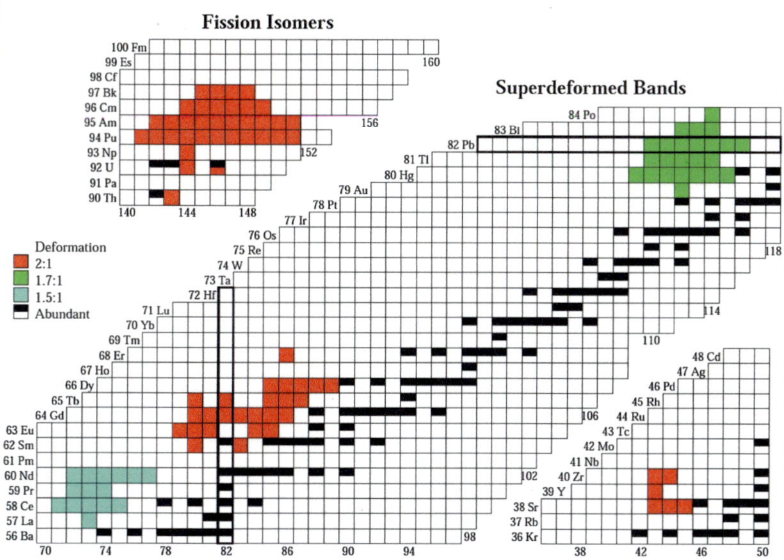

Fig. 7.10 The distribution of superdeformed states, taken from [24]

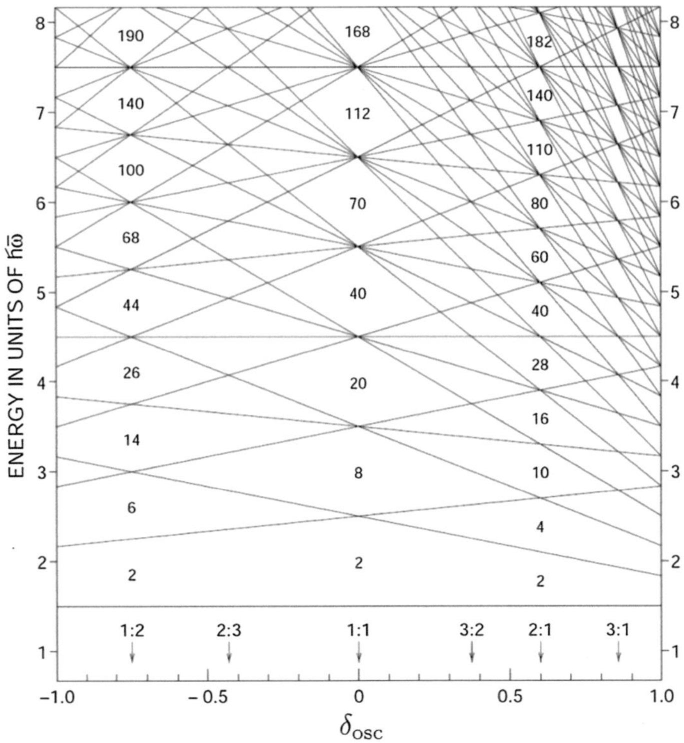

Fig. 7.11 The energy levels in the deformed harmonic oscillator model. Taken from [25]

References

1. A.B. Balantekin, N. Takigawa, Rev. Mod. Phys. **70**, 77 (1998)
2. M. Dasgupta, D.J. Hinde, N. Rowley, A.M. Stefanini, Annu. Rev. Nucl. Part. Sci. **48**, 401 (1998)
3. K. Hagino, N. Rowley, A.T. Kruppa, Comput. Phys. Commun. **123**, 143 (1999)
4. H. Esbensen, Nucl. Phys. A **352**, 147 (1981)
5. K. Hagino, N. Takigawa, Butsuri **57**, 588 (2002)
6. K. Hagino, N. Takigawa, Prog. Theor. Phys. **128**, 1001 (2012)
7. C.M. Lederer, V. Shirley, *Table of Isotopes*, 7th edn. (Wiley, New York, 1978)
8. K. Heyde, J.L. Wood, Rev. Mod. Phys. **83**, 1467 (2011)
9. A. Messiah, *Quantum Mechanics*, vol. II (Elsevier, Amsterdam, 1962)
10. A. Bohr, B.R. Mottelson, *Nuclear Structure*, vol. I (Benjamin, New York, 1969)
11. F. Iachello, A. Arima, *The Interacting Boson Model* (Cambridge University Press, Cambridge, 1987)

12. G.E. Brown, *Unified Theory of Nuclear Models and Forces* (North-Holland, Amsterdam, 1967)
13. A. Bohr, B.R. Mottelson, *Nuclear Structure*, vol. II (Benjamin, New York, 1975)
14. S.G. Nilsson, Mat. Fys. Medd. Dan. Vid. Selsk. **29**(16), (1955)
15. C. Gustafson, I.L. Lam, B. Nilsson, S.G. Nilsson, Ark. Fys. **36**, 613 (1967)
16. D.L. Hendrie et al., Phys. Lett. B **26**, 127 (1968)
17. V.M. Strutinsky, Yad. Fiz. **3**, 614 (1966); Sov. J. Nucl. Phys. **3**, 449 (1966); Nucl. Phys. A **95**, 420 (1967); Nucl. Phys. A **122**, 1 (1968)
18. P. Möller, J.R. Nix, W.D. Myers, W.J. Swiatecki, At. Data Nucl. Data Tables **59**, 185 (1995)
19. P. Ring, P. Schuck, *The Nuclear Many-Body problem* (Springer, Berlin, 1980)
20. S. Yoshida, N. Takigawa, Phys. Rev. C **55**, 1255 (1997)
21. P.-G. Reinhard, M. Rufa, J. Maruhn, W. Greiner, J. Friedrich, Z. Phys. A **323**, 13 (1986)
22. P. Möller, J.R. Nix, Nucl. Phys. A **536**, 20 (1992)
23. P.J. Nolan, P.J. Twin, Annu. Rev. Nucl. Part. Sci. **38**, 533 (1988); R.V.F. Janssens, T.L. Khoo, Annu. Rev. Nucl. Part. Sci. **41**, 321 (1991)
24. B. Singh, R.B. Firestone, S.Y. Frank Chu, Table of Superdeformed Nuclear Bands and Fission Isomers, WWW Edition, Updated June 1997; An updated version of the figure can be found in B. Singh, R. Zywina, R.B. Firestone. Nucl. Data Sheets **97**, 241 (2002)
25. I. Ragnarsson, S.G. Nilsson, R.K. Sheline, Phys. Rep. **45**, 1 (1987)

Chapter 8
Nuclear Decay and Radioactivity

Abstract Many nuclei with finite lifetimes are produced by nuclear reactions and/or nuclear decays, though the number of stable nuclei is limited to less than 300 if we call those nuclei whose lifetime is longer than or nearly equal to the age of the universe, i.e., about 13.8×10^9 years, as stable nuclei. Also, all the excited states of nuclei eventually decay into other nuclei via α or β decay or fission, or make transitions to lower energy levels of the same nucleus by emitting γ rays. The decay called cluster decay or heavy particle decay or cluster radioactivity, where a nucleus decays by emitting a nucleus heavier than α particle such as ^{14}C, has also been actively studied since 1980s (see Rose and Jones, Nature, 307: 245 (1984), [1]). In 2000s, the study of one-proton and two-proton radioactivities of proton-rich nuclei has also become active (see Blank and Borge, Prog Part Nucl Phys, 60: 403 (2008); Blank and Płoszajczak Rep Prog Phys, 71: 046301 (2008), [2]). In this chapter we learn the α-decay and electromagnetic transitions among the decay of nuclei and radioactivity. We also briefly discuss recent developments concerning fission.

8.1 Alpha Decay

Unstable nuclei with large atomic number decay primarily by emitting an α particle (α-*decay*).

$$^{210}_{84}\mathrm{Po}\left(0^+_{\mathrm{g.s.}}\right) \to {}^{206}_{82}\mathrm{Pb} + \alpha \quad (\tau_{1/2} = 138.38\,\mathrm{days})\ , \tag{8.1}$$

$$^{237}_{93}\mathrm{Np}\left(5/2^+_{\mathrm{g.s.}}\right) \to {}^{233}_{91}\mathrm{Pa} + \alpha \quad (\tau_{1/2} = 2.14 \times 10^6\,\mathrm{years})\ , \tag{8.2}$$

are the examples. Each of the total numbers of protons and neutrons is preserved in the decay. The nucleus before and after the decay is called the parent and the daughter nucleus, respectively. ^{210}Po and ^{206}Pb are the parent and daughter nuclei, respectively, in the decay of ^{210}Po. Both of the states of the parent nucleus prior to the decay and of the daughter nucleus after the decay are not necessarily the ground states, and are also not necessarily unique. The decay of ^{210}Po mentioned above represents the decay of the ground state of ^{210}Po. The half-life of this state is 138.38

N. Takigawa and K. Washiyama, *Fundamentals of Nuclear Physics*,
DOI 10.1007/978-4-431-55378-6_8

days, and the decay occurs to the ground state of ^{206}Pb ($0^+_{\text{g.s.}}$) with probability of more than 99% and the remaining part decays to the first excited state (2^+_1 state) of ^{206}Pb. In the case of the α-decay of the ground state of ^{237}Np, the branching ratios to many different states of the daughter nucleus are nearly equal and the final states diverge.

The amount of energy given by

$$Q_\alpha = B(A-4, Z-2) + B(4,2) - B(A,Z) \tag{8.3}$$

is released in the decay from the ground state of the parent nucleus to the ground state of the daughter nucleus. Q_α is called the Q-value of the α-decay between the ground states.

The α-decay takes place only when Q_α is positive. Using Eq. (8.3), the condition can be approximately represented as $\frac{B}{A} < \frac{B(4,2)}{4} - A\frac{d}{dA}(\frac{B}{A})$. Figure 2.6 suggests that $\frac{d}{dA}(\frac{B}{A})$ is about -7.5×10^{-3} MeV. On the other hand, the binding energy of α particle is 28.3 MeV. Hence the condition for the α-decay to occur becomes

$$\frac{B}{A} < 7.08 + 7.5 \times 10^{-3} A \text{ MeV}. \tag{8.4}$$

By adding the right-hand side of Eq. (8.4) to Fig. 2.6, we find that the Q-value of the α-decay gets positive, and hence the nucleus becomes in principle unstable with respect to α-decay, when the mass number exceeds about 150.

8.1.1 Decay Width

There exist various theoretical methods to estimate the *α-decay width*. Here, we introduce some of them.[1]

8.1.1.1 The Method to Obtain the Resonance Formula from the Pole of the S-Matrix

One of the simple methods is the potential model. In this method, one considers the α-decay to be the process that a metastable state in a potential field $V_{D\alpha}$, which we write simply as V in the following, of the daughter nucleus for the α particle decays with time by a quantum tunneling. The metastable state corresponds to the state of the parent nucleus. V consists of the nuclear and Coulomb potentials[2] and will behave as shown in Fig. 8.1. Figure 8.2 is its simplification. The abscissa r denotes the distance

[1] As a related topic, the study of resonance states by using the so-called *complex scaling method* is actively going on [3]. It is related also to the study of unstable nuclei.

[2] Here, we consider the decay of the s-wave.

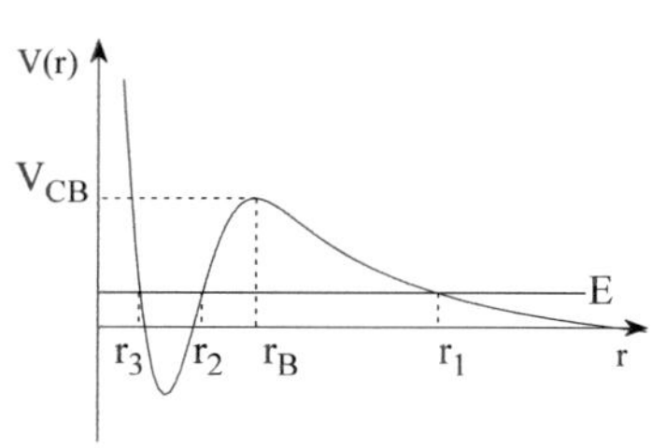

Fig. 8.1 The interaction potential between α particle and nucleus

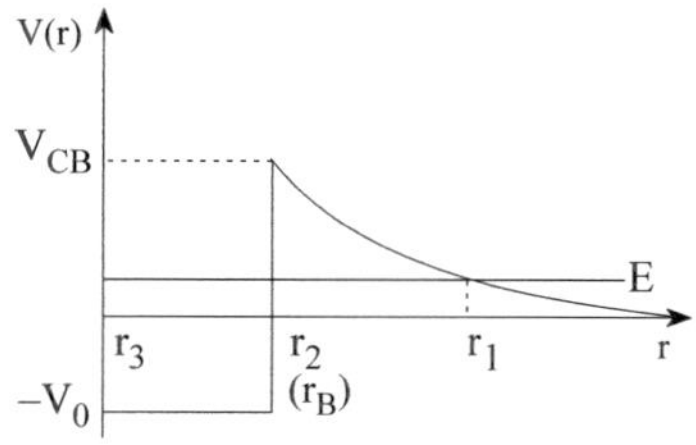

Fig. 8.2 The interaction potential between α particle and nucleus. The Gamow model

between the centers of the α particle and the daughter nucleus. For simplicity, we assumed that the daughter nucleus is spherical and correspondingly that V is a central potential.

The information on the metastable state can be obtained by studying the pole of the S-matrix for the scattering if one takes the view that the α-decay is the inverse process of the resonance scattering between the α particle and the daughter nucleus by the potential V. Although numerical calculations are needed in general to that end, one can derive formulae to determine the energy E, the width Γ of the metastable state and hence the lifetime of the α-decay $\tau = \hbar/\Gamma$ with intuitive physical understanding if one uses a semi-classical theory. In that connection, Fig. 8.1 shows the position of the metastable state, the Q-value of the α-decay ($E = Q_\alpha$) and the three classical turning points r_1, r_2, r_3 where the potential energy coincides with the Q-value of the α-decay: $V(r_i) = E (i = 1, 2, 3)$. Leaving the details of the derivation of the formulae to [4–6], here we quote the results:

$$\int_{r_3}^{r_2} k(r)\mathrm{d}r = \left(n + \frac{1}{2}\right)\pi + \frac{1}{2}\mathrm{arg}N \sim \left(n + \frac{1}{2}\right)\pi \quad (n: \text{ integer}), \tag{8.5}$$

$$\frac{\Gamma}{\hbar} = \frac{1}{T}\exp\left(-2\int_{r_2}^{r_1} \kappa(r)\mathrm{d}r\right) = \frac{1}{T}t^{(\mathrm{WKB})}, \tag{8.6}$$

$$k(r) = \sqrt{2\mu(E - V(r))}/\hbar, \quad \kappa(r) = \sqrt{2\mu(V(r) - E)}/\hbar, \tag{8.7}$$

$$T = 2\int_{r_3}^{r_2} \frac{\mathrm{d}r}{\sqrt{2\mu(E - V(r))/\mu}} = 2\int_{r_3}^{r_2} \frac{\mathrm{d}r}{v(r)}. \tag{8.8}$$

μ is the reduced mass for the relative motion between the α particle and the daughter nucleus. Equation (8.5) is the quantization condition to determine the position E of the metastable state and E has to agree with the Q value of the α-decay. The N is a factor related to the quantum tunneling. The resonance position reflects the properties of the potential barrier through the argument of the factor N. However, one can set arg$N$ to be 0 provided the α-decay takes place at an energy well below the potential barrier. The exponential factor of the second term of Eq. (8.6) is nothing but the tunneling probability $t^{(\mathrm{WKB})}$ estimated in the WKB approximation. On the other hand, as the definition of T, Eq. (8.8), suggests, the factor $1/T$ in front of the exponential function is the attempt frequency per unit time with which the α particle tries to decay.

8.1.1.2 Gamow Factor

The potential $V(r)$ is often assumed to be given by

$$V(r) = \begin{cases} -V_0 & \text{for } r < r_B \\ \frac{Z_D Z_\alpha e^2}{r} & \text{for } r \geq r_B, \end{cases} \tag{8.9}$$

as shown in Fig. 8.2 for the discussion of α-decay. Z_D and Z_α are the atomic numbers of the daughter nucleus and the α particle, respectively. In the following, we call the model which assumes Eq. (8.9) the Gamow model. Since the condition for the applicability of the WKB approximation is not satisfied in this case, one cannot use Eq. (8.6) for the estimate of the decay width as we learn later (see Eqs. (8.13)–(8.17)). However, it is useful for various discussions to evaluate the exponential factor $t^{(\mathrm{WKB})} = \exp(-2\int_{r_2}^{r_1} \kappa(r)\mathrm{d}r)$ for this case by denoting r_2 and r_1 as R and r_e, where r_e is the external classical turning point of the tunneling region. With the substitution $r = r_e \cos^2\theta$ the integral can be easily evaluated to give

$$t_{\mathrm{GM}}^{(\mathrm{WKB})} \equiv \exp\left(-2\int_{r_2}^{r_1} \kappa(r)\mathrm{d}r\right) = \exp\left[-4\eta\left(\cos^{-1}\sqrt{\frac{R}{r_e}} - \sqrt{\frac{R}{r_e}}\sqrt{1-\frac{R}{r_e}}\right)\right], \tag{8.10}$$

where η is called the *Sommerfeld parameter*, and is defined by $\eta \equiv Z_D Z_\alpha e^2/\hbar v$, where v is the speed in the asymptotic region.

Suppose that the Q-value is much smaller than the height of the potential barrier $V_{\mathrm{CB}} = Z_D Z_\alpha e^2/R$. Since $r_e \gg R$ in this case, the exponential factor becomes

$$t_{\mathrm{GM}}^{(\mathrm{WKB})} \sim F_{\mathrm{G}}\, \mathrm{e}^{4\sqrt{2\eta k R}}, \tag{8.11}$$

$$F_{\mathrm{G}} \equiv \mathrm{e}^{-2\pi\eta}, \tag{8.12}$$

where k is the wave number in the asymptotic region. The F_{G} is called the *Gamow factor*, and plays the principal role in governing the probability of quantum tunneling when the decay energy is small. The contribution of the second factor $\mathrm{e}^{4\sqrt{2\eta kR}}$ is much smaller than that of the Gamow factor. However, it cannot be ignored or set equal to 1 for the α-decay of heavy nuclei in order to keep the quantitative accuracy. In fact, it is larger than the pre-exponential factor which will be defined later.

8.1.1.3 Direct Method Based on the Gamow Model

In the case of the Gamow model, the wave functions are given by the spherical Bessel functions of the first kind in the potential region and by the Coulomb wave functions in the external region. Hence one can derive the formulae for the decay width by using them and the asymptotic form of the Coulomb wave function G_0 [7]. Leaving the derivation to Appendix A.8, here we quote the result for the case when $E \ll V_{\mathrm{CB}}$. Letting $T = 2R/v'$, where $v' = \hbar K/\mu$ is the speed of the α particle inside the potential well, i.e., at $r < r_B$, we have

$$\frac{\Gamma}{\hbar} = \frac{1}{T} t_{\mathrm{GM}}^{(\mathrm{D})} , \tag{8.13}$$

$$t_{\mathrm{GM}}^{(\mathrm{D})} \sim 4 \frac{k}{K} \frac{1}{G_0^2(\eta, kR)} , \tag{8.14}$$

$$\frac{1}{G_0^2(\eta, kR)} \sim \left(\frac{2\eta}{kR}\right)^{1/2} \mathrm{e}^{-2\pi\eta}\, \mathrm{e}^{4\sqrt{2\eta kR}} \sim \left(\frac{2\eta}{kR}\right)^{1/2} t_{\mathrm{GM}}^{(\mathrm{WKB})} , \tag{8.15}$$

$$t_{\mathrm{GM}}^{(\mathrm{D})} = 4 \frac{k}{K} \left(\frac{r_e}{R}\right)^{1/2} t_{\mathrm{GM}}^{(\mathrm{WKB})} = A_{\mathrm{GM}}^{(\mathrm{D})}\, t_{\mathrm{GM}}^{(\mathrm{WKB})} , \tag{8.16}$$

$$K \equiv \sqrt{2\mu(E + V_0)}/\hbar, \quad k = \sqrt{2\mu E}/\hbar . \tag{8.17}$$

$t_{\mathrm{GM}}^{(\mathrm{D})}$ is the probability of the tunneling effect given by Eq. (A.127). Note that it differs from the value given by Eq. (8.11), which is obtained by simply applying the WKB approximation, by the factor $A_{\mathrm{GM}}^{(\mathrm{D})}$.

8.1.1.4 Time-Dependent Perturbation Theory Based on the Gamow Model: Two-Potential Approach

The two-potential approach [8] is also often used for the calculation of the α-decay width. In this method, one introduces the potential $V_1(r)$, which has an infinitely thick potential barrier, i.e., $V_1(r) = -V_0$ for $r < R$ and $V_1(r) = V_{\mathrm{CB}}$ for $r \geq R$, and the potential $V_2(r)$, which has only the Coulomb potential, i.e., $V_2(r) = V_C(r) = \frac{Z_D Z_\alpha e^2}{r}$. The Gamow potential can be decomposed as

$$V(r) = \begin{cases} V_1(r) & \text{for } r < R \\ V_2(r) & \text{for } r \geq R \,, \end{cases} \tag{8.18}$$

by using these potentials. The α-decay width can be obtained by calculating the transition probability from a bound state $\phi_n(r)$ with energy E_n in the potential $V_1(r)$ to a continuum state $\phi_E(r)$ with energy E close to E_n in the potential $V_2(r)$. Since the perturbation Hamiltonian is $V(r) - V_1(r)$, the transition probability per unit time is given by

$$\frac{\Gamma}{\hbar} = \frac{2\pi}{\hbar} \left| \int \phi_E^*(r)[V(r) - V_1(r)]\phi_n(r)\mathrm{d}r \right|^2 . \tag{8.19}$$

One finally obtains[3]

$$\frac{\Gamma}{\hbar} = \frac{\hbar k_1}{\mu} D^2 \exp\left(-2\int_R^{r_e} \kappa(r)\mathrm{d}r\right) = \frac{\hbar k_1}{\mu} D^2 \mathrm{e}^{-2\pi\eta} \mathrm{e}^{4\sqrt{2\eta k R}} \sim \frac{\hbar k_1}{\mu} D^2 t_{\mathrm{GM}}^{(\mathrm{WKB})} , \tag{8.20}$$

$$\frac{\hbar k_1}{\mu} D^2 = \frac{2}{R}\left(\frac{2}{\mu}\right)^{1/2} (V_{\mathrm{CB}} - E)^{1/2} \frac{E + V_0}{V_{\mathrm{CB}} + V_0} , \tag{8.21}$$

$$D \equiv \left(\frac{2}{R}\right)^{1/2} \frac{K}{\sqrt{k_1^2 + K^2}} ; \quad k_1 = \sqrt{2\mu(V_{\mathrm{CB}} - E)}/\hbar . \tag{8.22}$$

According to [8], the pre-exponential factor $(\frac{\hbar k_1}{\mu})D^2$ represents the probability that the α particle appears at the nuclear surface per unit time.

8.1.1.5 Comparison of Three Methods

The formulae for the decay width in the three methods discussed in Sects. 8.1.1.1, 8.1.1.3 and 8.1.1.4 contain the Gamow factor in common. However, the pre-exponential factor depends on the way of derivation of the formulae in each method.

The formulae introduced in Sect. 8.1.1.1 are superior in the sense that they can be applied to smooth realistic potentials, since they have been derived based on the WKB approximation. However, one has to be careful when they are applied to the Gamow model as they are, and when the tunneling probability $t^{(\mathrm{WKB})}$ in Eq. (8.6) is replaced by the Gamow factor $\mathrm{e}^{-2\pi\eta}$ or by the formula multiplied with the correction factor Eq. (8.11). Since the discussions in both Sect. 8.1.1.3 and 8.1.1.4 are based on the Gamow model, their formulae must yield the correct result within the Gamow model, and should in principle give the same decay width.

As an example, let us consider the α-decay of ^{210}Po. As we discuss in Sect. A.8, there exists an ambiguity for the potential describing the α-decay, and there are

[3] See [8] for the details of the derivation and physical interpretation. See also [9].

various potentials ranging from a shallow to a deep potentials.[4] Here, we assume that the wave function of the relative motion between the daughter nucleus and the α particle has 11 nodes by considering the Pauli exclusion principle. In this case, the potential becomes a deep potential, and $V_0 = 108\,\text{MeV}$, $R = 7.92\,\text{fm}$ as we show in Sect. A.8. V_{CB} is about 30 MeV, while Q-value is 5.4 MeV. Hence it holds that

$$V_0 \gg V_{\text{CB}} \gg E_\alpha \,. \tag{8.24}$$

In this case, the pre-exponential factor $\frac{\hbar k_1}{\mu} D^2$ in Eq. (8.20) approximately becomes $\frac{1}{T} 4 \frac{k_\alpha}{K} (\frac{r_e}{R})^{1/2}$. The result of Sect. 8.1.1.3 given by Eqs. (8.13), (8.15), and (8.20) in Sect. 8.1.1.4 thus become practically equivalent for the model which assumes a deep potential for the potential between the daughter nucleus and the α particle.[5]

---

[4]Let us take a view of cluster model that a nucleus consists of two constituent clusters A and B. In the case of α-decay, the nucleus corresponds to the parent nucleus, and the clusters A and B to the daughter nucleus and the emitted α particle, respectively. Accordingly, we first represent the total wave function as the product of the intrinsic wave functions of the clusters A and B and the wave function for their relative motion. We then antisymmetrize with respect to the whole nucleons. This is called the *resonating group method (RGM)*. If we represent the intrinsic wave functions of two clusters in the harmonic oscillator model with the same oscillator parameter, then there exist in general special states called either the Pauli forbidden states or the redundant states for the wave function of the relative motion because of the constraint that the total wave function has to be antisymmetric with respect to the exchange of any two nucleons. They are the states, for which the total wave function vanishes as the result of antisymmetrization. Consequently, there appears ambiguity in the wave function of the relative motion concerning the admixture of the Pauli forbidden states, and it turns out that there exist several equivalent wave functions with different number of nodes, which we denote by $n_r = 0, 1, \ldots$ by excluding the node at the origin. Correspondingly, there appears an ambiguity for the potential describing the relative motion ranging from shallow to deep potentials. For example, if we represent ^{8}Be as the product of two α particles and their relative motion, then the 0s and 1s states are the Pauli forbidden states for the relative motion with angular momentum 0. By counting the quantum number in the harmonic oscillator shell model, one can write the condition for the Pauli forbidden states as

$$2n_r + \ell < N_{\text{C}}^{(\text{SM})} - N_{\text{A}}^{(\text{SM})} - N_{\text{B}}^{(\text{SM})} \,, \tag{8.23}$$

when the angular momentum of the relative motion is ℓ. Here, $N_{\text{C}}^{(\text{SM})}$, $N_{\text{A}}^{(\text{SM})}$ and $N_{\text{B}}^{(\text{SM})}$ are the total quantum number of the whole nucleus (C) and of the constituent clusters (A) and (B) when they are represented by the harmonic oscillator shell model. For example, in the case of α–α scattering or in the case of the α-decay of ^{8}Be, $N_{\text{A}}^{(\text{SM})} = N_{\text{B}}^{(\text{SM})} = 0$, $N_{\text{C}}^{(\text{SM})} = 4$. Equation (8.23) is called the *Wildermuth condition*. In the case of the α-decay of ^{210}Po, the right-hand side of Eq. (8.23) is $(5 + 6) \times 2$, so that the states with $n_r = 11$ and/or larger n_r are the allowed states. Incidentally, the existence of the Pauli forbidden states is also the origin of the repulsive core in the potential between complex nuclei. In the α–α scattering, the 0s and 1s states are the redundant states for $\ell = 0$ and the 0d state is the redundant state for $\ell = 2$. Hence there exists repulsive core in the potential for these partial waves. On the other hand, there exists no redundant state for $\ell = 4$ and higher partial waves, so that there is no repulsive core in the potential for those high partial waves.

[5]Various methods to calculate the decay width have been numerically compared for the proton decay [10].

8.1.1.6 Preformation Factor—*R*-Matrix Theory

All the three methods so far discussed presume that the α particle with the same structure as that of the α particle in nature exists inside the parent nucleus in advance of the α-decay. Let us call such a point of view the potential model. On the other hand, one can make a detailed description of the state of the parent nucleus using the wave functions of nucleons based on, e.g., the shell model or the cluster model. In such calculations the α particle is not presumed to exist in advance inside the nucleus with the same structure as that in nature. For example, a simple shell model is based on the thought that each nucleon inside a nucleus moves independently from each other in the mean field except for the constraint due to the Pauli exclusion principle. In order to evaluate the α-decay width based on the study of nuclear structure by a many-body theory such as the shell model, it is necessary to estimate the formation probability of α particle inside the nucleus prior to the decay. The formalism based on the *R-matrix theory*[6] is one of the representatives of such approaches which takes into account the preformation probability, where the decay width is given by the following formulae:

$$\Gamma = 2kr_c\gamma_L^2(r_c)v_L(r_c) = 2P_L(r_c)\gamma_L^2(r_c) , \tag{8.25}$$

$$v_L(r_c) \equiv \frac{1}{F_L^2(r_c)+G_L^2(r_c)} \sim \frac{1}{G_L^2(r_c)} , \tag{8.26}$$

$$\gamma_L(r) = \left(\frac{\hbar^2}{2\mu r}\right)^{1/2} \mathscr{Y}_L(r) , \tag{8.27}$$

$$\frac{\mathscr{Y}_L(r)}{r} = \langle \phi_\alpha \phi_D Y_{LM}(\hat{\mathbf{r}}) | \psi_P \rangle_{int,\hat{\mathbf{r}}} . \tag{8.28}$$

The F_L and G_L are the Coulomb wave functions. r_c is chosen such that the effect of short-range force between the daughter nucleus and the emitted particle, i.e., the α particle in the case of alpha decay, is absent in the region $r > r_c$, and is called the channel radius. $v_L(r_c)$ is called the *penetration factor*, where the index L shows that the decay takes place for the partial wave L.[7,8] The ψ_P is the wave function of the parent nucleus, and ϕ_D, ϕ_α are the intrinsic wave functions of the daughter nucleus and the α particle, respectively. They are determined based on appropriate

[6] The R-matrix theory is a theory to describe resonances in nuclear reactions. It describes the reaction by dividing the space into the internal region and the external region, i.e., the asymptotic region where the nuclear force between the emitted nucleus and the remaining nucleus can be ignored. See [11–15].

[7] As the asymptotic formula $G_L + \mathrm{i}F_L \sim \exp[\mathrm{i}(\rho - \eta\ln(2\rho) - L\pi/2 + \sigma_L)]$, where $\rho = kr$ and σ_L being the Coulomb phase shift, indicates, the wave function which becomes the outgoing wave of unit magnitude is given by $\psi_L(r) = F_L(r) + \mathrm{i}G_L(r)$ in the region $r \geq r_c$ for the Gamow model. Hence the $v_L(r_c) \equiv \frac{1}{F_L^2(r_c)+G_L^2(r_c)}$ gives the magnitude of the penetration probability from r_c to the outside of the potential barrier.

[8] As it was shown by Eq. (8.14) in Sect. 8.1.1.3, v_0 and the transmission coefficient t in the Gamow model are related as $t_{\mathrm{GM}}^{(\mathrm{D})} \sim 4\frac{k}{K}\frac{1}{G_0^2(kr_c)} \sim 4\frac{k}{K}v_0(kr_c)$.

many-body theories such as the shell model, cluster model or mean-field theory. The lower index *int*, $\hat{\mathbf{r}}$ of Eq. (8.28) means to integrate over the intrinsic coordinates and the angular coordinates of the relative motion. The $\mathscr{Y}_L(r)/r$ is therefore the radial wave function of the relative motion between the daughter nucleus and the α particle. $\gamma_L^2(r)$ has the dimension of energy and called the *reduced width*. It gives a measure of formation probability of the α particle at the surface ($r = r_c$) of the parent nucleus. It is often expressed as

$$\gamma_L^2(r_c) \equiv \theta^2(r_c)\gamma_W^2(r_c)\ , \quad \gamma_W^2(r_c) = \frac{3\hbar^2}{2\mu r_c^2}\ . \tag{8.29}$$

$\gamma_W^2(r_c)$ is the value of $\gamma_L^2(r_c)$ in the case when the initial state of the α-decay in the parent nucleus has such a structure that an α particle is clusterized and the radial wave function of the relative motion $Y_L(r) = \mathscr{Y}_L(r)/r$ distributes uniformly in the region $0 \leq r \leq r_c$. It is called the *Wigner limit* [16].[9] θ^2 is the reduced width in units of the Wigner limit.

Exercise 8.1 Noting that $G_0(\rho = 0) = 1/C_0(\eta)$ and $C_0^2(\eta) = 2\pi\eta(e^{2\pi\eta} - 1)^{-1}$, confirm that $1/G_0^2$ is proportional to the Gamow factor, which is the principal factor to govern the tunneling probability, if $\eta \gg 1$.

Let us investigate the relation between the formulae for the α-decay width discussed in Sect. 8.1.1.1, 8.1.1.3, and 8.1.1.4, and Eq. (8.25). If we denote the momentum for $\mathscr{Y}(r)$ in the region $r < r_c$ by p_{in}, and estimate it as $p_{in}r_c \sim \hbar$ based on the uncertainty relation, we have

$$\frac{\gamma_W^2}{\hbar} = \frac{3}{2}\frac{\hbar^2}{\mu}\frac{1}{r_c}\frac{p_{in}}{\hbar^2} \sim \frac{3}{T}\ . \tag{8.30}$$

Hence Eq. (8.25) can be expressed as

$$\frac{\Gamma}{\hbar} = \frac{1}{T} t f_P \tag{8.31}$$

by writing the penetration probability as t, which is tentatively identified with $P_L(r_c)$. $f_P \propto \theta^2$ represents the probability that the α particle exists inside the parent nucleus in advance of the decay. Let us therefore refer to it as the *preformation factor*.

The value of θ^2 provides important information on the extent of clusterization in light nuclei [18]. A large value of the reduced width for the α-decay corresponds to a large preformation factor of the α particle, and suggests that the picture of the alpha cluster structure holds. The preformation factor is ignored in some analyses of the α-decay. In that case, the effect of the preformation factor is effectively incorporated in the parameters such as the depth V_0 and the range parameter R of the potential.[10] Reference [20] is an example of the calculation of θ^2 based on the shell model.

[9]See [17] for the connection to the Teichmann–Wigner sum rule.

[10]Here, we quote [19] as references for the analysis of α-decay.

Incidentally, for heavy particle decay such as the ^{14}C emission, the possibility that the species of the emitted particle and its structure change on the way of decay due to the channel coupling effects has also been discussed.

8.1.2 *The Geiger–Nuttal Rule*

The α-decay width strongly depends on the Q-value of the decay and the atomic number of the parent or daughter nuclei through the tunneling probability. Their principal effects are expressed by the Gamow factor. Let us therefore express the decay width, more precisely, the decay rate per unit time as

$$\frac{\Gamma}{\hbar} = A e^{-2\pi\eta} \tag{8.32}$$

by separating out the Gamow factor. Taking the logarithm of both sides and using $\eta \equiv Z_D Z_\alpha e^2/\hbar v$, we obtain

$$\log_{10}\frac{\Gamma}{\hbar} = \log_{10} A - C Z_D E_\alpha^{-1/2}\,, \tag{8.33}$$

$$C \equiv 2\pi \frac{e^2}{\hbar c}\sqrt{\frac{\mu c^2}{2}} Z_\alpha \log_{10} \mathrm{e} \approx 1.72\,. \tag{8.34}$$

The factor A does not depend so much on the decaying system as long as we consider nuclei in a region of similar mass number. Hence Eq. (8.33) suggests that the experimental data distribute almost along a straight line if we plot them by taking $\log_{10}\Gamma/\hbar$ as the ordinate and $Z_D E_\alpha^{-1/2}$ as the abscissa [12, 21]. This property is called the *Geiger–Nuttal rule*.

Incidentally, detailed information can be obtained from the analysis of the factor A, since it contains the information on the system such as the potential and the preformation factor.

8.2 Fission

In Chap. 2 we studied fission somewhat in detail with respect to nuclear instability. Here, we briefly discuss recent developments of the study of fission. As shown in Fig. 2.19 in a simplified fashion, fission is, in a general term, one of the decay processes of a metastable state, and occurs by quantum tunneling if the temperature of the system is 0 or low, while occurs thermally, i.e., by thermal hopping, if the temperature becomes sufficiently high. Figure 8.3 schematically shows this situation. The critical temperature T_C which divides the quantum decay and the thermal decay is given by

Fig. 8.3 The temperature or energy dependence of the decay mechanism and the width of a metastable state

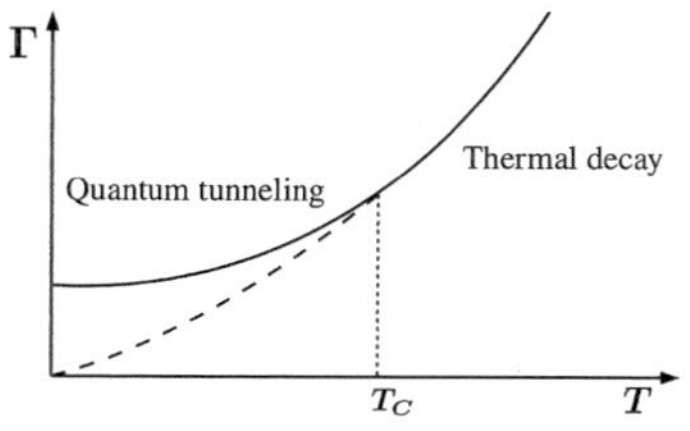

$$T_C \sim \frac{\hbar\Omega}{2\pi}\,, \tag{8.35}$$

if we denote the curvature of the potential barrier by $\hbar\Omega$ (Here, we consider in terms of the corresponding energy) [22, 23]. If we take a simple view of a one-dimensional process and express the curvature of the potential, in which the metastable state is made, around the potential minimum by $\hbar\Omega_0$, then the decay width, more precisely, the number of decay per unit time is given by

$$\Gamma \sim \begin{cases} \frac{\Omega_0}{2\pi}\frac{1}{1+\exp(\frac{2\pi}{\hbar\Omega}V_f)} & \text{for } T < T_C;\ \text{Quantum decay} \\ \frac{\Omega_0}{2\pi}\mathrm{e}^{-V_f/T} & \text{for } T \geq T_C;\ \text{Thermal decay.} \end{cases} \tag{8.36}$$

In each equation, the first factor $\frac{\Omega_0}{2\pi}$ represents the frequency of the metastable state to attempt decay and the second factor the success probability. V_f, which corresponds to E_f in Sect. 2.3.4, is the height of the fission barrier. We used the formula of the uniform approximation for the tunneling probability and approximated the potential barrier by a quadratic function. $\mathrm{e}^{-V_f/T}$ is a characteristic factor for thermal decay, and is called the *Arrhenius factor*.

The spontaneous fission is an example of the decay by quantum tunneling. A characteristic is that it is a quantum tunneling in a multi-dimensional space. Many experimental data such as those of the mass and energy distributions of the fission fragments have been accumulated. On the other hand, detailed theoretical calculations of the potential energy in the multi-dimensional space, where the mass ratio and the deformation parameters of the fission fragments are variables, have been performed by using, e.g., the macroscopic–microscopic method, and the search for the decay path and the calculation of the fission lifetime with, e.g., the *multi-dimensional escape path method* [24] have been developed [25, 26].[11]

---

[11] Not necessarily limited to the spontaneous fission, it has been experimentally reported that there are cases where there exist multiple decay paths with different barrier heights for the fission of the same nucleus. They are called *bimodal fission*. The bimodal fission appears as the coexistence of the symmetric and asymmetric fission for, e.g., ^{226}Ra(^{3}He,df). On the other hand, both are symmetric fission, but differ in the width of the mass distribution and the distribution of the kinetic energy of fission fragments in the case of the spontaneous fission of some of the actinides with large mass number such as ^{258}Fm.

The induced fission corresponds to the thermal decay if the excitation energy is high as it is the case for the fission induced by heavy-ion collisions. The formula for the thermal decay given by the lower line of Eq. (8.36) is known as the formula of the *transition state theory* by Bohr and Wheeler [27]. The *Kramers formula* [28] which takes into account the effect of friction,

$$\Gamma \sim \begin{cases} \gamma \frac{V_f}{T} \mathrm{e}^{-V_f/T} & \text{when } \frac{\gamma}{2\Omega} < \frac{T}{V_f} \text{ (the case of weak friction)} \\ \frac{\Omega_0}{2\pi} K \mathrm{e}^{-V_f/T} & \text{when } \frac{\gamma}{2\Omega} > \frac{T}{V_f} \text{ (the case of strong friction)}, \end{cases} \tag{8.37}$$

where

$$K = \frac{1}{\Omega}\left[\sqrt{\Omega^2 + \frac{1}{4}\gamma^2} - \frac{\gamma}{2}\right] \tag{8.38}$$

is also often used as the basic formula for the decay width of the thermal fission in calculating the competition between fission and the emission of particles such as neutrons by viewing the fission as a kind of Brownian motion [29, 30] in the external field and hence by formulating it in terms of the Fokker–Planck equation in the phase space. γ is the friction or viscosity coefficient in the Fokker–Planck equation. K is the factor which takes into account the effect of friction on the decay rate and is called the *Kramers factor*.[12]

The studies of the reaction mechanism of heavy-ion reactions and of various physical quantities governing the collision have significantly advanced since 1970s. In this procedure, various concepts have been introduced. They are, e.g., the *complete fusion reaction*, where the colliding nuclei completely fuse and decay after a long time, and the *quasi-fission*, which resembles the complete fusion concerning the mass distribution of the reaction products, but differs in the angular distribution and the *deep inelastic collision (DIC)*, which is supposed to have a short reaction time because of the feature that the masses of the reaction products stay near the masses of the projectile and the target nuclei even though a large energy loss from the relative motion is involved. Many experimental data have been accumulated in these studies concerning the mass and kinetic energy distributions of the reaction products such as the fission fragments, and the number of emitted neutrons associated with fission. The theoretical approach which follows the time evolution of the system based on the Langevin equation by simulating the reaction as a *random walk* (or stochastic process, or drunker's walk or Wiener process) in a multi-dimensional space has been developed as a powerful method to analyze these data by paying attention to the dissipation and associated fluctuation properties [35] (see also [36]).

---

[12]There appears in addition a multiplicative correction factor which takes into account the quantum fluctuation around the classical path if we relate the decay width to the imaginary part of the Helmholtz free energy [31] and evaluate the partition function by using the path integral method [32–34].

8.3 Electromagnetic Transitions

A nucleus in the ground state makes transitions to excited states by absorbing photon, while a nucleus in an excited state makes transition to the ground state or to lower excited states by emitting photon. One can extract much information on the structure of the nucleus such as the excitation energy, the spin and parity of each energy level, the nuclear shape and the collectivity of each excitation motion through these transitions.

8.3.1 Multipole Transition, Reduced Transition Probability

The probability of the electromagnetic transition can be given to a good accuracy by the Fermi golden rule in the first order perturbation theory,

$$T_{fi} = \frac{2\pi}{\hbar} |\langle f|\hat{H}_{\text{int}}|i\rangle|^2 g(E_f) \, . \tag{8.39}$$

$\hat{H}_{\text{int}}$ is the interaction between the nucleus and the electromagnetic field, and is given by Eq. (A.173) as will be described in Appendix A.9. If we denote the initial and the final states of the nucleus by $|\psi_i\rangle$ and $|\psi_f\rangle$, respectively, and express the state of the radiation field in the occupation number representation (see Appendix A.9.3), then the change of the state can be schematically written as

$$|i\rangle = |\psi_i\rangle|\cdots n_\lambda \cdots\rangle \rightarrow \begin{cases} |f\rangle = |\psi_f\rangle|\cdots n_\lambda - 1 \cdots\rangle & \text{for absorption,} \\ |f\rangle = |\psi_f\rangle|\cdots n_\lambda + 1 \cdots\rangle & \text{for emission.} \end{cases} \tag{8.40}$$

Also, it holds that $\hbar kc = |E_f - E_i|$ because of the energy conservation if we denote the wave length of the photon by $\lambda = 2\pi/k$, and the energy of the initial and final states of the nucleus by E_i and E_f, respectively.

$g(E_f)$ is the number of final states of the radiation field per unit energy. If we consider the magnetic radiation as an example, we have

$$0 = j_l(k_n R) \sim \frac{1}{k_n R} \sin\left(k_n R - \frac{\pi}{2} l\right) \tag{8.41}$$

from the boundary condition given by Eq. (A.167). Here, we considered $R \rightarrow \infty$. Hence it follows that

$$k_n R = \frac{\pi}{2} l + n\pi, \quad n = 0, \pm 1, \pm 2, \cdots \, . \tag{8.42}$$

The density of the final state of the radiation field g is thus given by

$$g(k) = \frac{1}{\Delta E(k)} = \frac{1}{\hbar(\Delta k)c} = \frac{1}{\hbar\frac{\pi}{R}c} = \frac{R}{\hbar c\pi} \ . \tag{8.43}$$

Using the explicit expressions of the vector fields, Eqs. (A.164) and (A.165), the expression of the normalization constant Eq. (A.169), and the formula for the level density Eq. (8.43), we can write the transition probability per unit time by emitting one photon of the (λ, k, I, M) type as

$$T_{fi}(E, kIM) = \frac{8\pi k}{\hbar}\left|\left\langle f\left|\frac{1}{ck}\int \mathbf{j}\cdot\nabla\times j_I(kr)\mathbf{Y}_{IIM}d\mathbf{r}\right|i\right\rangle\right|^2 , \tag{8.44}$$

$$T_{fi}(M, kIM) = \frac{8\pi k}{\hbar}\left|\left\langle f\left|\frac{1}{c}\int \mathbf{j}\cdot j_I(kr)\mathbf{Y}_{IIM}d\mathbf{r}\right|i\right\rangle\right|^2 . \tag{8.45}$$

Let us write the T_{fi}, in general, as

$$T_{fi}(\lambda, kIM) = \frac{8\pi(I+1)}{\hbar I[(2I+1)!!]^2}\left(\frac{E_\gamma}{\hbar c}\right)^{2I+1}\left|\langle f|\hat{\mathscr{M}}(\lambda, kIM)|i\rangle\right|^2 \tag{8.46}$$

by using the multipole transition operator $\hat{\mathscr{M}}(\lambda, kIM)$.

In nuclear physics, the long wave length approximation usually well applies. Namely, it well holds that

$$kR \ll 1 \tag{8.47}$$

if we denote the nuclear radius by R. Equation (8.47) is equivalent to[13]

$$E_\gamma = \hbar kc \ll \frac{\hbar c}{R} = \frac{\hbar c}{r_0}A^{-1/3} \approx 197A^{-1/3}\ \text{MeV} \ . \tag{8.48}$$

In the long wave length approximation, using

$$j_I(kr) \sim \frac{(kr)^I}{(2I+1)!!}\left[1 - \frac{1}{2}\frac{(kr)^2}{2I+3} + \cdots\right] , \tag{8.49}$$

we can show that the multipole transition operators are nothing but the electric multipole and the magnetic multipole operators introduced in Sect. 4.2 as

$$\hat{\mathscr{M}}(E, kIM) \approx \int \rho r^I Y_{IM}d\mathbf{r} = \hat{Q}_{IM} , \tag{8.50}$$

$$\hat{\mathscr{M}}(M, kIM) \approx \frac{1}{c(I+1)}\int (\mathbf{r}\times\mathbf{j})\cdot\nabla(r^I Y_{IM})d\mathbf{r} = \hat{M}_{IM} \ . \tag{8.51}$$

[13]Particle decay usually occurs with a higher probability when a high energy γ ray is emitted.

Exercise 8.2 Show that the wave length of photons of 1 MeV is about 400π fm.

One usually does not distinguish the difference of the orientation of the angular momentum of the nucleus except for the experiments related to polarization. By taking average with respect to the initial value M_i of the z component of the angular momentum of the nucleus, and sum over the value M_f in the final state, the total electric or magnetic transition probability with multipolarity I is given by

$$T_{fi}(\lambda, I) = \frac{1}{2I_i + 1} \sum_{M_i, M_f, M} T_{fi}(\lambda, kIM) , \tag{8.52}$$

where I_i is the angular momentum of the initial state of the nucleus. Performing the sum in Eq. (8.52) by using the Wigner–Eckart theorem and the properties of the $3j$ symbol or the Clebsch–Gordan coefficients, we obtain

$$T_{fi}(\lambda, I) = \frac{8\pi(I+1)}{\hbar I[(2I+1)!!]^2} \left(\frac{E_\gamma}{\hbar c}\right)^{2I+1} B(\lambda I; I_i \to I_f) . \tag{8.53}$$

Here, $B(\lambda)$ is called the *reduced transition probability*, and is given by

$$B(EI; I_i \to I_f) = \frac{1}{2I_i + 1} |\langle f \| \hat{Q}_I \| i \rangle|^2 \tag{8.54}$$

and

$$B(MI; I_i \to I_f) = \frac{1}{2I_i + 1} |\langle f \| \hat{M}_I \| i \rangle|^2 \tag{8.55}$$

for the electric and magnetic transitions, respectively.

8.3.2 *General Consideration of the Selection Rule and the Magnitude*

If one denotes the angular momentum and parity of the initial and final states by I_i, Π_i and I_f, Π_f, then the selection rules,

$$|I_i - I_f| \leq I \leq I_i + I_f , \quad \Pi_i \Pi_{\lambda I} \Pi_f = 1 , \tag{8.56}$$

hold due to the structure of the transition matrices $\langle I_f M_f | \hat{Q}_{IM} | I_i M_i \rangle$, $\langle I_f M_f | \hat{M}_{IM} | I_i M_i \rangle$. $\Pi_{\lambda I}$ is the parity of the multipole radiation, and is given by

$$\Pi_{EI} = (-1)^I, \quad \Pi_{MI} = (-1)^{I+1} , \tag{8.57}$$

because the interaction with the radiation field is given by $\hat{H}_{\rm int} \propto \mathbf{A} \cdot \mathbf{j}$.

Table 8.1 Summary of the electromagnetic transition probability. The unit of each quantity is: $B(E\lambda)$: e^2 fm$^{2\lambda}$; $B(M\lambda)$: μ_N^2 fm$^{2\lambda-2}$; E_γ: MeV; T: s^{-1}. The "yes" and "no" in the column of "Parity change" mean that the transition can take place only when the parities of the initial and final states, Π_i and Π_f, are different or the same, respectively. The γ ray transition is absolutely forbidden if $I_i = I_f = 0$, since the intrinsic spin of photon is 1

Type	Angular momentum selection rule	Parity change	Transition probability $T(\mathrm{s}^{-1})$
E1	$\lvert I_i - I_f\rvert \le 1 \le I_i + I_f$	yes ($\Pi_i \ne \Pi_f$)	$T(E1) = 1.59 \times 10^{15} E_\gamma^3\, B(E1)$
E2	$\lvert I_i - I_f\rvert \le 2 \le I_i + I_f$	no ($\Pi_i = \Pi_f$)	$T(E2) = 1.22 \times 10^{9}\, E_\gamma^5\, B(E2)$
E3	$\lvert I_i - I_f\rvert \le 3 \le I_i + I_f$	yes ($\Pi_i \ne \Pi_f$)	$T(E3) = 5.67 \times 10^{2}\, E_\gamma^7\, B(E3)$
M1	$\lvert I_i - I_f\rvert \le 1 \le I_i + I_f$	no ($\Pi_i = \Pi_f$)	$T(M1) = 1.76 \times 10^{13} E_\gamma^3 B(M1)$
M2	$\lvert I_i - I_f\rvert \le 2 \le I_i + I_f$	yes ($\Pi_i \ne \Pi_f$)	$T(M2) = 1.35 \times 10^{7}\, E_\gamma^5\, B(M2)$
M3	$\lvert I_i - I_f\rvert \le 3 \le I_i + I_f$	no ($\Pi_i = \Pi_f$)	$T(M3) = 6.28 \times 10^{0}\, E_\gamma^7\, B(M3)$

The transition probability T_{fi} is proportional to E_γ^{2l+1} as Eq. (8.53) shows. Also, the radiation of the γ ray is more strongly hindered with increasing multipolarity l. Table 8.1 summarizes the properties of the electromagnetic transition.

8.3.3 Single-Particle Model Estimate: Weisskopf Units and Experimental Values

The electromagnetic transitions give information on the lifetime of the excited states and the strength of the photo-absorption as the inverse process. They provide information also on the nuclear structure such as the nuclear shape, and also on the motion of nucleons inside a nucleus, especially on collective motions. In this connection, it is convenient to have an estimate of the value for the case when the single-particle model holds as a standard value to be compared.

As an example, let us consider the electric transition probability of an odd nucleus. The reduced transition probability is given by

$$B(E\lambda; I_i \to I_f) = \frac{1}{2I_i+1} \left| \left\langle \psi_f(I_f) \left\| \int \rho r^\lambda Y_\lambda \mathbf{dr} \right\| \psi_i(I_i) \right\rangle \right|^2 \tag{8.58}$$

$$= \frac{1}{2I_i+1} \left| \left\langle \psi_f(I_f) \left\| \sum_{k=1}^{A} e_k r_k^\lambda Y_\lambda(\theta_k, \varphi_k) \right\| \psi_i(I_i) \right\rangle \right|^2 . \tag{8.59}$$

In the single-particle model, one considers that the electromagnetic transition occurs by the transition of the valence nucleon from the point of view of the shell model. If we denote the initial and final states by the quantum numbers of the shell model $(n\ell j)_{i,f}$, then we have

$$B_{\rm sp}(E\lambda; I_i \to I_f) = \frac{1}{2I_i+1}\left|\langle (n\ell j)_f \| e_{\rm eff} r^\lambda Y_\lambda \| (n\ell j)_i \rangle\right|^2 , \tag{8.60}$$

which becomes

$$B_{\rm sp}(E\lambda; I_i \to I_f) = e_{\rm eff}^2 \frac{2\lambda+1}{4\pi}\left|\left\langle j_i \lambda \frac{1}{2} 0 \middle| j_f \frac{1}{2}\right\rangle\right|^2 \\ \times \left|\langle (n\ell j)_f | r^\lambda | (n\ell j)_i \rangle\right|^2 \frac{1+(-1)^{\ell_i+\ell_f+\lambda}}{2} \tag{8.61}$$

after some calculations. The $e_{\rm eff}$, named the *effective charge*, effectively takes into account the effect of the nucleons other than the valence nucleon due to, e.g., the core polarization. Let us consider especially the case when $j_i = I + 1/2, j_f = 1/2$, and approximate the Clebsch–Gordan coefficient by the asymptotic value:

$$\left\langle j_i \lambda \frac{1}{2} 0 \middle| j_f \frac{1}{2}\right\rangle = (-1)^I \frac{1}{\sqrt{2I+1}}\sqrt{\frac{2(I+1)}{2I+1}} \sim (-1)^I \frac{1}{\sqrt{2I+1}} , \tag{8.62}$$

and furthermore estimate $\langle f | r^I | i \rangle$ as

$$\langle f | r^I | i \rangle = \int_0^\infty R_f(r) r^I R_i(r) r^2 {\rm d}r = \frac{\int_0^{R_0} r^I r^2 {\rm d}r}{\int_0^{R_0} r^2 {\rm d}r} = \frac{3}{I+3} R_0^I , \tag{8.63}$$

where $R_f(r)$ and $R_i(r)$ are the radial wave functions of the final and initial states, by assuming that the nucleon uniformly distributes inside the region of radius R_0. We then obtain

$$B_W(EI) = e^2 \frac{1}{4\pi}\left(\frac{3}{I+3}\right)^2 R_0^{2I} . \tag{8.64}$$

We have set $e_{\rm eff}$ to the elementary charge e in order to have a measure of the single-particle transition. By introducing similar approximations, and further approximating $I^2(g_s - 2(I+1)^{-1} g_\ell)^2 \approx 10$ [37], we obtain

$$B_W(MI) = \frac{10}{\pi}\left(\frac{3}{I+3}\right)^2 R_0^{2I-2} \mu_N^2 \tag{8.65}$$

for the magnetic transition [37]. The values given by Eqs. (8.64) and (8.65) are called the *Weisskopf unit* and play an important role in judging the properties of the nucleus, especially concerning whether a certain motion of the nucleus is of single-particle nature or of a collective nature.

For even–even nuclei as well, by taking the initial and final states to be

$$|0^+_{\rm g.s.}\rangle = |(n\ell j)^2_\pi; 0^+\rangle, \quad |2^+_1\rangle = |(n\ell j)^2_\pi; 2^+_1\rangle , \tag{8.66}$$

Table 8.2 $E2$ transition probability $B(E2)$ in units of e^2 fm^4

Nucleus	Initial configuration	Final configuration	$B_{exp.}(E2)$	$B_{sp}(E2)$	Effective charge e_{eff}/e
$^{17}_{8}$O	$s_{1/2}$	$d_{5/2}$	6.3	35	0.42
$^{17}_{9}$F	$s_{1/2}$	$d_{5/2}$	64	43	1.2
$^{41}_{20}$Ca	$p_{3/2}$	$f_{7/2}$	66	40	1.3
$^{41}_{21}$Sc	$p_{3/2}$	$f_{7/2}$	110	40	1.7
$^{207}_{82}$Pb	$f_{5/2}^{-1}$	$p_{1/2}^{-1}$	70	81	0.9
$^{209}_{83}$Bi	$f_{7/2}$	$h_{9/2}$	40 ± 20	2.3	4 ± 1.5

we can derive

$$B\left(E2; 0^+_{\text{g.s.}} \to 2^+_1\right) \sim \frac{5}{4\pi} e^2 \left(\frac{3}{5} R_0^2\right)^2 \tag{8.67}$$

as an estimate based on the simple shell model. The index π in Eq. (8.66) refers to proton. Like the case for odd nuclei, we assumed that j is large, and used $\langle j|r^2|j\rangle = 3R_0^2/5$.

Incidentally, note that in general

$$B(E2; 2^+_1 \to 0^+_{\text{g.s.}}) = \frac{1}{5} B(E2; 0^+_{\text{g.s.}} \to 2^+_1) \ . \tag{8.68}$$

Table 8.2 compares the experimental values of the quadrupole transition probability from the first excited state to the ground state for several doubly magic ± 1 nuclei with the single particle estimate given by Eq. (8.61). In the latter, we supposed the configurations for the valence nucleon in the initial and final states to be those given in the second and third columns, respectively, from the experimentally known spin and parity and the shell model. The last column is the effective charge which is required to reproduce the experimental value with the single-particle model. The table shows that the single-particle model well reproduces the experimental value. However, a large effective charge is needed for Bi.

On the other hand, Fig. 8.4 shows the experimental values of the deformation parameter β, which have been deduced from the experimental values of the quadrupole transition probability $B(E2)$ from the ground state to the first excited 2^+ state based on the collective model (Eq. (8.76), which will be discussed later), in the ratio to the estimate of the single-particle model β_{sp} for many even–even nuclei in the wide range of the atomic number [38]. The equivalent radius has been assumed to be $R_0 = 1.2A^{1/3}$ fm in estimating β. Also, the formula $\beta_{sp} = 1.59/Z$, which is obtained by replacing $B(E2)$ in Eq. (8.76) with the single-particle estimate Eq. (8.67), was used for β_{sp}. In contrast to the case for doubly magic± 1 nuclei shown in Table 8.2, the experimental values for the nuclei far from the magic numbers significantly deviate from the estimate of the single-particle model. The experimental values are much more enhanced than the estimates of the single-particle model especially for the nuclei in the region of mass number between 150 and 200 such as rare-earth nuclei, and for the nuclei whose atomic number exceeds 86 such as Rn, U and Th.

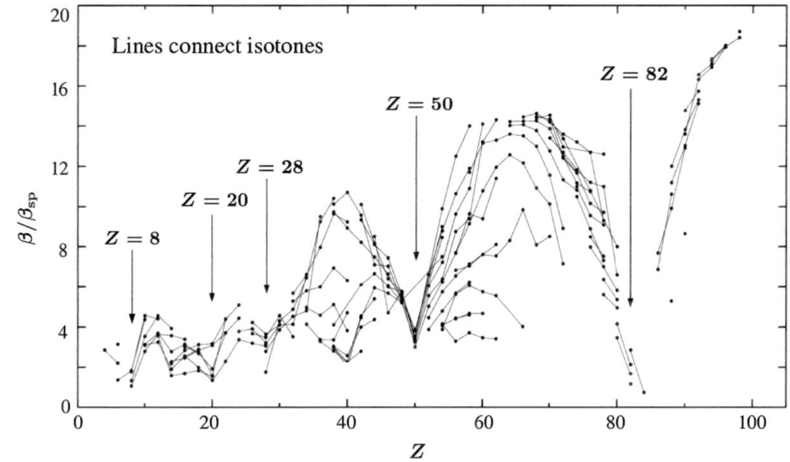

Fig. 8.4 The atomic number dependence of the deformation of even–even nuclei

8.3.4 Connection Between Electromagnetic Transitions and the Shapes and Collective Motions of Nuclei

The nuclei with large deformation, i.e., with large quadrupole transition probability, which have been pointed out regarding Fig. 8.4, well coincide with those nuclei which are expected to be statically deformed in their ground state and to be easily excited to the rotational excited states from the energy level structure in the vicinity of the ground state. In addition, the experimental values of the quadrupole transition probability systematically show significantly larger values than the prediction of the single-particle model for the nuclei whose mass number is around 120 such as Cd isotopes with the atomic number 48 as well. The structure of the energy spectrum suggests that they are the nuclei which are expected to be easily excited to the vibrational excited states of the quadrupole type, although they are spherical in the ground state. These nuclei can be more naturally described in the collective model than in the single-particle model.

Here, we mention some basic results concerning the collective model, i.e., the geometrical collective model, which is often called the Bohr–Mottelson model.

The geometrical collective model uses the deformation parameters $\alpha_{\lambda\mu}$ introduced by Eq. (7.2) as dynamical variables. Correspondingly, the density distribution of protons is assumed to be given by

$$\hat{\rho}_{\rm p}(\mathbf{r}) = \rho_{\rm p}^{(0)}\Theta[\hat{R}(\theta,\varphi) - r]$$
$$= \rho_{\rm p}^{(0)}\Theta\left[R_0\left(1 + \hat{\alpha}_{00} + \sum_{\lambda=2}^{\infty}\sum_{\mu=-\lambda}^{\lambda}\hat{\alpha}^*_{\lambda\mu}Y_{\lambda\mu}(\theta,\varphi)\right) - r\right] \qquad (8.69)$$

using the step function $\Theta(x)$.[14] The operator of the electric transition is then given by

$$\hat{Q}_{\lambda\mu} = e\int_V \hat{\rho}_{\rm p}(\mathbf{r})r^{\lambda}Y_{\lambda\mu}(\theta,\varphi)\mathrm{d}^3r \sim eZ\frac{3}{4\pi}R_0^{\lambda}\hat{\alpha}_{\lambda\mu}\,. \qquad (8.70)$$

Here, we truncated the sum by the first order of the operator of the deformation parameter.[15]

For spherical nuclei, if we approximate the surface vibration by the harmonic oscillator model, then the quadrupole transition probability from the first excited state to the ground state is given by

$$B(E2;2_1^+ \to 0_{\rm g.s.}^+) = \left(eZ\frac{3}{4\pi}R_0^2\right)^2\frac{\hbar}{2B_2\Omega_2}\,, \qquad (8.71)$$

if the first excited state is the one phonon state with the angular momentum and parity 2^+. The $\hbar\Omega_2$ and B_2 represent the excitation energy and the mass parameter of the quadrupole vibration, respectively.[16]

Exercise 8.3 Derive the expressions of $B(E2)$ from the two phonon states $0_2^+, 2_2^+, 4_1^+$ of the quadrupole vibration to the one phonon state 2_1^+ and discuss their ratios to $B(E2;2_1^+ \to 0_{\rm g.s.}^+)$.

On the other hand, for axially symmetric even–even nuclei with quadrupole deformation, the Euler angles $\omega = \{\phi,\theta,\psi\}$ which specify the directions of principal axes become the variables for the rotational motion and the total wave function is given by

$$|\Psi_{IMK}\rangle = \left(\frac{2I+1}{16\pi^2(1+\delta_{K0})}\right)^{1/2}\left[\mathscr{D}^I_{MK}(\omega)\Phi_K(q) + (-1)^{I+K}\mathscr{D}^I_{M-K}\Phi_{\bar{K}}(q)\right]\,, \qquad (8.72)$$

where $\mathscr{D}^I_{MK}$ is the Wigner $\mathscr{D}$ function, and K is the component of the angular momentum along the direction of the symmetry axis. $\Phi_K(q)$ is the wave function of the intrinsic motions, and $\Phi_{\bar{K}}(q)$ is the wave function which is obtained by rotating

[14] $\Theta(x) = 1\ (x \geq 0)$, $\Theta(x) = 0\ (x < 0)$.

[15] See p.13 of [39] for the expression including higher order terms.

[16] The values of $\hbar\Omega_2$ and B_2 can be theoretically estimated in the liquid-drop model which views the nucleus as an irrotational incompressible fluid. However, the resulting values are quantitatively in significant disagreement with the experimental data. Instead, more reliable estimates can be obtained phenomenologically from the experimental values of the excitation energy of the first excited 2^+ state E_2 and of $B(E2;2_1^+ \to 0_{\rm g.s.}^+)$.

$\Phi_K(q)$ by 180° around the second axis when the symmetry axis is taken to be the third axis. The operator of the electric quadrupole moment and the $E2$ transition is given by

$$\hat{Q}_{\lambda\mu} \sim \left(\frac{5}{16\pi}\right)^{1/2} eQ_0 \mathscr{D}^{\lambda}_{\mu 0} , \tag{8.73}$$

$$Q_0 = \left(\frac{16\pi}{5}\right)^{1/2} Z\frac{3}{4\pi} R_0^2 \beta , \tag{8.74}$$

where $\lambda = 2$. Q_0 is the intrinsic quadrupole moment.

Using Eqs. (8.72)–(8.74) and the properties of the $\mathscr{D}$ function, one can show that the electric quadrupole transition probability within a rotational band with the quantum number K is given by

$$B(E2; KI_1 \rightarrow KI_2) = \frac{5}{16\pi} e^2 Q_0^2 \langle I_1 2K0 | I_2 K \rangle^2 . \tag{8.75}$$

Especially, the value of $B(E2)$ from the first excited 2_1^+ state to the ground state $0^+_{\text{g.s.}}$ of the ground state rotational band with $K = 0$ of even–even nuclei is given by

$$B(E2; 2_1^+ \rightarrow 0^+_{\text{g.s.}}) = \frac{1}{5}\left(\frac{3}{4\pi} ZeR_0^2\right)^2 \beta^2 . \tag{8.76}$$

Equation (8.76) shows that the magnitude of $B(E2)$ increases proportionally to the square of the deformation β.[17,18] Note that the electromagnetic transition probabili-

[17]One can estimate the deformation parameter β, more precisely βR_0^2, from the experimental data of $B(E2)$ by using Eq. (8.76). However, $B(E2)$ gives only the magnitude of the deformation parameter. One of the standard methods to determine the sign is to use the so-called *reorientation effect* in the Coulomb excitation. Recently, it has been attempted to precisely determine the deformation including the sign through the analyses of heavy-ion fusion reactions at energies below the Coulomb barrier [40–45].

[18]The transition from high excited states to lower states within a nucleus can take place not only by the electromagnetic transition which emits a photon, but by ejecting an electron with the energy released in the decay. The latter process is called the *internal conversion* (see [46] for details). Hence the total transition probability is given by

$$w = w_\gamma + w_e = (1 + \alpha) w_\gamma , \tag{8.77}$$

$$\alpha \equiv w_e / w_\gamma , \tag{8.78}$$

if we denote the transition probability by emitting photon by w_γ, and the transition probability by the internal conversion process by w_e. α is called the *internal conversion coefficient*. If the transition energy is large to some extent, then α is small, so that one can ignore the contribution of the internal conversion. On the other hand, the internal conversion becomes the main decay process if the excitation energy of the first excited 2_1^+ state is small as in the case of $^{238}_{92}$U, where it is about 45 keV. Equation (8.76) holds in this case as well. However, care should be taken when one tries to estimate the magnitude of β from the lifetime of the 2_1^+ state.

ties between different states within a rotational band are related to each other through the Clebsch–Gordan coefficients as is given by Eq. (8.75) in the case of ideal rotational motion, where the deformation does not change with the angular momentum, namely, where the intrinsic state does not change in the ground and excited states.[19]

References

1. H.J. Rose, G.A. Jones, Nature **307**, 245 (1984)
2. B. Blank, M.J.G. Borge, Prog. Part. Nucl. Phys. **60**, 403 (2008); B. Blank, M. Płoszajczak. Rep. Prog. Phys. **71**, 046301 (2008)
3. K. Kato, K. Ikeda, Butsuri **61**, 814 (2006); T. Myo, Y. Kikuchi, H. Masui, K. Kato. Prog. Part. Nucl. Phys. **79**, 1 (2014)
4. D.M. Brink, N. Takigawa, Nucl. Phys. A **279**, 159 (1977)
5. S.Y. Lee, N. Takigawa, Nucl. Phys. A **308**, 189 (1978)
6. S.Y. Lee, N. Takigawa, C. Marty, Nucl. Phys. A **308**, 161 (1978)
7. M. Abramowitz, I.A. Stegun, *Handbook of Mathematical Functions with Formulas, Graphs, and Mathematical Tables* (Dover, London, 1978)
8. M. Nogami, *Nuclear Physics* (Shoukabou, Tokyo, 1973). Japanese edn.
9. D.F. Jackson, M. Rhoades-Brown, Ann. Phys. (NY) **105**, 151 (1977); S. A. Gurvitz, G. Kalbermann, Phys. Rev. Lett. **59**, 262 (1987); S. A. Gurvitz. Phys. Rev. A **38**, 1747 (1988)
10. S. Åberg, P.B. Semmes, W. Nazarewicz, Phys. Rev. C **56**, 1762 (1997)
11. J.M. Blatt, V.F. Weiskopf, *Theoretical Nuclear Physics* (Wiley, New York, 1952)
12. M.A. Preston, *Physics of the Nucleus* (Addison-Wesley, London, 1962)
13. K. Kikuchi, M. Kawai, *Nuclear Matter and Nuclear Reactions* (North-Holland, Amsterdam, 1968)
14. S. Yoshida, M. Kawai, *Nuclear Reactions* (Asakura, Tokyo, 2002). Japanese edn.
15. A.M. Lane, R.G. Thomas, Rev. Mod. Phys. **30**, 257 (1958)
16. E.P. Wigner, Phys. Rev. **70**, 606 (1946); E.P. Wigner, L. Eisenbud. Phys. Rev. **72**, 29 (1947)
17. T. Teichman, E.P. Wigner, Phys. Rev. **87**, 123 (1952)

---

[19]Here, we add comments on the liquid-drop model. The experimental data show systematics given by

$$E_{2_1^+}\, B(E2; 2_1^+ \to 0_{\rm g.s.}^+) \approx (25 \pm 8)\frac{Z^2}{A}\left[\mathrm{MeV}\, e^2\, \mathrm{fm}^4\right] \propto Z^2/A\,. \tag{8.79}$$

On the other hand, the liquid-drop model leads to

$$\hbar\Omega_2 B(E2) \propto Z^2 R_0^4/B_2 \propto Z^2 R_0^4/AR_0^2 \propto Z^2 A^{2/3}/A \sim Z^2 A^{-1/3} \tag{8.80}$$

for vibrational excitations.

The prediction of the liquid-drop model Eq. (8.80) has a different mass number dependence from that of the experimental data given by Eq. (8.79). One reason of the discrepancy is the quantum mechanical effect, which plays important roles in nuclear physics. It is necessary to treat nucleus as a quantum liquid in order to handle nucleus in the liquid-drop model in a quantitatively proper way. For example, the classical liquid-drop model evaluates the potential energy by taking into account the effect of deformation only in the ordinary space in deriving the Hamiltonian for collective motions. Accurately, however, one needs to consider the effects of deformation in the momentum space as well as suggested by the uncertainty principle in quantum mechanics. This effect shows up, for example, in the vibrational excited states such as the giant quadrupole resonance (GQR) [39]. The discrepancy between the experimental data and the prediction of the liquid-drop model appears also regarding the moment of inertia of deformed nuclei.

18. A. Arima, H. Horiuchi, K. Kubodera, N. Takigawa, Adv. Nucl. Phys. **5**, 345 (1972)
19. B. Buck, A.C. Merchant, S.M. Perez, Phys. Rev. C **45**, 2247 (1992); B. Buck, J. C. Johnson, A. C. Merchant, S. M. Perez. Phys. Rev. C **53**, 2841 (1996)
20. I. Tonozuka, A. Arima, Nucl. Phys. A **323**, 45 (1979)
21. C.M. Lederer, V. Shirley, *Table of Isotopes*, 7th edn. (Wiley, New York, 1978)
22. P. Hänggi, P. Talkner, M. Borkovec, Rev. Mod. Phys. **62**, 251 (1990). and references therein
23. U. Weiss, *Quantum Dissipative Systems* (World Scientific, Singapore, 1999)
24. P. L. Kapur, R. Peierls, Proc. Roy. Soc. London A **163**, 606 (1937); A. Schmid, Ann. Phys. (NY) **170**, 333 (1986)
25. T. Kindo, A. Iwamoto, Phys. Lett. B **225**, 203 (1989)
26. N. Takigawa, K. Hagino, M. Abe, Phys. Rev. C **51**, 187 (1995)
27. N. Bohr, J.A. Wheeler, Phys. Rev. **56**, 426 (1939)
28. H.A. Kramers, Physica **7**(284), 284 (1940)
29. R. Kubo, M. Toda, N. Hashitsume, *Statistical Physics II, Non-equilibrium Statistical Mechanics* (Springer, New York, 1985)
30. H. Risken, *The Fokker-Planck Equation* (Springer, Berlin, 1989)
31. J.S. Langer, Ann. Phys. (NY) **41**, 374 (1967); I. K. Affleck. Phys. Rev. Lett. **46**, 388 (1981)
32. R.P. Feynman, A.R. Hibbs, *Quantum Mechanics and Path Integrals* (McGraw-Hill, New York, 1965)
33. H. Grabert, P. Olschowski, U. Weiss, Phys. Rev. B **36**, 1931 (1987)
34. N. Takigawa, M. Abe, Phys. Rev. C **41**, 2451 (1990)
35. Y. Abe, C. Gregoire, H. Delagrange, J. Phys. **47**, C4 (1986); T. Wada, Y. Abe, N. Carjan, Phys. Rev. Lett. **70**, 3538 (1993); Y. Abe, S. Ayik, P.-G. Reinhard, E. Suraud, Phys. Rep. **275**, 49 (1996)
36. P. Fröbrich, R. Lipperheide, *Theory of Nuclear Reactions* (Clarendon Press, Oxford, 1996)
37. Aage Bohr, Ben R. Mottelson, *Nuclear Structure*, vol. I (Benjamin, New York, 1969)
38. S. Raman, C.W. Nestor Jr., P. Tikkanen, At. Data Nucl. Data Tables **78**(1), 1 (2001)
39. P. Ring, P. Schuck, *The Nuclear Many-Body problem* (Springer, Berlin, 1980)
40. A.B. Balantekin, N. Takigawa, Rev. Mod. Phys. **70**, 77 (1998)
41. M. Dasgupta, D.J. Hinde, N. Rowley, A.M. Stefanini, Ann. Rev. Nucl. Part. Sci. **48**, 401 (1998)
42. K. Hagino, N. Rowley, A.T. Kruppa, Comp. Phys. Commun. **123**, 143 (1999)
43. H. Esbensen, Nucl. Phys. A **352**, 147 (1981)
44. K. Hagino, N. Takigawa, Butsuri **57**, 588 (2002)
45. K. Hagino, N. Takigawa, Prog. Theor. Phys. **128**, 1001 (2012)
46. K. Yagi, *Nuclear Physics* (Asakura, Tokyo, 1971). Japanese edn.

Chapter 9
Synthesis of Elements

Abstract It is a long-standing intriguing problem to explore the synthesis of elements, and extensive studies are going on (See Clayton (Principles of Stellar Evolution and Nucleosynthesis, The University of Chicago Press, Chicago, 1968, [1]), Rolfs and Rodney (Cauldrons in the Cosmos, The University of Chicago Press, Chicago, 1988, [2]), Thompson and Nunes (Nuclear Reactions for Astrophysics: Principles, Calculation and Applications of Low-Energy Reactions, Cambridge University Press, Cambridge, 2009, [3]), VHS Element Genesis—Solving the Mystery, 2001, [4]). Roughly speaking, the synthesis of elements can be divided into the *primordial nucleosynthesis*, which is also called the *Big Bang nucleosynthesis*, and the *stellar nucleosynthesis*. The light elements such as deuterons, He and Li have been synthesized by nuclear reactions within 3–15 min after the Big Bang. This is the Big Bang nucleosynthesis. It terminates at the elements of mass number 7, since there is no stable nucleus of mass number 8. Stars started to be formed about one billion years later. Then, thermal nuclear reactions took place inside stars, and nuclei up to Fe, which has the largest binding energy per nucleon, have been successively synthesized depending on the mass of each star. (Precisely speaking, the nucleus which has the largest binding energy per nucleon is $^{62}_{28}$Ni as remarked in Chap. 2.) Nuclei beyond Fe are synthesized either slowly by the neutron capture reactions called slow process (s-process) inside red giant stars, or synthesized by the explosive astrophysical phenomenon called rapid process (r-process). Nuclei with extremely large mass number such as U are thought to be synthesized at the supernovae explosion, which is one of the last stages of stars. In this chapter we learn some basics concerning the nuclear reactions related to nucleosynthesis (See Clayton (Principles of Stellar Evolution and Nucleosynthesis, 1968, [1]), Rolfs and Rodney (Cauldrons in the Cosmos, 1988, [2]), Thompson and Nunes (Nuclear Reactions for Astrophysics: Principles, Calculation and Applications of Low-Energy Reactions, Cambridge University Press, Cambridge, 2009, [3]) for details).

9.1 The Astrophysical S-Factor and Gamow Factor

The nuclear reactions between two charged particles in stars occur with the collision energy much lower than the Coulomb barrier like the proton–proton reactions in the

N. Takigawa and K. Washiyama, *Fundamentals of Nuclear Physics*,
DOI 10.1007/978-4-431-55378-6_9

Sun discussed in Sect. 1.4. Hence it is customary to discuss not the cross section $\sigma(E)$, but $S(E)$, which is defined by

$$\sigma(E) = \frac{S(E)}{E} \exp(-2\pi\eta) , \tag{9.1}$$

$$\eta = \frac{Z_1 Z_2 e^2}{\hbar v} = \frac{Z_1 Z_2 e^2 \sqrt{\mu}}{\sqrt{2}\hbar} \frac{1}{\sqrt{E}} , \tag{9.2}$$

by taking out the trivial exponential factor (Gamow factor) $\exp(-2\pi\eta)$, where η is the Sommerfeld parameter. $S(E)$ is called the *astrophysical S-factor.*

Since the Gamow factor has been taken out, $S(E)$ is expected to vary gently as a function of energy unless resonance phenomena are involved. The energy of the Gamow peak (see the next section), which plays a central role in nuclear reactions in stars, is much lower than the lowest energy where experiments can be performed in many cases. The cross section at the Gamow peak is therefore often estimated by extrapolating the experimental data at high energies, where experiments can be performed. The gentle energy dependence of $S(E)$ is convenient in this connection, too.

Figure 9.1 shows the experimental data of $S(E)$ factor for the ^{3}He(^{3}He,2p)^{4}He reaction. It is remarkable that the experiments have been done down to energies as low as the Gamow peak in the Sun for this reaction. The figure shows that the experimental value of $S(E)$ is systematically larger than the value obtained by the extrapolation of high energy data. This situation is not special to this reaction, but is common to many other systems so far studied, and the enhancement of the cross

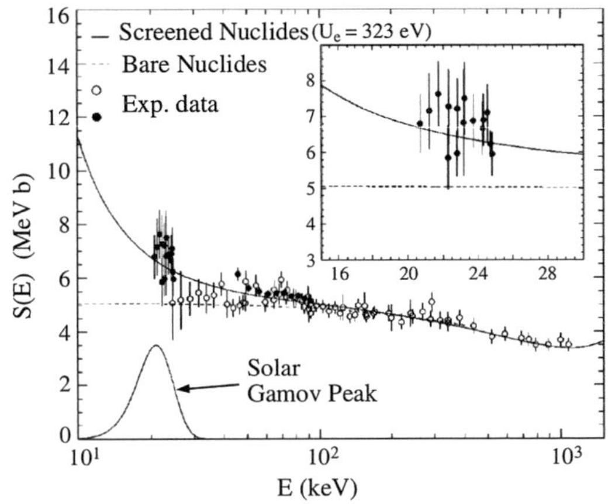

Fig. 9.1 The $S(E)$ factor for the ^{3}He(^{3}He,2p)^{4}He reaction. U_0 is the screening energy. Taken from [5]

section has been attempted to be explained in terms of the screening effect due to, e.g., the electrons of the target nucleus.

9.2 Gamow Peak

If we denote the reaction rate per pair of colliding nuclei by λ,[1] then it is given by

$$\lambda = \langle \sigma v \rangle = \int \sigma(E) v(E) \phi(v) \mathrm{d}\mathbf{v} \,. \tag{9.3}$$

Here, $\phi(v)$ is the normalized Maxwell–Boltzmann distribution, and is given by

$$\phi(v)\mathrm{d}\mathbf{v} = \frac{2}{\sqrt{\pi}} \beta^{3/2} \exp(-\beta E) \sqrt{E} \mathrm{d}E \,, \tag{9.4}$$

with $\beta = 1/kT$, where T and k are the temperature of the star and the Boltzmann constant, respectively.

Inserting Eqs. (9.1) and (9.4) into Eq. (9.3), and using the definition of the Sommerfeld parameter Eq. (9.2), we obtain

$$\lambda = \sqrt{\frac{8}{\mu \pi}} \beta^{3/2} \int_0^\infty S(E) \exp\left(-\beta E - b E^{-1/2}\right) \mathrm{d}E \,, \tag{9.5}$$

$$b = \frac{\pi Z_1 Z_2 e^2 \sqrt{2\mu}}{\hbar} \,. \tag{9.6}$$

b^2 is called the Gamow energy.

The exponential factor in the integrand of Eq. (9.5) is sharply peaked at an energy because of the competition between the Gamow factor and the Maxwell distribution. Since $S(E)$ is a gentle function of energy unless resonances are involved, one can apply the method of steepest descent to calculate the integral with little error. From the condition

$$\frac{\mathrm{d}}{\mathrm{d}E} \left(\beta E + b E^{-1/2}\right) = 0 \,, \tag{9.7}$$

the saddle point energy can be obtained as

$$E_0 = \left(\frac{b}{2\beta}\right)^{2/3} \,. \tag{9.8}$$

E_0 is called the *Gamow peak*. Finally, the reaction rate is given by

$$\lambda \approx \frac{4\sqrt{2}}{\sqrt{3}} \frac{1}{\sqrt{\mu}} E_0^{1/2} \beta S_0 \mathrm{e}^{-3\beta E_0} \,, \tag{9.9}$$

where $S_0 = S(E_0)$.

[1] The actual reaction rate per pair is given by multiplying the pair density.

Exercise 9.1 Prove Eq. (9.9) based on the formula of the method of steepest descent

$$\int g(x)\mathrm{e}^{-f(x)}\mathrm{d}x \approx \sqrt{\frac{2\pi}{f''(x_0)}}g(x_0)\mathrm{e}^{-f(x_0)},$$

where x_0 and f'' are the position of the saddle point and the second derivative of f, respectively.

9.3 Neutron Capture Cross Section

We show in this section that the capture cross section of s-wave neutrons, i.e., the cross section of low-energy reactions such as the reactions by thermal neutrons, is proportional to $1/v$ ($1/v$-*law*), where v is the velocity of the neutron relative to the target nucleus, and that the v dependence changes as the energy increases.

To make the argument simple, we approximate the cross section of the capture reaction by the absorption cross section by a strongly absorbing square well potential. Furthermore, we represent the strong absorption effect by the incoming wave boundary condition at the potential surface at $r = a$. If we express the wave function of the relative motion as $R(r) = u(r)/r$, and name the region of $r \leq a$ the region I, and that of $r > a$ the region II, then

$$u_I(r) = A\mathrm{e}^{-\mathrm{i}Kr} + B\mathrm{e}^{\mathrm{i}Kr}\ , \tag{9.10}$$

$$u_{II}(r) = \mathrm{e}^{-\mathrm{i}kr} - S\mathrm{e}^{\mathrm{i}kr}\ , \tag{9.11}$$

$$K = \sqrt{\frac{2\mu(E+V_0)}{\hbar^2}}\ , \tag{9.12}$$

$$k = \sqrt{\frac{2\mu E}{\hbar^2}}\ . \tag{9.13}$$

V_0 is the depth of the potential, which is assumed to be a real number.

$B = 0$ if we assume the strong absorption and impose the incoming wave boundary condition. In this case, the continuity conditions of the wave function and of its derivative at $r = a$ lead to

$$S = \frac{K-k}{K+k}\mathrm{e}^{-2\mathrm{i}ka}\ . \tag{9.14}$$

Hence the cross section of the capture reaction is given by

$$\begin{aligned}\sigma &= \frac{\pi}{k^2}(1-|S|^2)\\ &= \frac{\pi}{k^2}\frac{1}{|K+k|^2}4kK\ .\end{aligned} \tag{9.15}$$

Since $K \approx K_0 \equiv \sqrt{2\mu V_0/\hbar^2}$ at low energies, we have

$$\sigma \sim \frac{\pi}{k^2}\frac{1}{K_0^2}4kK_0 \propto \frac{1}{k} . \tag{9.16}$$

Equation (9.15) shows also that the v dependence of the capture cross section varies with the energy of collision (see Ref. [1]).

Exercise 9.2 Discuss the v dependence of the capture cross section of p-wave neutrons by heavy nuclei.

9.4 Synthesis of Heavy Elements: S-Process and R-Process

As mentioned at the beginning of this chapter, nuclei with large mass number beyond Fe are believed to be synthesized by either the so-called s-process (slow process) or by the r-process (rapid process).

In the abundance curve shown in Fig. 1.7, the right peak (indicated by "s") in each double peaks at $A \sim 80$, 130, and 200 is produced by the s-process. This peak corresponds to nuclei with a neutron magic number ($N = 50$, 82, or 126). The left peak (indicated by "r") is produced by the r-process.

Figure 9.2 conceptually shows the paths of the synthesis of elements by the s- and r-processes. It shows how heavy nuclei are built up by successive neutron captures, interspersed with β-decays. While the s-process nucleosynthesis proceeds almost along the stability line, the r-process nucleosynthesis rapidly proceeds via the region of neutron-rich unstable nuclei.[2] The study of r-process is therefore extensively going on in these days in connection with the study of unstable nuclei. For the r-process, it is suggested that the neutron number stays unchanged for a while when it coincides with one of the magic numbers, and nuclei with various atomic numbers with the fixed neutron number are synthesized.[3]

---

[2]Large neutron fluxes are expected in supernovae explosions. Consequently, neutron-rich nuclei are successively synthesized by (n,γ) reaction. The binding energy of neutron becomes small if the number of excess neutrons gets too large. The (n,γ) reaction then becomes balanced with the inverse (γ,n) reaction, and the synthesis does not proceed further. Eventually, the unstable neutron-rich nucleus either proceeds towards stability line via β-decays, or repeats (n,γ) reactions after increasing the atomic number via β-decays to synthesize nuclei with still larger proton and neutron numbers. The r-process path is thus determined.

[3]The regions where the neutron number does not change for a while in the r-process nucleosynthesis are called *waiting points* and the corresponding nuclei are called *waiting point nuclei*. It is considered that the peaks on the left side in Fig. 1.7 appear reflecting the waiting points.

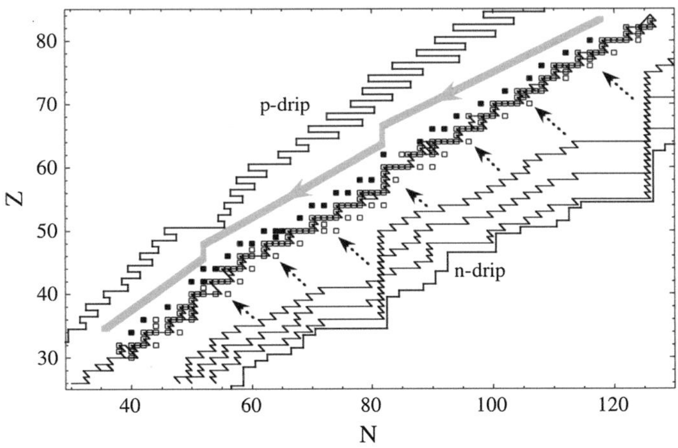

Fig. 9.2 Conceptual illustration of the synthesis of elements by s- and r-processes. Taken from [6]

9.5 Column: Overview of the Synthesis of Elements

Figure 9.3 summarizes the various paths of the synthesis of elements hitherto described. The rp-process, which has been added in the figure, is the process, where proton-rich nuclei with the mass number up to $\sim$ 100 are synthesized by (p,γ) and (α,p) reactions.[4] In addition, there exists the p-process, called sometimes γ process, where neutron-deficient nuclei with the mass number larger than that for the nuclei made by the rp-process are synthesized by photo-disintegration such as (γ,n), (γ,p), and (γ, α) reactions [3]. Nucleosynthesis by neutrinos associated with supernovae explosions is also one of the current topics [7].

In parallel with the research to uncover the synthesis of elements, various studies of astrophysical phenomena such as the physics of neutron stars from the points of view of nuclear physics are very actively ongoing. It forms an important branch of current nuclear physics.

[4]Recent research on neutron stars suggests that the rp-process ends at $^{105}_{52}$Te.

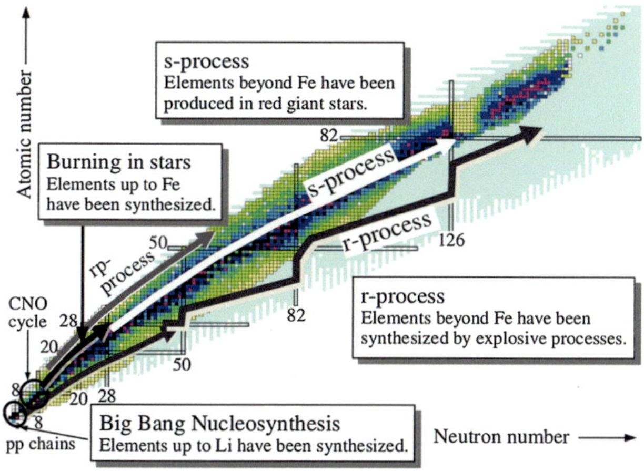

Fig. 9.3 Paths of nucleosynthesis. After [4] with some modifications

References

1. D.D. Clayton, *Principles of Stellar Evolution and Nucleosynthesis* (The University of Chicago Press, Chicago, 1968)
2. C.E. Rolfs, W.S. Rodney, *Cauldrons in the Cosmos* (The University of Chicago Press, Chicago, 1988)
3. I.J. Thompson, F.M. Nunes, *Nuclear Reactions for Astrophysics: Principles, Calculation and Applications of Low-Energy Reactions* (Cambridge University Press, Cambridge, 2009)
4. VHS *Element Genesis—Solving the Mystery*, Sci. Eds. Y. Motizuki, I. Tanihata, Y. Yano, R. Boyd (RIKEN & Image Science, Inc., 2001)
5. M. Junker et al., Phys. Rev. C **57**, 2700 (1998)
6. M. Arnould, S. Goriely, K. Takahashi, Phys. Rep. **450**, 97 (2007)
7. J.N. Bahcall, *Neutrino Astrophysics* (Cambridge University Press, Cambridge, 1989)

Appendix A
Important Formulae and Their Derivation

Abstract This chapter collects appendices to the previous chapters.

A.1 Basics of Scattering Problems

Here we summarize some basics for scattering problems such as the partial wave expansion of the scattering amplitude.[1]

A.1.1 Partial Wave Expansion

(1) The asymptotic form of the radial wave function for the partial wave ℓ (the case of a short range force; k is the incident wave number, S is the scattering matrix, i.e., the S-matrix, and δ is the phase shift)

$$R_\ell(r) \sim \frac{1}{kr}\left[e^{-i(kr-\frac{1}{2}\ell\pi)} - S(\ell)e^{i(kr-\frac{1}{2}\ell\pi)}\right] \tag{A.1}$$

$$\sim \frac{1}{kr}\sin\left(kr - \frac{1}{2}\ell\pi + \delta_\ell\right) . \tag{A.2}$$

(2) Partial wave expansion of the elastic scattering amplitude[2] (θ is the scattering angle, and α is the index of each state)

[1] We consider the scattering of particles without spin. See [1] for the scattering of particles with spin.

[2] When one takes the partial wave sum in the case of the scattering problem involving Coulomb force, it is traditional to separate out the Rutherford scattering amplitude.

N. Takigawa and K. Washiyama, *Fundamentals of Nuclear Physics*,
DOI 10.1007/978-4-431-55378-6

$$f^{(\mathrm{el})}(\theta) = \frac{1}{2ik}\sum_{\ell=0}^{\infty}(2\ell+1)\left(S_{\alpha,\alpha}(\ell)-1\right)P_\ell(\cos\theta)\ , \tag{A.3}$$

$$S_{\alpha,\alpha}(\ell) = \mathrm{e}^{2\mathrm{i}\delta_\ell}\ . \tag{A.4}$$

(3) Partial wave expansion of the inelastic scattering amplitude (α and β are the state indices)

$$f^{(\mathrm{inel})}_{\beta\neq\alpha}(\theta) = \frac{1}{2ik}\sum_{\ell}(2\ell+1)\,S_{\beta,\alpha}(\ell)P_\ell(\cos\theta)\ . \tag{A.5}$$

(4) Cross sections of the elastic, inelastic and total inelastic scattering

$$\sigma^{(\mathrm{el})} = \sum_{\ell}\sigma^{(\mathrm{el})}_{\ell} = \frac{\pi}{k_\alpha^2}\sum_{\ell}(2\ell+1)|S_{\alpha,\alpha}(\ell)-1|^2\ , \tag{A.6}$$

$$\sigma^{(\mathrm{inel})}_{\beta\neq\alpha} = \sum_{\ell}\sigma^{(\mathrm{inel})}_{\beta\neq\alpha}(\ell) = \frac{\pi}{k_\alpha^2}\sum_{\ell}(2\ell+1)|S_{\beta,\alpha}(\ell)|^2\ , \tag{A.7}$$

$$\sigma^{(\mathrm{inel})} = \sum_{\beta\neq\alpha}\sigma^{(\mathrm{inel})}_{\beta\neq\alpha} = \frac{\pi}{k_\alpha^2}\sum_{\ell}(2\ell+1)\left[1-|S_{\alpha,\alpha}(\ell)|^2\right]\ , \tag{A.8}$$

(5) Unitarity condition

$$\sum_{\beta}|S_{\beta,\alpha}(\ell)|^2 = 1\ . \tag{A.9}$$

(6) Total cross section (ReS is the real part of the S-matrix)

$$\sigma^{(\mathrm{total})} = \sigma^{(\mathrm{el})} + \sigma^{(\mathrm{inel})} = \frac{2\pi}{k_\alpha^2}\sum_{\ell}(2\ell+1)(1-\mathrm{Re}S)\ . \tag{A.10}$$

(7) Optical theorem (Im$f^{(\mathrm{el})}$ is the imaginary part of $f^{(\mathrm{el})}$)

$$\sigma^{(\mathrm{total})} = \frac{4\pi}{k_\alpha}\mathrm{Im}f^{(\mathrm{el})}(\theta=0)\ . \tag{A.11}$$

A.1.2 Sommerfeld–Watson Transformation

For high energy scattering, it is more convenient to use the integral representation of the scattering amplitude instead of the partial wave expansion. The integral representation sometimes becomes useful for scattering at low energies as well. Starting from the integral representation of the scattering amplitude and deriving characteristic closed form expression through some suitable approximation scheme such as the stationary phase approximation or the method of steepest descent, one can often

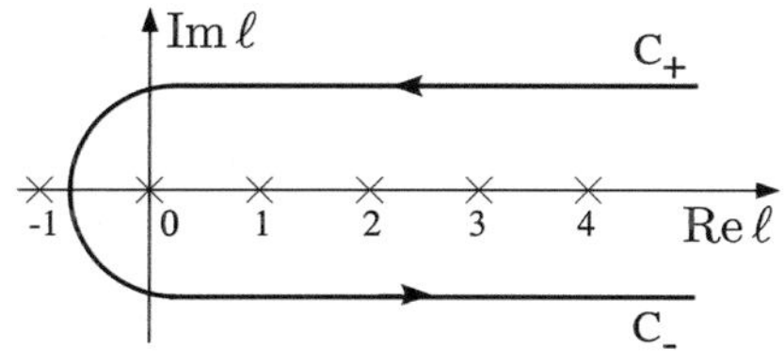

Fig. A.1 The contour used for the Sommerfeld–Watson transformation

obtain intuitive understanding for complicated patterns appearing in the differential cross section in terms of, e.g., the interference of various scattering waves or the caustics such as the rainbow scattering and the glory scattering [2–7].

The transformation from the partial wave sum to the integral representation can be achieved by using the knowledge of the contour integral in the complex plane. We first represent the scattering amplitude as

$$f(\theta) = -\frac{1}{2k}\int_C d\ell \, e^{i\pi\ell} \frac{1}{\sin\pi\ell}\left(\ell + \frac{1}{2}\right)[S(k,\ell) - 1]\, P_\ell(\cos\theta) \tag{A.12}$$

by using the integral along the contour C in the complex ℓ plane shown in Fig. A.1. The transformation from Eq. (A.3) to Eq. (A.12) is known as the *Sommerfeld–Watson transformation*. One can easily confirm that Eq. (A.12) reduces to Eq. (A.3) by noting that the integrand has a pole of order one at each integer value of ℓ and that the corresponding residue is given by

$$\lim_{\ell\to\ell_0} \frac{\ell - \ell_0}{\sin\pi\ell} = \frac{1}{\pi}(-1)^{\ell_0} \quad (\ell_0 : \text{integer}). \tag{A.13}$$

A.1.3 *Poisson Sum Formula*

The integral in Eq. (A.12) is a contour integral in the complex ℓ plane. Let us attempt to convert it into an integral along the real axis. To that end, we first rewrite Eq. (A.12) as

$$f(\theta) = -\frac{i}{k}\int_C d\ell \frac{e^{i\pi\ell}}{e^{i\pi\ell} - e^{-i\pi\ell}}\left(\ell + \frac{1}{2}\right)[S(k,\ell) - 1]\, P_\ell(\cos\theta)\,. \tag{A.14}$$

We then remark that

$$\frac{e^{i\pi\ell}}{e^{i\pi\ell} - e^{-i\pi\ell}} = \lim_{\varepsilon\to 0^+} \frac{e^{i\pi(\ell+i\varepsilon)}}{e^{i\pi(\ell+i\varepsilon)} - e^{-i\pi(\ell+i\varepsilon)}} \tag{A.15}$$

$$= -e^{2i\pi\ell}\left(1 + e^{2i\pi\ell} + e^{4i\pi\ell} + \cdots\right) \tag{A.16}$$

on the contour C_+ above the real axis, and that

$$\frac{e^{i\pi\ell}}{e^{i\pi\ell}-e^{-i\pi\ell}} = \lim_{\varepsilon\to 0^+}\frac{e^{i\pi(\ell-i\varepsilon)}}{e^{i\pi(\ell-i\varepsilon)}-e^{-i\pi(\ell-i\varepsilon)}} \tag{A.17}$$

$$= 1+e^{-2i\pi\ell}+e^{-4i\pi\ell}+\cdots \tag{A.18}$$

on the contour C_- below the real axis, hence

$$\int_C d\ell \frac{e^{i\pi\ell}}{e^{i\pi\ell}-e^{-i\pi\ell}}h(\ell) = \left(\int_{C^+}+\int_{C^-}\right) d\ell \frac{e^{i\pi\ell}}{e^{i\pi\ell}-e^{-i\pi\ell}}h(\ell)$$

$$= \sum_{m=-\infty}^{\infty}\int_{-1/2}^{\infty} d\ell\, e^{2i\pi m\ell}h(\ell)\,, \tag{A.19}$$

for a function $h(\ell)$. We finally obtain

$$f(k,\theta) = -\frac{i}{k}\sum_{m=-\infty}^{\infty}(-1)^m\int_0^{\infty} d\lambda e^{2\pi i m\lambda}\lambda\,[S(k,\lambda-1/2)-1]\,P_{\lambda-1/2}(\cos\theta) \tag{A.20}$$

by introducing the *Langer replacement* $\ell+1/2=\lambda$, which is referred to also as the *Langer correction* or the *Langer modification*. Equation (A.20) is called the *Poisson sum formula*. In the semi-classical theory for scattering [8], m represents the number of times which the incident particle circulates the target particle. The $m=0$ term governs the scattering at high energies or if the attractive force between the projectile and target particles is weak. However, the $m\geq 1$ terms also play important roles in the scattering at low energies if a strong attractive force is involved [6].

We obtain the following Poisson sum formula for the cross section of the total inelastic scattering by applying the similar procedure to Eq. (A.8),

$$\sigma^{(\text{inel})} = \frac{2\pi}{k_\alpha^2}\sum_{m=-\infty}^{\infty}(-1)^m\int_0^{\infty} d\lambda e^{2i\pi m\lambda}\lambda\left[1-|S_{\alpha,\alpha}|^2\right] \tag{A.21}$$

$$= 2\pi\sum_{m=-\infty}^{\infty}(-1)^m\int_0^{\infty} db\, b\, e^{2i\pi mkb}\left[1-|S_{\alpha,\alpha}(b)|^2\right] \tag{A.22}$$

$$\approx 2\pi\int_0^{\infty} db\, b\left[1-|S_{\alpha,\alpha}(b)|^2\right] \quad \text{(for high energy)}. \tag{A.23}$$

The relation between the angular momentum and the impact parameter $\lambda=kb$ has been used in transforming Eqs. (A.21) to (A.22).

A.2 Basics of Semi-classical Theory I: WKB Approximation

Here, let us first study the WKB approximation[3] among various semi-classical methods. The WKB approximation is the method to approximately obtain the solution of the Schrödinger equation in the limit of taking the Planck constant $\hbar$ to be zero. Physically, it is valid when the change of the potential energy within a wave length can be ignored compared with the magnitude of the local kinetic energy. Hence, it can be said to be a short wave length approximation. In this approximation, the wave function and the phase shift as well as the scattering matrix for scattering problems are given in terms of the action integrals in classical mechanics.

A.2.1 Wave Function

As an example, let us consider a problem where the potential is given by $V_n(r) + V_C(r)$, the former and the latter being the nuclear and the Coulomb potentials, respectively, and approximately obtain $\psi_\ell(r)$ when the radial wave function $R_\ell(r)$ of an eigenstate of energy or the scattering problem with the orbital angular momentum ℓ is expressed as $R_\ell(r) \equiv \psi_\ell(r)/r$. The $\psi_\ell(r)$ obeys the following equation:

$$\left[-\frac{\hbar^2}{2\mu}\frac{\mathrm{d}^2}{\mathrm{d}r^2} + V(r) - E\right]\psi_\ell(r) = 0\,, \tag{A.24}$$

$$V(r) = V_n(r) + V_C(r) + V_\ell(r)\,, \tag{A.25}$$

$$V_\ell(r) = \frac{\hbar^2}{2\mu}\frac{(\ell+1/2)^2}{r^2} = \frac{\hbar^2}{2\mu}\frac{\lambda^2}{r^2}\,. \tag{A.26}$$

We introduced the Langer replacement $\ell(\ell+1) \to (\ell+1/2)^2 \equiv \lambda^2$ in Eq. (A.26).

In the WKB approximation, we express the wave function in the form of the product of the amplitude and the phase factor, and write

$$\psi(r) = A(r)\mathrm{e}^{\mathrm{i}S(r)/\hbar}\,. \tag{A.27}$$

We leave out the index ℓ in Eq. (A.27) and throughout this section. Making the substitution Eq. (A.27) in Eq. (A.24) and treating $A(r)$, $S(r)$ and $V(r)$ as quantities independent of $\hbar$ and collecting the terms with the same powers of $\hbar$, we obtain

[3]Wenzel–Kramers–Brillion approximation. It is also called the JWKB approximation by including the name of Jefferey.

$$\left[\frac{1}{2\mu}\left(\frac{\mathrm{d}S}{\mathrm{d}r}\right)^2 + V(r) - E\right] A\hbar^0 - \frac{\mathrm{i}}{2\mu}\left[2\frac{\mathrm{d}A}{\mathrm{d}r}\frac{\mathrm{d}S}{\mathrm{d}r} + A\frac{\mathrm{d}^2 S}{\mathrm{d}r^2}\right]\hbar^1 - \frac{1}{2\mu}\frac{\mathrm{d}^2 A}{\mathrm{d}r^2}\hbar^2 = 0\,. \tag{A.28}$$

By considering the limit of $\hbar \to 0$, and setting the coefficients of the zeroth and the first order power of $\hbar$ to be zero separately, we obtain

$$\left[\frac{1}{2\mu}\left(\frac{\mathrm{d}S}{\mathrm{d}r}\right)^2 + V(r) - E\right] = 0\,, \tag{A.29}$$

$$\frac{\mathrm{d}}{\mathrm{d}r}\left(A^2\frac{\mathrm{d}S}{\mathrm{d}r}\right) = 0\,. \tag{A.30}$$

Equation (A.29) corresponds to the Hamilton–Jacobi equation in classical mechanics, and suggests that S is given by the action integral in classical mechanics

$$S = \pm\int\sqrt{2\mu(E - V(r))}\mathrm{d}r = \pm\hbar\int k(r)\mathrm{d}r\,. \tag{A.31}$$

Equation (A.30) denotes the conservation of the current density. Together with Eq. (A.31), it shows that the amplitude A is inversely proportional to the square root of the wave number, i.e.,

$$A \propto \frac{1}{\sqrt{k(r)}}\,. \tag{A.32}$$

Finally, the wave function is given by

$$\psi(r) = A_+\frac{1}{\sqrt{k(r)}}\exp\left(\mathrm{i}\int_{r_0}^{r} k(r')\mathrm{d}r'\right) + A_-\frac{1}{\sqrt{k(r)}}\exp\left(-\mathrm{i}\int_{r_0}^{r} k(r')\mathrm{d}r'\right)\,, \tag{A.33}$$

where A_+ and A_- are constants. In this way, the wave function in the WKB approximation is given by replacing the wave number times displacement in the arguments of the exponential functions for a free particle with the action integral using the local wave number. This is a natural consequence, since the WKB approximation is based on the assumption that the spatial change of momentum is small.

The $\hbar^2$ term in Eq. (A.28) has to be sufficiently small compared with the other terms in order for the WKB approximation to be applicable. Equations (A.31) and (A.32) lead to

$$\frac{\mathrm{d}^2 A}{\mathrm{d}r^2}\Big/\left[\frac{A}{\hbar^2}\left(\frac{\mathrm{d}S}{\mathrm{d}r}\right)^2\right] \sim \frac{3}{16}\frac{1}{(\hbar^2 k^2(r)/2\mu)^2}\frac{1}{k^2(r)}\left(\frac{\mathrm{d}V}{\mathrm{d}r}\right)^2 + \cdots\,. \tag{A.34}$$

Thus, the condition for the WKB approximation to be valid is that the amount of change of the potential energy within a wave length is much smaller than the kinetic energy at that position.

A.3 Basics of Semi-Classical Theory II: Method of Comparison Equation

In this section, we formulate the semi-classical theory, i.e., the WKB method, using the method of comparison equation, which is different from the standard method described in Sect. A.2.

A.3.1 *Principle of the Method of Comparison Equation*

Turning points play important roles in describing the reflection of a particle in classical mechanics. Since the turning points are defined as the points where the energy of the particle matches the value of the potential energy, namely, as the points where $E = V(r)$, they become in general complex numbers if the potential is a complex potential as in the case of scattering by an optical potential. There exist many turning points in the complex r plane even if $V(r)$ is a real potential. For example, when the potential between the scattering nuclei has the form shown in Fig. 8.1, and when the scattering energy is slightly above the potential barrier, there appears a turning point, whose real part is the barrier position, at each side of, i.e., above and below, the real axis in the complex plane.

Exercise A.1 Let us express the potential near the barrier position r_B as $V(r) = V_{\rm CB} - \mu\Omega^2(r - r_B)^2/2$ when the potential looks like Fig. 8.1. Discuss the change of the positions of the classical turning points in the complex r plane as the scattering energy E varies from below to above the barrier height $V_{\rm CB}$.

We therefore make an analytic continuation of the Schrödinger equation to complex plane and seek the solution of the second order differential equation of a complex variable r:

$$\left[\frac{\mathrm{d}^2}{\mathrm{d}r^2} + \chi(r)\right]\psi(r) = 0\ , \tag{A.35}$$

$$\chi(r) = \frac{2\mu}{\hbar^2}(E - V(r))\ . \tag{A.36}$$

$V(r)$ can have in general both the real and the imaginary parts; $V(r) = V_R(r) + \mathrm{i}V_I(r)$, where both $V_R(r)$ and $V_I(r)$ are analytic functions and are real functions when r is real.

In the method of comparison equation, we introduce another space of a complex number, which we denote by σ, besides the complex r space, and introduce a subsidiary second order differential equation in that space:

$$\left[\frac{\mathrm{d}^2}{\mathrm{d}\sigma^2} + \Gamma(\sigma)\right]\phi(\sigma) = 0\ , \tag{A.37}$$

which is called the comparison equation.

The function $\Gamma(\sigma)$ is chosen to have the same topological structure as that of $\chi(r)$. In simpler words, it is chosen to have the same number of important zero points, that is, the same number of turning points. Moreover, it is so chosen that the exact solutions of Eq. (A.37) are functions whose properties are well known. The basic idea of the method of comparison equation is to obtain an approximate solution $\psi(r)$ of the original Eq. (A.35) in terms of the solutions $\phi(\sigma)$ of Eq. (A.37), whose properties are well known, by introducing suitable comparison equation.

The mapping between the two spaces is made in the following way. We first map the important zeros in the r space onto the zeros in the σ space by one to one mapping, then define the mapping between the two spaces so that

$$\frac{\mathrm{d}\sigma}{\mathrm{d}r} = \left[\frac{\chi(r)}{\Gamma(\sigma)}\right]^{1/2} . \tag{A.38}$$

Since the integral form of Eq. (A.38) reads

$$\int_{r_0}^{r} [\chi(r)]^{1/2}\,\mathrm{d}r = \int_{\sigma_0}^{\sigma} [\Gamma(\sigma)]^{1/2}\,\mathrm{d}\sigma\ , \tag{A.39}$$

Equation (A.38) means that the mapping is made to conserve the action integral in two spaces.

One can show that in this case it holds

$$\psi(r) \approx \left(\frac{\mathrm{d}\sigma}{\mathrm{d}r}\right)^{-1/2} \phi(\sigma(r)) \tag{A.40}$$

in the limit of $\hbar \to 0$.

In fact, by changing Eq. (A.40) into an identity equation, and inserting it into Eq. (A.35), one obtains

$$\left(\frac{\mathrm{d}\sigma}{\mathrm{d}r}\right)^{3/2}\left[\frac{\mathrm{d}^2\phi}{\mathrm{d}\sigma^2} + \frac{1}{\hbar^2}Q(r)\left(\frac{\mathrm{d}\sigma}{\mathrm{d}r}\right)^{-2}\phi(\sigma)\right] + f(\sigma)\phi(\sigma) = 0\ , \tag{A.41}$$

where

$$f(\sigma) = -\frac{1}{2}\frac{\mathrm{d}}{\mathrm{d}r}\left[\left(\frac{\mathrm{d}\sigma}{\mathrm{d}r}\right)^{-3/2}\frac{\mathrm{d}^2\sigma}{\mathrm{d}r^2}\right] . \tag{A.42}$$

In Eq. (A.41), we expressed as $\chi(r) = Q(r)/\hbar^2$ in order to make the role of $\hbar$ clear. Equation (A.41) shows that Eq. (A.37) is obtained in the limit of $\hbar \to 0$ if one defines the mapping by Eq. (A.38). The error is given by $(\frac{\mathrm{d}\sigma}{\mathrm{d}r})^{-3/2} f(\sigma)\phi(\sigma)/\frac{\mathrm{d}^2\phi(\sigma)}{\mathrm{d}\sigma^2}$.

A.3.2 Derivation of the WKB Wave Function

As a simple example, let us obtain the wave function at a place far from any classical turning points. In this case, the Γ can be chosen as

$$\Gamma(\sigma) = 1 \, . \tag{A.43}$$

The solution of the corresponding comparison Eq. (A.37) is given by

$$\phi(\sigma) = A\mathrm{e}^{\mathrm{i}\sigma} + B\mathrm{e}^{-\mathrm{i}\sigma} \, . \tag{A.44}$$

On the other hand, Eq. (A.39) leads to

$$\sigma = \int \chi^{1/2}\mathrm{d}r = \int k(r)\mathrm{d}r \, . \tag{A.45}$$

Since $\chi(r) = k^2(r)$, we finally obtain

$$\psi(r) \sim [\chi(r)]^{-1/4}\left[A\exp\left(\mathrm{i}\int^r k(r')\mathrm{d}r'\right) + B\exp\left(-\mathrm{i}\int^r k(r')\mathrm{d}r'\right)\right] \, . \tag{A.46}$$

The right-hand side agrees with Eq. (A.33) obtained in Sect. A.2, and is nothing but the wave function in the WKB approximation.

Incidentally, the WKB formula for the tunneling probability can be derived by separately treating the two turning points on both sides of the potential barrier as an isolated turning point, and using σ as Γ for each region, and connecting the wave functions on both sides by using the asymptotic form of the Airy function, which is the corresponding ϕ. On the other hand, the formula of the barrier transmission probability in the uniform approximation (2.77) can be obtained by considering the potential barrier basically as a parabolic barrier and hence by taking $\Gamma = \sigma^2/4 - \varepsilon$ and by studying the asymptotic behaviours of the corresponding eigenfunctions ϕ, which are the Whittaker functions [9, 10]. The ε is determined so as to conserve the action integral between the two turning points at the potential barrier in the r and σ spaces. Also, the S matrix, which is the basis of Eqs. (8.5)–(8.8) can be derived by mapping the inner region including r_3 (see Fig. 8.1) by a linear function and the barrier region by a quadratic function, determining approximate wave function for each region and by connecting them using their asymptotic forms. In the connection, it is crucial to use the correct asymptotic forms of the Whittaker and Airy functions depending on the phase of the arguments.

A.4 Eikonal Approximation

A.4.1 Scattering Amplitude

The scattering amplitude by a potential V at high energies can be well given in the eikonal approximation as [11]

$$f(k,\theta) = -\mathrm{i}k \int_0^{\infty} \mathrm{d}b\, b J_0(kb\theta) \left[\mathrm{e}^{2\mathrm{i}\Delta(b)} - 1 \right] , \tag{A.47}$$

$$\Delta(b) = \delta_{\mathrm{eikonal}}(\ell) = \frac{-m}{2k\hbar^2} \int_{-\infty}^{+\infty} V(\sqrt{b^2+z^2})\mathrm{d}z , \tag{A.48}$$

and the cross section as

$$\sigma^{(\mathrm{inel})}_{\mathrm{eikonal}} \approx 2\pi \int_0^{\infty} \mathrm{d}b\, b \left[1 - \mathrm{e}^{-4\Delta_I(b)} \right] \quad \text{(for high energy)}, \tag{A.49}$$

$$\Delta_I(b) = \mathrm{Im}\Delta(b) . \tag{A.50}$$

A.4.2 Glauber Theory

Here, we derive the formulae of the Glauber theory in the optical limit given by Eqs. (2.34)–(2.37).

The starting point is Eqs. (A.47)–(A.49) of the eikonal approximation. In order to apply these equations to the nucleon–nucleus and nucleus–nucleus scattering, let us microscopically represent the scattering potential in terms of the sum of the nuclear force between individual nucleons $v_{NN}(\mathbf{r}_{i_P} - \mathbf{r}_{j_T})$,

$$V(\mathbf{R}) = \sum_{i_P, j_T} v_{NN}(\mathbf{r}_{i_P} - \mathbf{r}_{j_T}) \tag{A.51}$$

$$= \iint \mathrm{d}\mathbf{r}_P \mathrm{d}\mathbf{r}_T v_{NN}(\mathbf{r}_P - \mathbf{r}_T + \mathbf{R}) \rho_P(\mathbf{r}_P) \rho_T(\mathbf{r}_T) . \tag{A.52}$$

In these equations, $\mathbf{R}$ is the coordinate of the relative motion, $\mathbf{R} \equiv \mathbf{R}_P - \mathbf{R}_T$, between the centers of mass of the projectile and target nuclei, $\mathbf{R}_P$, $\mathbf{R}_T$. $\mathbf{r}_{i_P}$ ($\mathbf{r}_{j_T}$) is the coordinate of the i-th nucleon in the projectile (j-th nucleon in the target) measured from the origin of the space fixed coordinate system. $\mathbf{r}_P$ and $\mathbf{r}_T$ are the intrinsic coordinates of the nucleons in the projectile and target nuclei measured from $\mathbf{R}_P$ and $\mathbf{R}_T$, respectively. $\rho_P(\mathbf{r}_P)$ and $\rho_T(\mathbf{r}_T)$ are the nucleon densities of the projectile and target nuclei at $\mathbf{r}_P$ and $\mathbf{r}_T$, respectively. In transforming from Eqs. (A.51) to (A.52), we used that the nucleon densities are given by $\rho_P(\mathbf{r}_P) = \sum_{i_P} \delta(\mathbf{r}_{i_P} - \mathbf{R}_P - \mathbf{r}_P)$, and $\rho_T(\mathbf{r}_T) = \sum_{j_T} \delta(\mathbf{r}_{j_T} - \mathbf{R}_T - \mathbf{r}_T)$, like Eq. (5.2). $v_{NN}(\mathbf{r})$ is the nuclear potential between two nucleons, which are separated by $\mathbf{r}$. The potential given by Eq. (A.52) is called the *double folding potential*, and is often used for the interaction between two nuclei.

Since the nuclear force is of short range, we approximate it as

$$v_{NN}(\mathbf{r}) = (v_{\rm Re} + {\rm i}v_{\rm Im})\delta(\mathbf{r}). \tag{A.53}$$

The imaginary part (i$v_{\rm Im}$) has been introduced to take into account the other processes than the elastic scattering. The potential is then given by

$$V(\mathbf{R}) = (v_{\rm Re} + {\rm i}v_{\rm Im}) \int {\rm d}\mathbf{r}\rho_P(\mathbf{r} - \mathbf{R})\rho_T(\mathbf{r}) \ . \tag{A.54}$$

By inserting the result of Eq. (A.54) into Eq. (A.48) and using Eq. (A.49), we obtain

$$\sigma^{(\rm inel)} \approx 2\pi \int_0^\infty {\rm d}b\, b \left[1 - \exp\left(\frac{2\mu_{nn}}{k_{nn}\hbar^2} v_{\rm Im} O_v(b)\right)\right] \quad \text{(for high energy)}\ , \tag{A.55}$$

where the overlap function O_v is given by Eq. (2.36). μ_{nn} and k_{nn} are the reduced mass and the wave number in the nucleus–nucleus scattering, respectively.

We next use the Born approximation and the optical theorem in order to relate the strength of the nuclear force to the total cross section in the nucleon–nucleon scattering. By applying the formula in the Born approximation,

$$f^{(1)}(\theta) = -\frac{1}{4\pi}\frac{2m}{\hbar^2} \int {\rm e}^{-{\rm i}\mathbf{q}\cdot\mathbf{r}} V(\mathbf{r}){\rm d}\mathbf{r} \ ,$$

to the nucleon–nucleon scattering, and assuming the limit of the short range force given by Eq. (A.53), we obtain

$$f_{NN} = -\frac{\mu_{NN}}{2\pi\hbar^2}(v_{\rm Re} + {\rm i}v_{\rm Im}) \tag{A.56}$$

for the scattering amplitude f_{NN}. $\mu_{NN} = M_N/2$ is the reduced mass in the nucleon–nucleon scattering. If we use here the optical theorem, we obtain

$$v_{\rm Im} = -\frac{2\pi\hbar^2}{\mu_{NN}}\frac{k_{NN}}{4\pi}\sigma_{NN}^{(\rm total)}(k_{NN}) \ . \tag{A.57}$$

k_{NN} is the wave number of the relative motion in the nucleon–nucleon scattering, $\sigma_{NN}^{(\rm total)}(k_{NN})$ is the total cross section when the nucleon–nucleon collision takes place at the wave number k_{NN}. By inserting the results of Eq. (A.57) into Eq. (A.55), we obtain

$$\sigma^{(\rm inel)} \approx 2\pi \int_0^\infty {\rm d}b\, b \left[1 - \exp\left(-\frac{\mu_{nn}k_{NN}}{k_{nn}\mu_{NN}}\sigma_{NN}^{(\rm total)}(k_{NN}) O_v(b)\right)\right] \ . \tag{A.58}$$

In order for Eq. (A.58) to agree with Eq. (2.34), it is required that

$$\frac{\mu_{nn}k_{NN}}{k_{nn}\mu_{NN}} = 1 \; . \tag{A.59}$$

This is equivalent to the condition (2.37). Equations (2.34)–(2.37) are derived in this way. Since the proof used the same reaction cross section irrespective of the nucleon species, it will be adequate to use the average cross section $\bar{\sigma}_{NN}$.

A.5 Non-local Potential

A characteristic of nuclear physics is that a non-local potential is often involved. The Fock potential in the Hartree–Fock approximation for nuclear structure, and the exchange potentials which appear when one microscopically describes the scattering between complex nuclei such as the scattering between α particle and nucleus and between two heavy nuclei by taking into account antisymmetrization among nucleons based on, e.g., the resonating group method are typical examples.

On the other hand, one usually assumes a local potential for phenomenological analyses of nuclear structure and nuclear reactions. In this section, we review some standard methods to convert the Schrödinger equation involving a non-local potential, hence an integro-differential equation, into an equivalent local differential equation, and interpret the characteristic effects due to non-locality in other physical terms, such as the effective mass, which are relevant to the description by using local potential, hence by a local differential equation [1].

A.5.1 Integro-Differential Equation

We consider a one-dimensional problem and represent the local and the non-local potentials by U and V, respectively. In this case, the wave function obeys the integro-differential equation:

$$-\frac{\hbar^2}{2m}\frac{\mathrm{d}^2\psi(x)}{\mathrm{d}x^2} + U(x)\psi(x) + \int V(x,x')\psi(x')\mathrm{d}x' = E\psi(x) \; . \tag{A.60}$$

A.5.2 Equivalent Effective Local Potential: WKB Approximation

One useful method to study the characteristic physical effects of a non-local potential is the method based on the WKB approximation. To that end, we first rewrite the term of the non-local potential into

$$\int V(x,x')\psi(x')\mathrm{d}x' = \int V(x,x')\mathrm{e}^{\frac{\mathrm{i}}{\hbar}(x'-x)\hat{p}_\psi}\mathrm{d}x'\ \psi(x)\ . \tag{A.61}$$

The subscript ψ of the momentum operator $\hat{p}$ has been attached in order to indicate that it operates on the wave function ψ. The essence of the WKB approximation is to represent the wave function in the form of a plane wave by using the local momentum. Also, it is based on the assumption that the spatial variation of the wave number is small. Hence,

$$\hat{p}_\psi\psi(x) \sim p(x)\psi(x)\ , \tag{A.62}$$

where $p(x)$ is the local momentum at the position x.

A.5.2.1 WKB Approximation I: Local Momentum Approximation, Energy Dependent Potential

By applying the approximation (A.62) to Eq. (A.61), one obtains the following equivalent effective local potential:

$$U_{\text{eff}}(E;x) = U(x) + \int V(x,x')\mathrm{e}^{\frac{\mathrm{i}}{\hbar}(x'-x)p(x)}\mathrm{d}x' \tag{A.63}$$

$$= U(x) + \int V(x,x')\mathrm{e}^{\frac{\mathrm{i}}{\hbar}(x'-x)\sqrt{2m(E-U_{\text{eff}}(E;x))}}\mathrm{d}x' \tag{A.64}$$

$$\sim U(x) + \int V(x,x')\mathrm{e}^{\frac{\mathrm{i}}{\hbar}(x'-x)\sqrt{2m(E-U(x))}}\mathrm{d}x'\ . \tag{A.65}$$

Equation (A.65) is an approximate equation which is valid when the contribution from the non-local potential is small compared with $U(x)$. As these formulae show, the equivalent effective local potential depends on the energy.

The energy dependence discussed here originates from non-locality. The potential has in addition another energy dependence due to the polarization or the channel-coupling effects such as the phonon excitation on the way of collision. The latter is often called the intrinsic energy dependence in order to distinguish from the energy dependence due to non-locality [1].

A.5.2.2 WKB Approximation II: Effective Mass

Let us next expand the translation operator $\exp[\frac{\mathrm{i}}{\hbar}(x'-x)\hat{p}_\psi]$ and keep up to the second order terms,

$$\exp\left[\frac{\mathrm{i}}{\hbar}(x'-x)\hat{p}_\psi\right] \approx 1 + \frac{\mathrm{i}}{\hbar}(x'-x)\hat{p}_\psi - \frac{1}{2!}\frac{1}{\hbar^2}(x'-x)^2\hat{p}_\psi^2\ . \tag{A.66}$$

Equation (A.60) is then approximated as

$$-\frac{\hbar^2}{2m_{\text{eff}}}\frac{\mathrm{d}^2\psi(x)}{\mathrm{d}x^2} + U_{\text{eff}}^{(0)}(x)\psi(x) + V^{(1)}(x)\frac{\mathrm{d}}{\mathrm{d}x}\psi(x) = E\psi(x) , \tag{A.67}$$

where

$$\frac{1}{m_{\text{eff}}} = \frac{1}{m} - \frac{1}{\hbar^2}\int (x'-x)^2 V(x,x')\mathrm{d}x' , \tag{A.68}$$

$$U_{\text{eff}}^{(0)}(x) \equiv U(x) + \int V(x,x')\mathrm{d}x' , \tag{A.69}$$

$$V^{(1)}(x) \equiv \int (x'-x)V(x,x')\mathrm{d}x' . \tag{A.70}$$

In particular, suppose that the non-local potential is given by a separable type such that [12]

$$V(x,x') = F\left(\frac{x+x'}{2}\right) G(x'-x) \tag{A.71}$$

$$= F\left(\frac{x+x'}{2}\right)\frac{1}{\sqrt{\pi}}\frac{1}{\lambda_{NL}}\mathrm{e}^{-[(x'-x)/\lambda_{NL}]^2} \tag{A.72}$$

$$\sim F(x)\frac{1}{\sqrt{\pi}}\frac{1}{\lambda_{NL}}\mathrm{e}^{-[(x'-x)/\lambda_{NL}]^2} . \tag{A.73}$$

λ_{NL} is a parameter to measure the degree of non-locality. The degree of non-locality is larger for larger values of λ_{NL}. Equation (A.73) leads to

$$\frac{1}{m_{\text{eff}}(x)} = \frac{1}{m} - \frac{1}{\hbar^2}F(x)\frac{\lambda_{NL}^2}{2} , \tag{A.74}$$

$$U_{\text{eff}}^{(0)}(x) \equiv U(x) + F(x) , \tag{A.75}$$

$$V^{(1)}(x) = 0 . \tag{A.76}$$

Equation (A.74) indicates that Eq. (A.67) is valid when the degree of non-locality is small, and that the mass parameter becomes position dependent due to the non-local potential. In addition, Eq. (A.75) shows that the static part of the potential is also modified by the non-local potential.

A.5.2.3 WKB Approximation III: Wigner Transformation

Here we derive an alternative expression of the equivalent local potential based on the Wigner transformation. Let us first rewrite the non-local potential term by using the Wigner transformation of the non-local potential as

$$\int V(x, x')\psi(x')\mathrm{d}x'$$
$$= \left[\frac{1}{2\pi\hbar}\int \mathrm{d}x' \int \mathrm{d}p \mathrm{e}^{-\mathrm{i}p(x'-x)/\hbar}\mathrm{e}^{\mathrm{i}(x'-x)\hat{p}_V/2\hbar}V_W(x, p)\mathrm{e}^{\mathrm{i}(x'-x)\hat{p}_\psi/\hbar}\right]\psi(x)\ . \tag{A.77}$$

$\hat{p}_V$ is an operator which operates on x in $V_W(x, p)$. If we assume

$$\mathrm{e}^{\mathrm{i}(x'-x)\hat{p}_V/2\hbar}V_W(x, p) \approx V_W(x, p)\ , \tag{A.78}$$

we obtain

$$\int V(x, x')\psi(x')\mathrm{d}x' = V_W(x, \hat{p}_\psi)\psi(x) \tag{A.79}$$
$$\sim V_W(x, p(x))\psi(x)\ . \tag{A.80}$$

The local momentum approximation (A.62), which is based on the WKB approximation, has been used in transforming from Eq. (A.79) to (A.80). Finally, we obtain

$$U_{\mathrm{eff}}^{(W)} = U(x) + V_W(x, p(x))\ , \tag{A.81}$$

for the equivalent effective local potential. The right-hand side of Eq. (A.81) agrees with the equivalent potential in the phase space which appears in the classical theory in the presence of a non-local potential [13].[4]

A.6 Tensor Algebra

A.6.1 Definition of Tensor Operators

The tensor property of a physical quantity or an operator represents their transformation property under a rotation of either the coordinate system or the object and is a unified generalization of scalar, vector and tensor by introducing the index named rank. The tensor property is one of the fundamental properties of each physical quantity like the spatial and/or temporal dimension related to units, and plays an important role, e.g., in evaluating transition matrices and in deriving selection rules.

[4]Thanks to the large effective mass and hence the short wave length property, the classical theory which uses the concept of classical trajectory is useful for heavy-ion collisions, and has been successfully applied to analysing fusion reactions and deep inelastic collisions involving a huge amount of energy dissipation from the relative motion to intrinsic motions and the associated fluctuations [14, 15]. In that connection, it is an interesting question how to incorporate the exchange effects of nucleons in the classical theory. The semi-classical treatment of a non-local potential has been discussed also in [16].

Let us consider a translation before we discuss a rotation. In that case, there are two alternative ways. In the one, one keeps the object as it is, and performs a translation of the coordinate system from the old one $\mathscr{K}(x, y, z)$ to a new one $\mathscr{K}'(x', y', z')$. In the other, one translates the object by keeping the coordinate system as it is. Here, we take the latter. Let us consider the translation of an object by a. If we express the wave function before and after the translation as $\psi(x)$ and $\psi'(x)$, respectively, then

$$\begin{aligned} \psi'(x) &= \psi(x-a) \\ &= \psi(x) - a\frac{\mathrm{d}}{\mathrm{d}x}\psi(x) + \frac{1}{2!}\frac{\mathrm{d}^2}{\mathrm{d}x^2}\psi(x) + \cdots \\ &= \mathrm{e}^{-\mathrm{i}a\frac{\hbar}{\mathrm{i}}\frac{\mathrm{d}}{\mathrm{d}x}/\hbar}\psi(x) = \mathrm{e}^{-\mathrm{i}a\hat{p}/\hbar}\psi(x)\ . \end{aligned} \tag{A.82}$$

Equation (A.82) shows that $\hat{p}$ is the generator of the infinitesimal translation in space. Equation (A.82) is expressed as

$$|\psi'\rangle = \hat{T}(a)|\psi\rangle\ , \tag{A.83}$$

$$\hat{T}(a) \equiv \mathrm{e}^{-\mathrm{i}a\hat{p}/\hbar}\ , \tag{A.84}$$

in the Dirac notation. $\hat{T}(a)$ given by Eq. (A.84) is the unitary operator which causes the translation by a.[5] Also, if we define the operator $\hat{O}'$ for an operator $\hat{O}$ representing some observable in such a way that the expectation value is conserved under transformation, namely, by

$$\langle\psi|\hat{O}|\psi\rangle \equiv \langle\psi'|\hat{O}'|\psi'\rangle\ , \tag{A.85}$$

then by rewriting the right-hand side of Eq. (A.85) as $\langle\psi'|\hat{O}'|\psi'\rangle = \langle\psi|\hat{T}^\dagger\hat{O}'\hat{T}|\psi\rangle$, we find that $\hat{O}'$ is given by

$$\hat{O}' = \hat{T}(a)\hat{O}\hat{T}^\dagger(a) = \hat{T}(a)\hat{O}\hat{T}^{-1}(a)\ . \tag{A.86}$$

Equation (A.86) shows that the operator is transformed by the same unitary operator as that for the state vector ψ.

Exercise A.2 Show

$$\hat{x}' \equiv \hat{T}(a)\hat{x}\hat{T}^\dagger(a) = \hat{x} - a\ , \tag{A.87}$$

$$\langle\psi'|\hat{x}|\psi'\rangle = \langle\psi|\hat{x}|\psi\rangle + a\ , \tag{A.88}$$

by using Eq. (A.84). Equation (A.88) agrees with that $|\psi'\rangle$ is the state, where the object is translated by a.

---

[5] It is the operator to cause the translation by $-a$ if one takes the point of view of fixing the object and transforming the coordinate system.

Analogously, by considering infinitesimal rotations, one can show that the generators of the infinitesimal rotations are the angular momentum operators $\hat{J}_\alpha$ ($\alpha = x, y, z$). For example, the rotation of an object through an angle α about the z-axis is given by the unitary operator $\hat{R}_z = \mathrm{e}^{-\mathrm{i}\alpha\hat{J}_z}$. By extending this result, the unitary operator for a rotation in the three-dimensional space is in general given either as

$$\hat{R}(\mathbf{w}) = \mathrm{e}^{-\mathrm{i}\mathbf{w}\cdot\hat{\mathbf{J}}} \tag{A.89}$$

by using the vector $\mathbf{w}$, which specifies the direction and the magnitude of rotation, or as

$$\hat{R}(\phi, \theta, \psi) = \mathrm{e}^{-\mathrm{i}\phi\hat{J}_z}\mathrm{e}^{-\mathrm{i}\theta\hat{J}_y}\mathrm{e}^{-\mathrm{i}\psi\hat{J}_z} \tag{A.90}$$

by using the Euler angles [17].

(i) First Definition of a Tensor Operator: When $2k + 1$ operators $\hat{T}_q^{(k)}$ ($q = -k, -k+1, \ldots, k-1, k$) transform to each other following

$$(\hat{T}_q^{(k)})' = \hat{R}\hat{T}_q^{(k)}\hat{R}^{-1} = \sum_{q'=-k}^{k} \hat{T}_{q'}^{(k)} R_{q'q}^{(k)}(\phi, \theta, \psi) \tag{A.91}$$

under a rotation, $\hat{T}^{(k)}$ is called an *irreducible tensor of rank* k, and $\hat{T}_q^{(k)}$ the q-component. The rotation matrix $R_{q'q}^{(k)}(\phi, \theta, \psi)$ is defined by

$$R_{q'q}^{(k)}(\phi, \theta, \psi) \equiv \langle kq'|\hat{R}(\phi, \theta, \psi)|kq\rangle\ . \tag{A.92}$$

[N.B.] Equation (A.92) obeys [17]. It is the same as D function in Ref. [18]: $R_{q'q}^{(k)}(\phi, \theta, \psi) = (D_{q'q}^{(k)}(\phi, \theta, \psi))_{\mathrm{BS}}$. It is related to the $\mathscr{D}$ function in Ref. [19] as $R_{q'q}^{(k)}(\phi, \theta, \psi) = (\mathscr{D}_{q'q}^{(k)*}(\phi, \theta, \psi))_{\mathrm{BM}}$. Since Ref. [20] transforms coordinate system rather than the object, the sign of the arguments of the D function is inverted: $R_{q'q}^{(k)}(\phi, \theta, \psi) = (\mathscr{D}_{q'q}^{(k)}(-\phi, -\theta, -\psi))_{\mathrm{RS}}$. In this book, the definition of $\mathscr{D}_{qq'}^{(k)}$ follows Ref. [19].

(ii) Second Definition of a Tensor Operator: Since the generators of the infinitesimal rotations are the angular momentum operators, one can alternatively define the irreducible tensor of rank k as the set of operators which satisfy the following commutation relations:

$$\left[\hat{J}_z, \hat{T}_q^{(k)}\right] = q\hat{T}_q^{(k)}\ , \tag{A.93}$$

$$\left[\hat{J}_\pm, \hat{T}_q^{(k)}\right] = \sqrt{(k \mp q)(k \pm q + 1)}\hat{T}_{q\pm1}^{(k)}\ . \tag{A.94}$$

In order to show the equivalence between the two definitions, we show here that Eqs. (A.93) and (A.94) can be derived from Eq. (A.91). Let us now assume that $\hat{R}$ is the infinitesimal rotation operator $(1 - \mathrm{i}\alpha\hat{J}_\lambda)$. In this case, from the definition

Eq. (A.92), we have

$$R^{(k)}_{q'q} = \langle kq'|(1 - \mathrm{i}\alpha \hat{J}_\lambda)|kq\rangle = \delta_{q'q} - \mathrm{i}\alpha\langle kq'|\hat{J}_\lambda|kq\rangle \ . \tag{A.95}$$

On the other hand,

$$\hat{R}\hat{T}^{(k)}_q\hat{R}^{-1} = (1 - \mathrm{i}\alpha\hat{J}_\lambda)\hat{T}^{(k)}_q(1 + \mathrm{i}\alpha\hat{J}_\lambda) \ . \tag{A.96}$$

Hence from Eq. (A.91)

$$(1 - \mathrm{i}\alpha\hat{J}_\lambda)\hat{T}^{(k)}_q(1 + \mathrm{i}\alpha\hat{J}_\lambda) = \left(\delta_{q'q} - \mathrm{i}\alpha\langle kq'|\hat{J}_\lambda|kq\rangle\right)\hat{T}^{(k)}_{q'} \ . \tag{A.97}$$

Thus it follows

$$[\hat{J}_\lambda, \hat{T}^{(k)}_q] = \sum_{q'}\langle kq'|\hat{J}_\lambda|kq\rangle\hat{T}^{(k)}_{q'} \ . \tag{A.98}$$

By putting here $\hat{J}_\lambda = \hat{J}_z$ and $\hat{J}_\lambda = \hat{J}_\pm$ and noting

$$\langle kq'|\hat{J}_z|kq\rangle = q\delta_{q'q} \ , \quad \langle k\, q \pm 1|\hat{J}_\pm|kq\rangle = \sqrt{(k \mp q)(k \pm q + 1)} \ , \tag{A.99}$$

we obtain Eqs. (A.93) and (A.94).

A.6.2 Examples of Irreducible Tensor

(i) Spherical Harmonics: $Y_{kq}(\theta, \varphi)$.

(ii) Tensor of Rank 1: Vector Operators A vector is an irreducible tensor of rank 1 and its $1, 0, -1$ and x, y, z components are related as [18],

$$\begin{aligned} V_{\pm 1} &= \mp\frac{1}{\sqrt{2}}(V_x \pm \mathrm{i}V_y) \ , \\ V_0 &= V_z \ . \end{aligned} \tag{A.100}$$

Equation (A.100) can be understood from the fact that the x, y and z components of the position vector $\mathbf{r}$ can be expressed as

$$\begin{aligned} rY_{1\pm 1} &= \mp\sqrt{\frac{3}{4\pi}}\frac{1}{\sqrt{2}}(x \pm \mathrm{i}y) \ , \\ rY_{10} &= \sqrt{\frac{3}{4\pi}}z \ , \end{aligned} \tag{A.101}$$

by using the explicit expressions of the irreducible tensor of rank 1, Y_{1q}.

A.6.3 Wigner–Eckart Theorem

The following formula holds[6]

$$\langle \alpha J M | T_q^{(k)} | \alpha' J' M' \rangle = \frac{1}{\sqrt{2J+1}} \langle J' k M' q | J M \rangle \langle \alpha J \| T^{(k)} \| \alpha' J' \rangle \,, \qquad \text{(A.102)}$$

for the q component of a tensor of rank k, $T_q^{(k)}$. The α and α' are the quantum numbers other than the angular momentum to specify states.

Equation (A.102) is called the *Wigner–Eckart theorem*, and shows that the matrix element of a tensor operator can factor out the dependence on the z-component of the angular momentum in terms of the geometrical factor given by the Clebsch–Gordan coefficient. The factor $\langle \alpha J \| T^{(k)} \| \alpha' J' \rangle$ is called the *reduced matrix element*. The Wigner–Eckart theorem allows to automatically determine all the matrix elements based on the Clebsch–Gordan coefficient if the reduced matrix element is once determined by calculating the matrix element for one combination of M', q, M for which the Clebsch–Gordan coefficient is finite.

Especially when T is 1 or the angular momentum operators or the spherical harmonics, the reduced matrix elements are given by

$$\langle \alpha J \| 1 \| \alpha' J' \rangle = \delta_{\alpha\alpha'} \delta_{JJ'} \sqrt{2J+1} \,, \qquad \text{(A.103)}$$

$$\langle \alpha J \| \mathbf{J} \| \alpha' J' \rangle = \delta_{\alpha\alpha'} \delta_{JJ'} \sqrt{J(J+1)(2J+1)} \qquad \text{(A.104)}$$

$$\begin{aligned} \langle \ell \| Y^{(k)} \| \ell' \rangle &= (-1)^{\ell-\ell'-k} \sqrt{\frac{(2k+1)(2\ell'+1)}{4\pi}} \langle k \ell' 0 0 | \ell 0 \rangle \\ &= (-1)^{\ell} \sqrt{\frac{(2k+1)(2\ell'+1)(2\ell+1)}{4\pi}} \begin{pmatrix} \ell & k & \ell' \\ 0 & 0 & 0 \end{pmatrix} \,, \end{aligned} \qquad \text{(A.105)}$$

where $\begin{pmatrix} \ell & k & \ell' \\ 0 & 0 & 0 \end{pmatrix}$ is a $3j$ symbol.

A.6.4 Projection Theorem

Using the Wigner–Eckart theorem, one can show that the matrix element of any vector operator $\mathbf{A}$ can be expressed as

$$\langle \alpha J M | \mathbf{A} | \alpha J M' \rangle = \frac{1}{J(J+1)} \langle \alpha J M | \mathbf{J} | \alpha J M' \rangle \langle \alpha J J | (\mathbf{A} \cdot \mathbf{J}) | \alpha J J \rangle \,. \qquad \text{(A.106)}$$

[6]There exists an alternative definition, which does not explicitly take out the factor $1/\sqrt{2J+1}$ on the right-hand side, but includes it in the factor $\langle \alpha J \| T \| \alpha' J' \rangle$.

Equation (A.106) is called the *projection theorem* or the *Landé formula*, and is useful, e.g., in deriving the Schmidt values for the magnetic dipole moment given in Sect. 5.4.1.

A.6.5 Relation Between the Scalar Product and the Tensor Product of Rank 0

One can define a combined tensor operator of two tensor operators by taking an appropriate sum of the product of their components using the Clebsch–Gordan coefficients as follows

$$V_Q^{(K)} \equiv \left[T^{(k_1)} \times U^{(k_2)}\right]_Q^{(K)} \equiv \sum_{q_1} \langle k_1 k_2 q_1 q_2 | K\, Q \rangle T_{q_1}^{(k_1)} U_{q_2}^{(k_2)} \,. \tag{A.107}$$

In the case of a combined tensor of rank 0, one can use the scalar product defined by

$$S \equiv (T^{(k)} \cdot U^{(k)}) \equiv \sum_q (-1)^q T_q^{(k)} U_{-q}^{(k)} \quad \text{(scalar product)}, \tag{A.108}$$

for $V_0^{(0)}$. Especially, the scalar product of two vectors is given by

$$S = (\mathbf{V}^{(1)} \cdot \mathbf{W}^{(1)}) = V_x W_x + V_y W_y + V_z W_z \,. \tag{A.109}$$

Exercise A.3 Show that the scalar product and the tensor of rank 0 are related as

$$S = (-1)^k \sqrt{2k+1}\, V_0^{(0)} \tag{A.110}$$

when one combines two tensors of rank k.

A.7 Relation Between the Quadrupole Moment and Intrinsic Quadrupole Moment

Here, we prove Eq. (4.47), which relates the experimentally observed quadrupole moment for a deformed nucleus to the intrinsic quadrupole moment.

We first remark from the general transformation formula of a tensor, Eq. (A.91), that the multipole moment operator $\hat{Q}_{\lambda\mu}^{\text{intr.}}$ given by the intrinsic coordinates and the corresponding operator $\hat{Q}_{\lambda\mu}^{\text{lab.}}$ which is related to the experimental measurement and is defined by the coordinates in the laboratory system are related as

$$\hat{Q}^{\text{intr.}}_{\lambda\mu} = \sum_{\mu'} \mathscr{D}^{\lambda *}_{\mu'\mu}(\omega)\hat{Q}^{\text{lab.}}_{\lambda\mu'} , \quad \hat{Q}^{\text{lab.}}_{\lambda\mu} = \sum_{\nu} \mathscr{D}^{\lambda}_{\mu\nu}(\omega)\hat{Q}^{\text{intr.}}_{\lambda\nu} , \tag{A.111}$$

through the Euler angles $\omega = (\phi, \theta, \psi)$ which specify the orientation of the principal axes. We used the properties of the $\mathscr{D}$ function in transforming from the first to the second equations. Since the wave function for an axial-symmetrically deformed state is given by Eq. (8.72), the matrix element of the operator is given in general by

$$\begin{aligned}
\left\langle \Psi_{I_1 M_1 K_1} \left| \hat{Q}^{\text{lab.}}_{\lambda\mu} \right| \Psi_{I_2 M_2 K_2} \right\rangle &= \sum_{\nu} \left(\frac{2I_1+1}{16\pi^2(1+\delta_{K_1 0})} \right)^{1/2} \left(\frac{2I_2+1}{16\pi^2(1+\delta_{K_2 0})} \right)^{1/2} \\
&\times \Big[\left\langle \mathscr{D}^{I_1}_{M_1 K_1} \left| \mathscr{D}^{\lambda}_{\mu\nu} \right| \mathscr{D}^{I_2}_{M_2 K_2} \right\rangle \left\langle \Phi_{K_1} \left| \hat{Q}^{\text{intr.}}_{\lambda\nu} \right| \Phi_{K_2} \right\rangle \\
&+ (-1)^{I_2+K_2} \left\langle \mathscr{D}^{I_1}_{M_1 K_1} \left| \mathscr{D}^{\lambda}_{\mu\nu} \right| \mathscr{D}^{I_2}_{M_2 -K_2} \right\rangle \left\langle \Phi_{K_1} \left| \hat{Q}^{\text{intr.}}_{\lambda\nu} \right| \Phi_{\bar{K}_2} \right\rangle \\
&+ (-1)^{I_1+K_1} \left\langle \mathscr{D}^{I_1}_{M_1 -K_1} \left| \mathscr{D}^{\lambda}_{\mu\nu} \right| \mathscr{D}^{I_2}_{M_2 K_2} \right\rangle \left\langle \Phi_{\bar{K}_1} \left| \hat{Q}^{\text{intr.}}_{\lambda\nu} \right| \Phi_{K_2} \right\rangle \\
&+ (-1)^{I_2+K_2+I_1+K_1} \left\langle \mathscr{D}^{I_1}_{M_1 -K_1} \left| \mathscr{D}^{\lambda}_{\mu\nu} \right| \mathscr{D}^{I_2}_{M_2 -K_2} \right\rangle \left\langle \Phi_{\bar{K}_1} \left| \hat{Q}^{\text{intr.}}_{\lambda\nu} \right| \Phi_{\bar{K}_2} \right\rangle \Big] .
\end{aligned} \tag{A.112}$$

In the following, we consider the quadrupole moment. Since the quadrupole moment is defined by $\langle \Psi_{IM=IK} | \hat{Q}^{\text{lab.}}_{\lambda=2,\mu=0} | \Psi_{IM=IK} \rangle$, one can easily show that the second and third terms on the right-hand side of Eq. (A.112) are zero for axial-symmetrically deformed even–even nuclei[7] and for odd nuclei in the strong coupling state with $I = K = 1/2$ or $K \geq 3/2$ by noting the properties of the $\mathscr{D}$ function, $|\nu| \leq 2$, the addition rule of two angular momenta and the fact that $K = 0, 2, 4 \cdots$ for collective motions of axial-symmetrically deformed even–even nuclei. In addition, the fourth and first terms become equal. It therefore follows

$$\langle \Psi_{IIK} | \hat{Q}^{\text{lab.}}_{20} | \Psi_{IIK} \rangle = \sum_{\nu} \frac{2I+1}{8\pi^2} \langle \mathscr{D}^{I}_{M=IK} | \mathscr{D}^{2}_{0\nu} | \mathscr{D}^{I}_{M=IK} \rangle \langle \Phi_K | \hat{Q}^{\text{intr.}}_{\lambda\nu} | \Phi_K \rangle \tag{A.113}$$

$$= \langle 2I0I | II \rangle \langle 2I0K | IK \rangle \langle \Phi_K | \hat{Q}^{\text{intr.}}_{20} | \Phi_K \rangle . \tag{A.114}$$

We then obtain Eq. (4.47) by using the explicit expression of the Clebsch–Gordan coefficient.

A.8 Derivation of the Formula for the α-Decay Width Based on the Gamow Model: Direct Method

Here, we derive the formulae (8.13)–(8.17) for the α-decay width. We first note that in the Gamow model the wave function is given by

[7] This excludes the case when $K_1 = K_2 = 0$. The second and third terms can be combined with the first and/or fourth terms when $K_1 = K_2 = 0$.

$$\phi(r) = \begin{cases} \phi_I(r) = N j_0(Kr) & \text{for } r < R \\ \phi_{II}(r) = (\mu/\hbar k)^{1/2} \left[G_0(\eta, kr) + \mathrm{i} F_0(\eta, kr) \right] / r & \text{for } r \geq R \;, \end{cases} \tag{A.115}$$

where K and $k = k_\alpha$ are given by Eq. (8.17). The continuity conditions at $r = R$ require that

$$\begin{aligned} N \sin KR &= \sqrt{\frac{\mu}{\hbar k}} K \left[G_0(\eta, kR) + \mathrm{i} F_0(\eta, kR) \right] \\ N \cos KR &= \sqrt{\frac{\mu}{\hbar k}} k \left[G_0'(\eta, kR) + \mathrm{i} F_0'(\eta, kR) \right] \;, \end{aligned} \tag{A.116}$$

where G_0' and F_0' are the derivative of $G_0(\eta, \rho)$ and $F_0(\eta, \rho)$ with respect to ρ, where $\rho = kr$. These equations lead to

$$\frac{k}{K} \tan KR = \frac{G_0(\eta, kR) + \mathrm{i} F_0(\eta, kR)}{G_0'(\eta, kR) + \mathrm{i} F_0'(\eta, kR)} \;. \tag{A.117}$$

On the other hand, the decay width Γ is given by the outgoing flux per unit time at the external turning point of the potential barrier in the case when the wave function inside the potential is normalized to be unity,

$$\frac{\Gamma}{\hbar} \equiv \frac{\text{flux at the external turning point}}{\text{probability to be inside the potential}} \;. \tag{A.118}$$

Since Eq. (A.115) normalizes the external flux to be unity, we have

$$\frac{\hbar}{\Gamma} = \int_0^R |\phi_I(r)|^2 r^2 \mathrm{d}r + \int_R^{r_e} |\phi_{II}(r)|^2 r^2 \mathrm{d}r \tag{A.119}$$

$$= \frac{\mu}{\hbar k^2} \left[\frac{1}{2} kR \frac{G_0^2(\eta, kR) + F_0^2(\eta, kR)}{\sin^2 KR} \left(1 - \frac{\sin 2KR}{2KR} \right) + \int_{kR}^{2\eta} (G_0^2(\eta, \rho) + F_0^2(\eta, \rho)) \mathrm{d}\rho \right] \;. \tag{A.120}$$

One can determine the energy E, i.e., the Q-value for the α-decay, and the decay width Γ by using Eqs. (A.117) and (A.120) if the depth of the potential V_0 and the range of the nuclear force R are known in advance. Conversely, one can determine the depth of the potential V_0 and the range of the nuclear force R for the α-decay from the Q-value and the decay width using Eqs. (A.117) and (A.120) if they are known experimentally. However, as discussed in Sect. 8.1.1, there remains an ambiguity of potential because of the presence of Pauli forbidden states, in other words, redundant states. In order to remove the ambiguity, one needs to specify the number of nodes in the wave function for the relative motion. As one example of such analyses, Table A.1

Table A.1 The potential for the α-decay $^{210}\mathrm{Po}(0^+_{\mathrm{g.s.}}) \rightarrow {}^{206}\mathrm{Pb}(0^+_{\mathrm{g.s.}}) + \alpha$ [21]

Number of nodes	R (fm)	V_0 (MeV)	KR/π
13	7.91	150	13.62
12	7.91	128	12.63
11	7.92	108	11.64
10	7.93	89.4	10.65
9	7.94	72.5	9.660
8	7.95	57.2	8.675
7	7.97	43.6	7.693
6	7.99	31.7	6.714
5	8.03	21.5	5.739
4	8.08	13.0	4.768
3	8.15	6.16	3.804
2	8.25	0.95	2.846
1	8.43	−2.65	1.894
0	8.77	−4.69	0.9461

shows the results of the analysis to determine the potential for the α-decay from the ground state of ^{210}Po to the ground state of ^{206}Pb given by Eq. (8.1) by using the experimental data for the Q-value and the decay width. As the Table shows, there exist a wide range of potentials from shallow to deep ones depending on the number of nodes in the radial wave function of the relative motion. The number of nodes is 11 for the α-decay from the ground state of ^{210}Po to the ground state of ^{206}Pb if we count the quantum number in the harmonic oscillator shell model.

The last column of Table A.1 gives the magnitude of the action integral inside the potential area in units of $\hbar\pi$. When the number of node n is large, including the case of $n = 11$, the value of the action integral inside the potential region becomes

$$KR \approx \left(n + \frac{1}{2}\right)\pi \,, \tag{A.121}$$

and well agrees with the quantization condition (8.5) derived based on the semi-classical theory.

In the following, suppose that the Q-value of the α-decay is small, so that $2\eta \gg \rho = kR$. In that case, the following asymptotic formulae hold for the Coulomb wave functions $F_0(\eta, kR)$, $F_0'(\eta, kR)$, $G_0(\eta, kR)$ and $G_0'(\eta, kR)$ [22]:

$$F_0 \sim \frac{1}{2}\beta e^{\alpha} \,; \quad F_0' \sim \frac{1}{2}\beta^{-1} e^{\alpha} \,; \quad G_0 \sim \beta e^{-\alpha} \,; \quad G_0' \sim -\beta^{-1} e^{-\alpha} \,, \tag{A.122}$$

$$\alpha = 2\sqrt{2\eta\rho} - \pi\eta \,, \quad \beta = (\rho/2\eta)^{1/4} \,. \tag{A.123}$$

Let us now relate the decay width to the tunneling probability by using Eq. (A.121). We set $\sin 2KR$ to 0 based on Eq. (A.121). Equations (A.122) and (A.123) suggest $|F_0(\eta, kR)| \ll |G_0(\eta, kR)|$, so that we can ignore F_0 in Eq. (A.120). Also, we ignore the integration term as well. The decay width is then approximately given by

$$\Gamma \sim \frac{2\hbar^2 k}{\mu R} \frac{1}{G_0^2(\eta, kR)} . \tag{A.124}$$

On the other hand, the tunneling probability t is given by the ratio of the outgoing flux $J_{II}^{(+)}$ to the flux $J_I^{(+)}$ which enters the potential barrier at $r = R$ in the outward direction from inside. Since $J_I^{(+)}$ and $J_{II}^{(+)}$ are given by

$$J_I^{(+)} = \int j_I^{(+)} \mathrm{d}S = \int \frac{1}{r^2} \frac{\hbar K}{\mu} \left| \frac{N}{2K} \right|^2 r^2 \mathrm{d}\Omega = 4\pi \frac{\hbar K}{\mu} \left| \frac{N}{2K} \right|^2 , \tag{A.125}$$

$$J_{II}^{(+)} = \int j_{II}^{(+)} \mathrm{d}S = \int \frac{1}{r^2} r^2 \mathrm{d}\Omega = 4\pi , \tag{A.126}$$

the tunneling probability becomes

$$t_{\mathrm{GM}}^{(\mathrm{D})} \equiv \frac{J_{II}^{(+)}}{J_I^{(+)}} \sim \frac{4k}{K} \frac{\sin^2 KR}{G_0^2(\eta, kR)} \sim \frac{4k}{K} \frac{1}{G_0^2(\eta, kR)} . \tag{A.127}$$

Equations (A.124) and (A.127) lead to Eq. (8.13).

Thus one can obtain Eqs. (8.13)–(8.17) by using the asymptotic forms of the Coulomb wave functions given by Eqs. (A.122) and (A.123) [22], which hold when $kR \ll 2\eta = kr_e$, i.e., when $R \ll r_e$ or when the Q-value is small.

A.9 Basics of Electromagnetic Transitions

A.9.1 Hamiltonian of the Total System

To discuss the electromagnetic transitions of a nucleus, it is necessary to treat the combined system consisting of the nucleus, i.e., a many-body system of nucleons, and the electromagnetic fields. The Hamiltonian of the total system is given by

$$\hat{H}_{\mathrm{tot}} = \hat{H}_{\mathrm{nucl}} + \hat{H}_{\mathrm{field}} + \hat{H}_{\mathrm{int}} . \tag{A.128}$$

$\hat{H}_{\mathrm{nucl}}$ is the Hamiltonian of an isolated nucleus, and consists of the kinetic energy of nucleons and the strong interaction among nucleons. Alternatively, it is expressed in an appropriate model such as the shell model or the collective model. We assume that the eigenstates and the corresponding eigenvalues are known and express as

$$\hat{H}_{\text{nucl}}\Psi_\alpha = E_\alpha \Psi_\alpha \ . \tag{A.129}$$

The Hamiltonian of the electromagnetic fields $\hat{H}_{\text{field}}$ and the interaction Hamiltonian between the nucleus and the electromagnetic fields $\hat{H}_{\text{int}}$ are given by

$$\hat{H}_{\text{field}} = \frac{1}{8\pi}\int \left[\mathbf{E}^2(\mathbf{r}, t) + \mathbf{B}^2(\mathbf{r}, t)\right] d\mathbf{r} \ , \tag{A.130}$$

$$\hat{H}_{\text{int}} = -\frac{1}{c}\int j_\mu A^\mu d\mathbf{r} \tag{A.131}$$

$$= \int \left[\rho(\mathbf{r}, t)\Phi(\mathbf{r}, t) - \frac{1}{c}\mathbf{j}(\mathbf{r}, t)\mathbf{A}(\mathbf{r}, t)\right] d\mathbf{r} \ , \tag{A.132}$$

where $\mathbf{E}$, $\mathbf{B}$, ρ and $\mathbf{j}$ are the electric field, the magnetic field, the charge and current densities of the nucleus, respectively.

From the point of view of the original many-body system of nucleons, ρ is given by

$$\rho(\mathbf{r}, t) = \sum_{i=1}^{A} e\left(\frac{1}{2} - t_3^{(i)}\right)\delta(\mathbf{r} - \mathbf{r}_i(t)) \ , \tag{A.133}$$

while $\mathbf{j}$ by

$$\mathbf{j}(\mathbf{r}, t) = c\nabla \times \boldsymbol{\mu}(\mathbf{r}, t) \tag{A.134}$$

using the magnetic moment $\boldsymbol{\mu}(\mathbf{r}, t)$. Since the magnetic moment originates from the orbital motion of protons, and also from the spins of protons and neutrons, $\mathbf{j}$ is given by a sum of two parts. If we denote $\mathbf{j}$ originating from the orbital motion of protons by $\mathbf{j}_c$, then

$$\mathbf{j}_c = \sum_{i=1}^{A} e\left(\frac{1}{2} - t_3^{(i)}\right)\frac{1}{2}\left[\mathbf{v}_i\delta(\mathbf{r} - \mathbf{r}_i(t)) + \text{h.c.}\right] \ . \tag{A.135}$$

$\mathbf{v}_i$ is given by $\mathbf{v}_i = \frac{\mathrm{i}}{\hbar}\left[\hat{H}_{\text{nucl}}, \mathbf{r}_i\right]$, and becomes $\mathbf{v}_i = \mathbf{p}_i/M_N$ if the potential in $\hat{H}_{\text{nucl}}$ does not depend on the momentum. On the other hand, if we denote the magnetic moment originating from the spin of nucleons by $\boldsymbol{\mu}_s$, then it is given by

$$\boldsymbol{\mu}_s(\mathbf{r}, t) = \sum_{i=1}^{A}\delta(\mathbf{r} - \mathbf{r}_i(t))\frac{e\hbar}{2M_N c}\left[\left(\frac{1}{2} - t_3^{(i)}\right)g_p + \left(\frac{1}{2} + t_3^{(i)}\right)g_n\right]\mathbf{s}_i \tag{A.136}$$

using the g-factors of protons and neutrons: $g_\text{p} = 2.792847 \times 2 \approx 5.586$, $g_\text{n} = -1.9130427 \times 2 \approx -3.826$.

A.9.2 *Wave Function of Photon: Vector Spherical Harmonics*

A.9.2.1 Coulomb Gauge

It is convenient to adopt the *Coulomb gauge*, known also as the *transverse gauge* or the *radiation gauge* [23]:

$$\nabla \cdot \mathbf{A} = 0 \tag{A.137}$$

in order to discuss the electromagnetic transitions. In this case, the Hamiltonian of the radiation field is given by

$$\hat{H}_{\text{field}} = \frac{1}{8\pi} \int \left(\frac{1}{c^2} \dot{\mathbf{A}}^2 + (\nabla \times \mathbf{A})^2 \right) \mathrm{d}\mathbf{r} \,, \tag{A.138}$$

and the vector potential $\mathbf{A}$ obeys the wave equation:

$$\Delta \mathbf{A} - \frac{1}{c^2} \frac{\partial^2}{\partial t^2} \mathbf{A} = 0 \,. \tag{A.139}$$

If we put the time dependence as e^{-ickt}, then we obtain the Helmholtz equation

$$(\Delta + k^2)\mathbf{A} = 0 \,. \tag{A.140}$$

Note that Eq. (A.137) indicates that the vector potential $\mathbf{A}$ is perpendicular to the direction of the propagation of the wave ($\mathbf{k} \perp \mathbf{A}$), i.e., that the wave is a transverse wave.

A.9.2.2 Rotational Transformation of Vector Wave Function and the Spin and Intrinsic Wave Function of Photon

When one uses the spherical polar coordinates and makes the partial wave expansion, the solution of the Helmholtz equation (A.140) for the partial wave ℓ is nothing but $j_\ell(kr)$ provided the field is a scalar field. In reality, photons are described by a vector field, so that they have spin 1 as we show later. It is therefore necessary to extend the wave function so as to be a simultaneous eigenfunction of $\hat{\mathbf{L}}^2$, $\hat{\mathbf{S}}^2$, $\hat{\mathbf{J}}^2$ and $\hat{J}_z$.

(i) The Transformation Property of a Vector Function: Let us first study how a vector function $\mathbf{V}(r, \theta, \varphi)$ is transformed under rotation. One needs to consider both the transformation of vector components and that of coordinates under rotation. For example, when one rotates the coordinate system through α about the z-axis, the vector components in the new coordinate system are given by

$$\begin{pmatrix} V'_x(r,\theta,\varphi) \\ V'_y(r,\theta,\varphi) \\ V'_z(r,\theta,\varphi) \end{pmatrix} = \begin{pmatrix} \cos\alpha & \sin\alpha & 0 \\ -\sin\alpha & \cos\alpha & 0 \\ 0 & 0 & 1 \end{pmatrix} \begin{pmatrix} V_x(r,\theta,\varphi+\alpha) \\ V_y(r,\theta,\varphi+\alpha) \\ V_z(r,\theta,\varphi+\alpha) \end{pmatrix} \,. \tag{A.141}$$

For small α, Eq. (A.141) can be represented as

$$\mathbf{V}' = (1 + \mathrm{i}\alpha \hat{J}_z)\mathbf{V} + \mathcal{O}(\alpha^2) , \tag{A.142}$$

where $\hat{J}_z$ is given by

$$\hat{J}_z = -\mathrm{i}\frac{\partial}{\partial \varphi} + \mathrm{i}\mathbf{e}_z \times \tag{A.143}$$

$$= -\mathrm{i}\left(x\frac{\partial}{\partial y} - y\frac{\partial}{\partial x}\right) + \mathrm{i}\mathbf{e}_z \times \tag{A.144}$$

$$= \hat{L}_z + \mathrm{i}\mathbf{e}_z \times \ . \tag{A.145}$$

$\mathbf{e}_z \times$ means to take the outer product between the unit vector in the z-axis $\mathbf{e}_z$ and the vector $\mathbf{V}$ which appears on the right side in Eq. (A.142).

Exercise A.4 Prove Eq. (A.143) by noting that Eq. (A.141) leads to

$$V_x'(r, \theta, \varphi) = V_x(r, \theta, \varphi) + \alpha\frac{\partial}{\partial \varphi}V_x + \alpha V_y , \tag{A.146}$$

$$V_z'(r, \theta, \varphi) = V_z(r, \theta, \varphi) + \alpha\frac{\partial}{\partial \varphi}V_z , \tag{A.147}$$

and that

$$\mathbf{e}_z \times \mathbf{V} = \mathbf{e}_y V_x - \mathbf{e}_x V_y . \tag{A.148}$$

(ii) Spin of a Photon: Equation (A.145) shows that the intrinsic spin operator of a photon is given by

$$\hat{S}_x = \mathrm{i}\mathbf{e}_x \times \ , \quad \hat{S}_y = \mathrm{i}\mathbf{e}_y \times \ , \quad \hat{S}_z = \mathrm{i}\mathbf{e}_z \times \ . \tag{A.149}$$

Also, the total angular momentum is given by

$$\hat{\mathbf{J}} = \hat{\mathbf{L}} + \hat{\mathbf{S}} . \tag{A.150}$$

Exercise A.5 Show that $\hat{\mathbf{S}}$ defined by Eq. (A.149) satisfies the commutation relations of the angular momentum operator.

(iii) The Eigenvectors of the Intrinsic Spin of Photon: If we define

$$\begin{aligned} \mathbf{e}_{+1} &= -\frac{1}{\sqrt{2}}(\mathbf{e}_x + \mathrm{i}\mathbf{e}_y) , \\ \mathbf{e}_0 &= \mathbf{e}_z , \\ \mathbf{e}_{-1} &= \frac{1}{\sqrt{2}}(\mathbf{e}_x - \mathrm{i}\mathbf{e}_y) , \end{aligned} \tag{A.151}$$

in terms of the unit vectors in the x, y, z direction by following the general rule that relates the $1, 0, -1$ components of a tensor of rank 1 to the x, y, z components of a vector, we can show that

$$\hat{\mathbf{S}}^2\mathbf{e}_q = 2\mathbf{e}_q, \quad \hat{\mathbf{S}}_z\mathbf{e}_q = q\mathbf{e}_q \quad (q = \pm 1, 0) \; . \tag{A.152}$$

Equation (A.152) shows that the magnitude of the intrinsic spin of photon is 1 and that the eigenvector corresponding to $S_z = q$ is $\mathbf{e}_q$.

A.9.2.3 Vector Spherical Harmonics

(i) Definition: From the above arguments, it follows that the wave function of photon, which is the simultaneous eigenfunction of $\hat{\mathbf{L}}^2$, $\hat{\mathbf{S}}^2$, $\hat{\mathbf{J}}^2$ and $\hat{J}_z$, is given by

$$\mathbf{Y}_{J\ell M}(\theta, \varphi) = \sum_{mq} \langle \ell 1\, mq | JM \rangle Y_{\ell m}(\theta, \varphi)\mathbf{e}_q \tag{A.153}$$

by using the addition rule of two angular momenta. $\mathbf{Y}_{J\ell M}(\theta, \varphi)$ are called vector spherical harmonics.

One can easily show that the vector spherical harmonics have the following properties,

$$\hat{\mathbf{J}}^2\mathbf{Y}_{J\ell M}(\theta, \varphi) = J(J+1)\mathbf{Y}_{J\ell M}(\theta, \varphi) \; , \tag{A.154}$$

$$\hat{J}_z\mathbf{Y}_{J\ell M}(\theta, \varphi) = M\mathbf{Y}_{J\ell M}(\theta, \varphi) \; , \tag{A.155}$$

$$\hat{\mathbf{L}}^2\mathbf{Y}_{J\ell M}(\theta, \varphi) = \ell(\ell+1)\mathbf{Y}_{J\ell M}(\theta, \varphi) \; , \tag{A.156}$$

$$\hat{\mathbf{S}}^2\mathbf{Y}_{J\ell M}(\theta, \varphi) = 2\mathbf{Y}_{J\ell M}(\theta, \varphi) \; , \tag{A.157}$$

and the orthogonality,

$$\int \mathrm{d}\Omega \mathbf{Y}^*_{J\ell M}(\theta, \varphi) \cdot \mathbf{Y}_{J\ell M}(\theta, \varphi) = \delta_{JJ'}\delta_{\ell\ell'}\delta_{MM'} \; . \tag{A.158}$$

Note that there exist three kinds of vector spherical harmonics $\mathbf{Y}_{JJM}(\theta, \varphi)$ and $\mathbf{Y}_{JJ\pm 1M}(\theta, \varphi)$, which have different values of the orbital angular momentum for each set of J, M. They can be classified into the following two groups depending on the difference of parity:

- $\mathbf{Y}_{JJM}(\theta, \varphi)$: Parity $(-1)^J$
  This corresponds to the magnetic field for the electric multipole radiation and to the electric field for the magnetic multipole radiation.

- $\mathbf{Y}_{JJ\pm 1M}(\theta, \varphi)$: Parity $(-1)^{J+1}$
  This corresponds to the electric field for the electric multipole radiation and to the magnetic field for the magnetic multipole radiation.

(ii) The Method to Construct from the Ordinary Scalar Spherical Harmonics: The vector spherical harmonics can be constructed from the ordinary scalar spherical harmonics by the following operations:

$$\boldsymbol{\ell} Y_{\ell m}(\theta, \varphi) = \sqrt{\ell(\ell+1)}\mathbf{Y}_{\ell\ell m}(\theta, \varphi) , \tag{A.159}$$

$$\frac{\mathbf{r}}{r} Y_{\ell m} = -\left(\frac{\ell+1}{2\ell+1}\right)^{1/2} \mathbf{Y}_{\ell\ell+1m}(\theta, \varphi) + \left(\frac{\ell}{2\ell+1}\right)^{1/2} \mathbf{Y}_{\ell\ell-1m}(\theta, \varphi) , \tag{A.160}$$

(iii) Gradient Formulae:

$$\begin{aligned} \nabla \Phi(r) Y_{\ell m} = & -\left(\frac{\ell+1}{2\ell+1}\right)^{1/2} \left(\frac{\mathrm{d}}{\mathrm{d}r} - \frac{\ell}{r}\right) \Phi(r) \mathbf{Y}_{\ell\ell+1m}(\theta, \varphi) \\ & + \left(\frac{\ell}{2\ell+1}\right)^{1/2} \left(\frac{\mathrm{d}}{\mathrm{d}r} + \frac{\ell+1}{r}\right) \Phi(r) \mathbf{Y}_{\ell\ell-1m}(\theta, \varphi) , \end{aligned} \tag{A.161}$$

$$\nabla \times (\Phi(r)\mathbf{Y}_{\ell\ell+1m}) = \mathrm{i}\left(\frac{\ell}{2\ell+1}\right)^{1/2} \left(\frac{\mathrm{d}}{\mathrm{d}r} + \frac{\ell+2}{r}\right) \Phi(r) \mathbf{Y}_{\ell\ell m}(\theta, \varphi) . \tag{A.162}$$

A.9.2.4 Eigenfunctions of Photon

(i) General form: As described in Sect. A.9.2.3, there exist three independent solutions of the Helmholtz equation which differ in the orbital angular momentum when the total angular momentum of photon is given, which we denote by I. They have to satisfy in addition the transversality condition $\nabla \cdot \mathbf{A} = 0$ in order to be the wave function of photon. Equation (A.159) shows that $j_I(kr)\mathbf{Y}_{IIM}(\theta, \varphi)$ satisfies this condition. The other wave functions, which are orthogonal to this wave function, are given by taking linear combinations of $j_{I+1}(kr)\mathbf{Y}_{II+1M}(\theta, \varphi)$ and $j_{I-1}(kr)\mathbf{Y}_{II-1M}(\theta, \varphi)$. Among them, the unique wave function which fulfills the transversality condition is

$$\frac{\mathrm{i}}{k} \nabla \times j_I(kr)\mathbf{Y}_{IIm} = \sqrt{\frac{I}{2I+1}} j_{I+1}(kr)\mathbf{Y}_{II+1m} - \sqrt{\frac{I+1}{2I+1}} j_{I-1}(kr)\mathbf{Y}_{II-1m} . \tag{A.163}$$

We distinguish them with the names of the magnetic and electric radiations, and represent as

$$\begin{aligned} \mathbf{A}_{MkIM}(\mathbf{r}) & = \mathscr{N} j_I(kr)\mathbf{Y}_{IIM}(\theta, \varphi) \\ & = \frac{\mathscr{N}}{\sqrt{I(I+1)}} \frac{1}{\mathrm{i}} (\mathbf{r} \times \nabla) j_I(kr) Y_{IM}(\theta, \varphi) , \end{aligned} \tag{A.164}$$

$$\mathbf{A}_{EkIM}(\mathbf{r}) = \frac{\mathrm{i}}{k}(\nabla \times \mathbf{A}_{MkIM}(\mathbf{r}))$$
$$= -\frac{\mathcal{N}}{\sqrt{I(I+1)}}\frac{1}{k}\left[\nabla\left(Y_{IM}(\theta,\varphi)\frac{\partial}{\partial r}rj_I(kr)\right) + k^2\mathbf{r}j_I(kr)Y_{IM}(\theta,\varphi)\right], \quad \text{(A.165)}$$

by using the indices M and E, respectively. $\mathcal{N}$ is the normalization constant.

(ii) Quantization of the Wave Number: The wave number is quantized by considering a perfectly conducting sphere of radius R,[8] and by imposing the following boundary conditions on the surface [23],

$$\mathbf{E}_{\parallel} = 0\,, \quad \mathbf{B}_{\perp} = 0\,. \quad \text{(A.166)}$$

Since $\mathbf{E} = -\frac{1}{c}\frac{\partial \mathbf{A}}{\partial t}$, $\mathbf{B} = \nabla \times \mathbf{A}$ in the Coulomb gauge, the magnetic and electric radiations are quantized respectively as

$$\begin{cases} j_I(k_nR) = 0 & \text{magnetic radiation,} \\ \frac{\partial}{\partial r}[rj_I(k_nr)]\big|_{r=R} = 0 & \text{electric radiation,} \end{cases} \quad \text{(A.167)}$$

based on Eqs. (A.164) and (A.165).

(iii) Normalization: The normalization constant is determined so that

$$\int_0^R r^2\mathrm{d}r \int \mathrm{d}\Omega\,\mathbf{A}^*_{\lambda'k'I'M'}(\mathbf{r})\cdot\mathbf{A}_{\lambda kIM}(\mathbf{r}) = \frac{2\pi\hbar c}{k}\delta_{\lambda\lambda'}\delta_{kk'}\delta_{II'}\delta_{MM'}\,, \quad \text{(A.168)}$$

where $\lambda = M, E$. We obtain

$$\mathcal{N} = \sqrt{\frac{4\pi\hbar ck}{R}} \quad \text{(A.169)}$$

by considering the asymptotic form of $j_\ell(kr)$ and using the boundary conditions (A.167) in the limit of $R \to \infty$.

A.9.3 Multipole Expansion and Quantization

The most general solution of the Helmholtz equation which satisfies the condition of a transverse wave is given by

$$\hat{\mathbf{A}}(\mathbf{r},t) = \sum_{\lambda kIM}\left[\mathbf{A}_{\lambda kIM}(\mathbf{r})\mathrm{e}^{-\mathrm{i}ckt}\hat{a}^\dagger_{\lambda kIM} + \mathbf{A}^*_{\lambda kIM}(\mathbf{r})\mathrm{e}^{\mathrm{i}ckt}\hat{a}_{\lambda kIM}\right]. \quad \text{(A.170)}$$

[8] R is introduced for the sake of quantization. It is taken to be much larger than the nuclear radius, and is eventually taken to be the limit of $R \to \infty$.

The coefficients $\hat{a}^\dagger$ and $\hat{a}$ are independent variables to describe the electromagnetic field, and are assumed to obey the following commutation relations of the Bose–Einstein statistics,

$$\left[\hat{a}_{\lambda kIM}, \hat{a}_{\lambda' k' I' M'}\right] = 0 \,, \quad \left[\hat{a}_{\lambda kIM}, \hat{a}^\dagger_{\lambda' k' I' M'}\right] = \delta_{\lambda\lambda'}\delta_{kk'}\delta_{II'}\delta_{MM'} \,. \qquad \text{(A.171)}$$

The Hamiltonian of the radiation field and that of the electromagnetic interaction are given by

$$\hat{H}_{\text{field}} = \sum_{\lambda kIM} \hbar kc \left(\hat{a}^\dagger_{\lambda kIM}\hat{a}_{\lambda kIM} + \frac{1}{2}\right) \,, \qquad \text{(A.172)}$$

$$\hat{H}_{\text{int}} = -\frac{1}{c}\sum_{\lambda kIM} \left(\hat{a}^\dagger_{\lambda kIM} \int d\mathbf{r}\, \mathbf{j}(\mathbf{r}, t) \cdot \mathbf{A}_{\lambda kIM}(\mathbf{r}) e^{-ikct} + \text{h.c.}\right) \,, \qquad \text{(A.173)}$$

in terms of $\hat{a}^\dagger$ and $\hat{a}$. The eigenstate of the radiation field is given as

$$|\cdots n_{\lambda kIM} \cdots\rangle \qquad \text{(A.174)}$$

in the occupation number representation. Since it holds that

$$|\langle\cdots n_{\lambda kIM} - 1 \cdots|\hat{a}_{\lambda kIM}|\cdots n_{\lambda kIM}\cdots\rangle|^2 = n_{\lambda kIM} \,, \qquad \text{(A.175)}$$

$$|\langle\cdots n_{\lambda kIM} + 1 \cdots|\hat{a}^\dagger_{\lambda kIM}|\cdots n_{\lambda kIM}\cdots\rangle|^2 = n_{\lambda kIM} + 1 \,, \qquad \text{(A.176)}$$

the creation and annihilation of a photon take place via $\hat{a}^\dagger$ and $\hat{a}$ in Eq. (A.173). On the other hand, the nucleus changes its state through $\mathbf{j}(\mathbf{r}, t)$.

A.10 Notation for Relativistic Equation of Motion and Dirac Equation

Here, we summarize the notations of the Dirac equation used in this book and their basic functions:

$$x^\mu = (ct,\ \mathbf{x}) \,, \quad x_\mu = (ct,\ -\mathbf{x}) \,, \qquad \text{(A.177)}$$

$$p^\mu = (E/c,\ \mathbf{p}) \,, \quad p_\mu = (E/c,\ -\mathbf{p}) \,, \qquad \text{(A.178)}$$

$$\partial^\mu \equiv \frac{\partial}{\partial x_\mu} = \left(\frac{\partial}{c\partial t},\ -\nabla\right) \,, \quad \partial_\mu \equiv \frac{\partial}{\partial x^\mu} = \left(\frac{\partial}{c\partial t},\ \nabla\right) \,, \qquad \text{(A.179)}$$

$$g^{\mu\nu} = g_{\mu\nu} = \begin{pmatrix} 1 & 0 & 0 & 0 \\ 0 & -1 & 0 & 0 \\ 0 & 0 & -1 & 0 \\ 0 & 0 & 0 & -1 \end{pmatrix} \,, \qquad \text{(A.180)}$$

$$a \cdot b = a_\mu b^\mu = a^\mu g_{\mu\nu} b^\nu = a^0 b^0 - \mathbf{a} \cdot \mathbf{b} \,, \tag{A.181}$$

$$\partial^\mu \partial_\mu = \frac{\partial^2}{c^2 \partial t^2} - \nabla^2 \,, \tag{A.182}$$

$$\boldsymbol{\alpha} = \begin{pmatrix} 0 & \boldsymbol{\sigma} \\ \boldsymbol{\sigma} & 0 \end{pmatrix} , \quad \beta = \begin{pmatrix} 1 & 0 \\ 0 & -1 \end{pmatrix} , \tag{A.183}$$

$$\gamma^0 = \beta \,, \quad \boldsymbol{\gamma} = \beta \boldsymbol{\alpha} = \begin{pmatrix} 0 & \boldsymbol{\sigma} \\ -\boldsymbol{\sigma} & 0 \end{pmatrix} , \quad \gamma^\mu = (\gamma^0, \boldsymbol{\gamma}) \,, \tag{A.184}$$

$$\bar{\psi} = \psi^\dagger \gamma^0 \,, \tag{A.185}$$

$$\hat{p}^\mu = \mathrm{i}\hbar \frac{\partial}{\partial x_\mu} = \left(\mathrm{i}\hbar \frac{\partial}{\partial (ct)}, \ \mathrm{i}\hbar \frac{\partial}{\partial x_1}, \ \mathrm{i}\hbar \frac{\partial}{\partial x_2}, \ \mathrm{i}\hbar \frac{\partial}{\partial x_3} \right) \tag{A.186}$$

$$= \mathrm{i}\hbar \nabla^\mu = \left(\mathrm{i}\hbar \frac{\partial}{\partial (ct)}, \ -\mathrm{i}\hbar \frac{\partial}{\partial x}, \ -\mathrm{i}\hbar \frac{\partial}{\partial y}, \ -\mathrm{i}\hbar \frac{\partial}{\partial z} \right) \tag{A.187}$$

$$= \mathrm{i}\hbar \left(\frac{\partial}{\partial (ct)}, \ -\nabla \right) . \tag{A.188}$$

A.11 Derivation of the Pairing Energy

In this section, we outline the derivation of Eqs. (5.67)–(5.69) for the pairing energy due to the residual interaction of the δ-function type given by Eq. (5.66).

We first remark that the antisymmetrized wave function of two-nucleon system is given by Eq. (5.63). We assume that each wave function $|(j_\alpha j_\beta)JM\rangle$ is normalized. The normalization constant N is then given by

$$N = \begin{cases} 1/\sqrt{2} & \text{for } (n_\alpha \ell_\alpha j_\alpha) \neq (n_\beta \ell_\beta j_\beta) \\ 1/2 & \text{for } (n_\alpha \ell_\alpha j_\alpha) = (n_\beta \ell_\beta j_\beta) \,. \end{cases} \tag{A.189}$$

Since the residual interaction is independent of spin, it is convenient to transform the wave function from the jj-coupling scheme to the LS-coupling scheme. It can be achieved by means of a recoupling coefficient as

$$|(j_\alpha j_\beta)JM\rangle = \big|(\ell_\alpha 1/2)j_\alpha, (\ell_\beta 1/2)j_\beta; JM\big\rangle \tag{A.190}$$

$$= \sum_{LS} \big|(\ell_\alpha \ell_\beta)L, (1/2\ 1/2)S; JM\big\rangle$$
$$\times \big\langle (\ell_\alpha \ell_\beta)L, (1/2\ 1/2)S; J \big| (\ell_\alpha 1/2)j_\alpha, (\ell_\beta 1/2)j_\beta; J \big\rangle \,, \tag{A.191}$$

where the recoupling coefficient is given by [24]

$$\begin{aligned}&\big\langle (j_1 j_2)J_{12}, (j_3 j_4)J_{34}; J \big| (j_1 j_3)J_{13}, (j_2 j_4)J_{24}; J\big\rangle \\ &\quad = [(2J_{12}+1)(2J_{34}+1)(2J_{13}+1)(2J_{24}+1)]^{1/2} \begin{Bmatrix} j_1 & j_2 & J_{12} \\ j_3 & j_4 & J_{34} \\ J_{13} & J_{24} & J \end{Bmatrix} . \end{aligned} \tag{A.192}$$

Similarly,

$$|(j_\beta j_\alpha)JM\rangle = \big|(\ell_\beta 1/2)j_\beta, (\ell_\alpha 1/2)j_\alpha; JM\big\rangle \tag{A.193}$$

$$\begin{aligned}&= \sum_{LS} \big|(\ell_\beta \ell_\alpha)L, (1/2\ 1/2)S; JM\big\rangle \\ &\quad \times \big\langle(\ell_\beta \ell_\alpha)L, (1/2\ 1/2)S; J \big| (\ell_\beta 1/2)j_\beta, (\ell_\alpha 1/2)j_\alpha; J\big\rangle . \end{aligned} \tag{A.194}$$

The recoupling coefficients in Eqs. (A.191) and (A.194) can be related to each other as

$$\begin{aligned}&\langle(\ell_\beta \ell_\alpha)L, (1/2\ 1/2)S; J|(\ell_\beta 1/2)j_\beta, (\ell_\alpha 1/2)j_\alpha; J\rangle \\ &\quad = [(2L+1)(2S+1)(2j_\beta+1)(2j_\alpha+1)]^{1/2} \begin{Bmatrix} \ell_\beta & \ell_\alpha & L \\ 1/2 & 1/2 & S \\ j_\beta & j_\alpha & J \end{Bmatrix} \end{aligned} \tag{A.195}$$

$$= (-1)^{\sigma_1} \big\langle(\ell_\alpha \ell_\beta)L, (1/2\ 1/2)S; J \big| (\ell_\alpha 1/2)j_\alpha, (\ell_\beta 1/2)j_\beta; J\big\rangle , \tag{A.196}$$

where

$$\sigma_1 = 1 + \ell_\alpha + \ell_\beta + j_\alpha + j_\beta + L + S + J \tag{A.197}$$

by using the following exchange property of the $9j$ symbol,

$$\begin{Bmatrix} j_1 & j_2 & J_{12} \\ j_3 & j_4 & J_{34} \\ J_{13} & J_{24} & J \end{Bmatrix} = (-1)^{\Sigma} \begin{Bmatrix} j_2 & j_1 & J_{12} \\ j_4 & j_3 & J_{34} \\ J_{24} & J_{13} & J \end{Bmatrix} , \tag{A.198}$$

where

$$\Sigma = j_1 + j_2 + j_3 + j_4 + J_{12} + J_{34} + J_{13} + J_{24} + J . \tag{A.199}$$

The antisymmetrized wave function is now given by

$$\begin{aligned}|(j_\alpha j_\beta)JM\rangle_a &= N \sum_{LS} \big\langle(\ell_\alpha \ell_\beta)L, (1/2\ 1/2)S; J \big| (\ell_\alpha 1/2)j_\alpha, (\ell_\beta 1/2)j_\beta; J\big\rangle \\ &\times \Big[\big|(\ell_\alpha \ell_\beta)L, (1/2\ 1/2)S; JM\big\rangle + (-1)^{\sigma} \big|(\ell_\beta \ell_\alpha)L, (1/2\ 1/2)S; JM\big\rangle\Big] , \end{aligned} \tag{A.200}$$

where

$$\sigma = \ell_\alpha + \ell_\beta + L + S\ . \tag{A.201}$$

The energy shift due to a residual interaction V is given by

$$\Delta E((n_\alpha \ell_\alpha j_\alpha, n_\beta \ell_\beta j_\beta)J) = {}_a\langle (j_\alpha j_\beta)J|V|(j_\alpha j_\beta)J\rangle_a \tag{A.202}$$

in the first order perturbation theory. Inserting Eq. (A.200), we have

$$\begin{aligned}
&\Delta E\left((n_\alpha \ell_\alpha j_\alpha, n_\beta \ell_\beta j_\beta)J\right) \\
&= 2N^2 \sum_{LS} \left\langle (\ell_\alpha \ell_\beta)L, (1/2\ 1/2)S;\ J\big|(\ell_\alpha 1/2)j_\alpha, (\ell_\beta 1/2)j_\beta;\ J\right\rangle^2 \\
&\quad \times \left[\left\langle (\ell_\alpha \ell_\beta)L, (1/2\ 1/2)S;\ J|V|(\ell_\alpha \ell_\beta)L, (1/2\ 1/2)S;\ J\right\rangle\right. \\
&\quad \left. + (-1)^\sigma \left\langle (\ell_\alpha \ell_\beta)L, (1/2\ 1/2)S;\ J|V|(\ell_\beta \ell_\alpha)L, (1/2\ 1/2)S;\ J\right\rangle\right]\ .
\end{aligned} \tag{A.203}$$

We now remark that

$$\delta(\mathbf{r}_1 - \mathbf{r}_2) = \frac{1}{r_1 r_2}\delta(r_1 - r_2)\delta(\cos\theta_1 - \cos\theta_2)\delta(\varphi_1 - \varphi_2) \tag{A.204}$$

for the residual interaction of the δ-function type, and that the two-body wave function is given by

$$\begin{aligned}
&|(\ell_\alpha \ell_\beta)L, (1/2\ 1/2)S;\ J\rangle \\
&= \sum_{M_L M_S m_\alpha m_\beta} \langle LSM_L M_S|JM\rangle\langle \ell_\alpha \ell_\beta m_\alpha m_\beta|LM_L\rangle|SM_S\rangle \\
&\qquad \times \frac{1}{r_1}u_{n_\alpha \ell_\alpha}(r_1)\frac{1}{r_2}u_{n_\beta \ell_\beta}(r_2)Y_{\ell_\alpha m_\alpha}(\Omega_1)Y_{\ell_\beta m_\beta}(\Omega_2)\ .
\end{aligned} \tag{A.205}$$

It is then straightforward to obtain

$$\begin{aligned}
&\left\langle (\ell_\alpha \ell_\beta)L', (1/2\ 1/2)S';\ JM\big|\delta(\mathbf{r}_1 - \mathbf{r}_2)\big|(\ell_\alpha \ell_\beta)L, (1/2\ 1/2)S;\ J\right\rangle \\
&= \delta_{SS'}\delta_{LL'}\frac{(2\ell_\alpha+1)(2\ell_\beta+1)}{4\pi}\begin{pmatrix}\ell_\alpha & \ell_\beta & L\\ 0 & 0 & 0\end{pmatrix}^2 \int \frac{1}{r^2}\left[u_{n_\alpha \ell_\alpha}(r)u_{n_\beta \ell_\beta}(r)\right]^2 \mathrm{d}r\ ,
\end{aligned} \tag{A.206}$$

and

$$\begin{aligned}
&\left\langle (\ell_\alpha \ell_\beta)L', (1/2\ 1/2)S';\ JM\big|\delta(\mathbf{r}_1 - \mathbf{r}_2)\big|(\ell_\beta \ell_\alpha)L, (1/2\ 1/2)S;\ J\right\rangle \\
&= \delta_{SS'}\delta_{LL'}(-1)^{\ell_\alpha+\ell_\beta+L}\frac{(2\ell_\alpha+1)(2\ell_\beta+1)}{4\pi}\begin{pmatrix}\ell_\alpha & \ell_\beta & L\\ 0 & 0 & 0\end{pmatrix}^2 \int \frac{1}{r^2}[u_{n_\alpha \ell_\alpha}(r)u_{n_\beta \ell_\beta}(r)]^2 \mathrm{d}r\ .
\end{aligned} \tag{A.207}$$

Equations (A.203), (A.206), and (A.207) lead to

$$\begin{aligned}
&\Delta E\left((n_\alpha \ell_\alpha j_\alpha, n_\beta \ell_\beta j_\beta)J\right) \\
&= -V_0 2N^2 \sum_{LS} \left\langle (\ell_\alpha \ell_\beta)L, (1/2\ 1/2)S; J \middle| (\ell_\alpha 1/2) j_\alpha, (\ell_\beta 1/2) j_\beta; J \right\rangle^2 \\
&\quad \times [1+(-1)^S] \frac{(2\ell_\alpha+1)(2\ell_\beta+1)}{4\pi} \begin{pmatrix} \ell_\alpha & \ell_\beta & L \\ 0 & 0 & 0 \end{pmatrix}^2 \int \frac{1}{r^2} \left[u_{n_\alpha \ell_\alpha}(r) u_{n_\beta \ell_\beta}(r)\right]^2 \mathrm{d}r\,.
\end{aligned} \tag{A.208}$$

The factor $[1+(-1)^S]$ shows that only $S=0$ state contributes. This is a consequence of the δ-force for the two-body system of identical nucleons. The isospin wave function for the system of identical nucleons is symmetric. On the other hand, the δ-force has a finite effect only in the positive parity states, i.e., spatially symmetric states. Consequently, the δ-force affects only the spin singlet state. Equation (A.208) now becomes

$$\begin{aligned}
&\Delta E\left((n_\alpha \ell_\alpha j_\alpha, n_\beta \ell_\beta j_\beta)J\right) \\
&= -V_0 4N^2 \left\langle (\ell_\alpha \ell_\beta)J, (1/2\ 1/2)0; J \middle| (\ell_\alpha 1/2) j_\alpha, (\ell_\beta 1/2) j_\beta; J \right\rangle^2 \\
&\quad \times \frac{(2\ell_\alpha+1)(2\ell_\beta+1)}{4\pi} \begin{pmatrix} \ell_\alpha & \ell_\beta & J \\ 0 & 0 & 0 \end{pmatrix}^2 \int \frac{1}{r^2} \left[u_{n_\alpha \ell_\alpha}(r) u_{n_\beta \ell_\beta}(r)\right]^2 \mathrm{d}r\,.
\end{aligned} \tag{A.209}$$

One has to use various properties of the $3j$, $6j$ and $9j$ symbols, which can be found in [17, 24], in order to reach the final result given by Eqs. (5.67)–(5.69). The relevant formulae in the present context are

$$\begin{Bmatrix} j_1 & j_2 & J \\ j_1' & j_2' & J \\ k & k & 0 \end{Bmatrix} = \frac{1}{\sqrt{(2J+1)(2k+1)}} (-1)^{-j_2-J-j_1'-k} \begin{Bmatrix} j_1 & j_2 & J \\ j_2' & j_1' & k \end{Bmatrix} \tag{A.210}$$

and

$$\begin{pmatrix} j_1 & j_2 & J \\ 1/2 & -1/2 & 0 \end{pmatrix} = -\sqrt{(2\ell_1+1)(2\ell_2+1)} \begin{pmatrix} \ell_1 & \ell_2 & J \\ 0 & 0 & 0 \end{pmatrix} \begin{Bmatrix} j_1 & j_2 & J \\ \ell_2 & \ell_1 & 1/2 \end{Bmatrix}\,. \tag{A.211}$$

Using these formulae, one can rewrite Eq. (A.209) into Eqs. (5.67)–(5.69).

A.12 Errata to the Japanese Edition

In Table A.2, we list the errata for the original Japanese edition.

Table A.2 Errata for the Japanese edition

Page	Place	Original	Correction
8	Sect. 1.5, 5th line	92	roughly speaking 92
20	Eq. (2.24)	$0.14\,\mathrm{fm}^{-3}$	$0.14–0.17\,\mathrm{fm}^{-3}$
21	Eq. (2.28)	$1.18\sim1.2\,\mathrm{fm}$	$1.1–1.2\,\mathrm{fm}$
45	Eq. (2.67)	$\frac{(n-1)(n+2)}{2(2n+1)}\alpha_n^2$	$\sum_{n=4,5...}\frac{(n-1)(n+2)}{2(2n+1)}\alpha_n^2$
45	Eq. (2.68)	$\frac{5(n-1)}{(2n+1)^2}\alpha_n^2$	$\sum_{n=4,5...}\frac{5(n-1)}{(2n+1)^2}\alpha_n^2$
72	Footnote	amnd	and
72	Footnote	potetial	potential
79	4th line	(3.87)	(3.86)
79	Eq. (3.95)	$\frac{4\pi}{3\mu^2}\mu c^2\frac{g_\pi^2}{4\pi\hbar c}$	$\frac{4\pi}{\mu^3}\frac{g_\pi^2}{3\hbar c}m_\pi c^2$
83	Eq. (3.103)	f	g_π
96	Below Eq. (4.38)	electromagnetic moment	electromagnetic multipole moments
104	Fig. 5.1	occupation number	degeneracy
104	Fig. 5.1	connection lines between H.O. and ISQ for 1h, 1i	from N = 5 and 6, respectively
120	Eq. (5.68)	$R_{n_\alpha\ell_\alpha}, R_{n_\beta\ell_\beta}$	$u_{n_\alpha\ell_\alpha}, u_{n_\beta\ell_\beta}$
120	Below Eq. (5.69)	$R_{n\ell}(r)$	$u_{n\ell}(r)$
131	Below Eq. (6.32)	Eq. (6.32)	Eq. (6.31)
134	2nd line	$0.14\,\mathrm{fm}^{-3}$	$0.17\,\mathrm{fm}^{-3}$
163	Eq. (7.18)	$\frac{2}{9}$	$\frac{1}{9}$
172	Fig. 7.12	superdeformed rotors	superdeformed rotational bands
174	1st line	13.7 billion years	13.8 billion years
176	Fig. 8.1	positions of r_2, r_3	turning point position for each
177	(8.11), below (8.12)	$e^{2\sqrt{2\eta k R}}$	$\mathrm{e}^{4\sqrt{2\eta k R}}$
178	Eqs. (8.15), (8.20)	$e^{2\sqrt{2\eta k R}}$	$\mathrm{e}^{4\sqrt{2\eta k R}}$
181	Eq. (8.30)	$\frac{3}{2}\frac{\hbar^2}{\mu}\frac{1}{r_c}\frac{p_{in}}{\hbar}$	$\frac{3}{2}\frac{\hbar^2}{\mu}\frac{1}{r_c}\frac{p_{in}}{\hbar^2}$
199	2nd line	60	100
205	Eq. (10.31)	k	$\hbar k$
222	Eq. (10.123)	$\alpha = \sqrt{2\eta\rho} - \pi\eta$	$\alpha = 2\sqrt{2\eta\rho} - \pi\eta$
222	Below Eq. (10.127)	$\alpha = \sqrt{2\eta\rho} - \pi\eta$	$\alpha = 2\sqrt{2\eta\rho} - \pi\eta$
222	Below Eq. (10.127)	(10.123)	(10.122), (10.123)
228	Eq. (10.167)	$rj_l(k_n R)$	$rj_l(k_n r)$
228	Eq. (10.168)	$\mathbf{A}^*_{\lambda k I M}(\mathbf{r})$	$\mathbf{A}_{\lambda k I M}(\mathbf{r})$

References

1. G.R. Satchler, *Direct Nuclear Reactions* (Clarendon Press, Oxford, 1983)
2. K.W. Ford, J.A. Wheeler, Ann. Phys. **7**:259 (1959); 287 (1959)
3. W.E. Frahn, Ann. Phys. **72**, 524 (1972); W.E. Frahn, M.S. Hussein, Nucl. Phys. A **346**, 237 (1980)
4. K.W. McVoy, Phys. Rev. C **3**, 1104 (1971)
5. N. Rowley, H.C. Doubre, C. Marty, Phys. Lett. B **69**, 147 (1977)
6. N. Takigawa, S.Y. Lee, Nucl. Phys. A **292**, 173 (1977)
7. M. Ueda, M.P. Pato, M.S. Hussein, N. Takigawa, Phys. Rev. Lett. **81**, 1809 (1998)
8. D.M. Brink, *Semi-Classical Methods for Nucleus–Nucleus Scattering* (Cambridge University Press, London, 1985)
9. D.M. Brink, N. Takigawa, Nucl. Phys. A **279**, 159 (1977)
10. S.Y. Lee, N. Takigawa, C. Marty, Nucl. Phys. A **308**, 161 (1978)
11. J.J. Sakurai, *Modern Quantum Mechanics* (Addison-Wesley, New York, 1985)
12. W.E. Frahn, R.H. Lemmer, Nuovo Cimento **5**, 1564 (1957); **6**, 664 (1957); F.G. Perey, B. Buck, Nucl. Phys. **32**, 353 (1962)
13. N. Takigawa, K. Hara, Z. Phys. A **276**, 79 (1976)
14. D.H.E. Gross, H. Kalinowski, Phys. Lett. B **48**, 302 (1974); Phys. Rep. **45**, 175 (1978)
15. P. Fröbrich, R. Lipperheide, *Theory of Nuclear Reactions* (Clarendon Press, Oxford, 1996)
16. H. Horiuchi, Prog. Theor. Phys. **63**, 725 (1980); **64**, 184 (1980)
17. A. Messiah, *Quantum Mechanics*, vol. II (Elsevier, Amsterdam, 1962)
18. D.M. Brink, G.R. Satchler, *Angular Momentum* (Oxford University Press, Oxford, 1962)
19. A. Bohr, B.R. Mottelson, Nucl. Struct. vol. I (Benjamin, New York, 1969)
20. P. Ring, P. Schuck, *The Nuclear Many-Body problem* (Springer, Berlin, 1980)
21. DSc thesis of Yoshihiro Nozawa, Tohoku University (2002)
22. M. Abramowitz, I.A. Stegun, *Handbook of Mathematical Functions with Formulas, Graphs, and Mathematical Tables* (Dover, London, 1978)
23. J.D. Jackson, *Classical Electrodynamics*, 3rd edn. (Wiley, New York, 1998)
24. A. de-Shalit, I. Talmi, *Nuclear Shell Theory* (Academic, New York, 1963)

Index

N. Takigawa and K. Washiyama, *Fundamentals of Nuclear Physics*,
DOI 10.1007/978-4-431-55378-6

F

G

H

I

T

U

Made in the USA
Columbia, SC
17 September 2026

27554b19-ccf5-4ea6-829c-c68c3f5b98ceR01